Barron's Regents Exams and Answers

Biology
The Living Environment

GABRIELLE I. EDWARDS
Former Science Consultant
Board of Education
City of New York

Former Assistant Principal Supervision
Science Department
Franklin D. Roosevelt High School
Brooklyn, New York

MARION CIMMINO
Former Teacher, Biology and Laboratory Techniques
Franklin D. Roosevelt High School
Brooklyn, New York

FRANK J. FODER
Teacher, Advanced Placement Biology
Franklin D. Roosevelt High School
Brooklyn, New York

G. SCOTT HUNTER
Former Teacher, Biology and Advanced Biology
Former Consultant, State Education Department Bureaus of Science and Testing
Former Central Office Administrator
Former Superintendent of Schools
Mexico and Chatham, New York

Barron's Educational Series, Inc.

All inquiries should be addressed to:
Barron's Educational Series, Inc.
250 Wireless Boulevard
Hauppauge, New York 11788
www.barronseduc.com

ISBN-13: 978-0-8120-3197-3
ISBN-10: 0-8120-3197-0
ISSN 1069-2940

PRINTED IN THE UNITED STATES OF AMERICA
9 8 7 6 5 4 3 2 1

Contents

How to Use This Book   **1**
Organization of the Book   1
How to Study   6

Test-Taking Tips—A Summary   **10**

New York State Regents Biology—The Living Environment: Core Curriculum and Assessment Map   **15**

Tips for Teachers   **23**

Study Questions and Answers   **26**
Questions on Standard 1, Key Idea 1—Purpose of Scientific Inquiry   26
Questions on Standard 1, Key Idea 2—Methods of Scientific Inquiry   32
Questions on Standard 1, Key Idea 3—Analysis in Scientific Inquiry   42
Questions on Standard 4, Key Idea 1—Application of Scientific Principles   60
Questions on Standard 4, Key Idea 2—Genetic Continuity   73
Questions on Standard 4, Key Idea 3—Organic Evolution   81
Questions on Standard 4, Key Idea 4—Reproductive Continuity   90
Questions on Standard 4, Key Idea 5—Dynamic Equilibrium and Homeostasis   99
Questions on Standard 4, Key Idea 6—Interdependence of Living Things   104
Questions on Standard 4, Key Idea 7—Human Impact on the Environment   112

Glossary   126

Prominent Scientists   126
Biological Terms   127

Regents Examinations, Answers, and Student Self-Appraisal Guides   161

Living Environment Regents August 2007 162
Living Environment Regents June 2008 241
Living Environment Regents August 2008 311
Living Environment Regents June 2009 390

How to Use This Book

ORGANIZATION OF THE BOOK

Study Questions

Section 1 of the book consists of 173 questions taken from past Regents examinations in biology. Most of the questions require that you select a correct response from four choices given. A few questions provide you with a list of words or phrases from which to select the one that best matches a given description. Others are constructed-response, graphical analysis, or reading comprehension questions. You should become familiar with the formats for questions that will appear on the new Living Environment Regents examination.

Each question in Section 1 of this book has a well-developed answer. Each answer provides the number of the correct response, the reason why the response is correct, and explanations of why the other choices are incorrect.

A useful feature of Section 1 is the student Self-Appraisal Guide. This device allows you to determine where your learning strengths and weaknesses lie in the major topics of each unit. For specific topics within the units, the numbers of related questions are given. As you attempt to answer each question in Section 1, you may wish to circle on the appraisal form the numbers of the questions that you are unable to answer. The circled items then help you to identify at a glance subject matter areas in which you need additional study.

New York State Learning Standards

COMMENCEMENT STANDARDS

There are several commencement standards required of students in New York State public schools regarding their performance in math, science, and technology. The Core Curriculum for The Living Environment addresses two of these standards:

Standard 1: **Students will use mathematical analysis, scientific inquiry, and engineering design, as appropriate, to pose questions, seek answers, and develop solutions.**

Standard 4: **Students will understand and apply scientific concepts, principles, and theories pertaining to the physical setting and living environment and recognize the historical development of ideas in science.**

The Core Curriculum for The Living Environment was built from these two commencement standards. It is important to recognize that the Core Curriculum is not a syllabus. It does not prescribe what will be taught and learned in any particular classroom. Instead, it defines the skills and understandings that you must master in order to achieve the commencement standards for life science.

Instead of memorizing a large number of details at the commencement level, then, you are expected to develop the skills needed to deal with science on the investigatory level, generating new knowledge from experimentation and sharpening your abilities in data analysis. You are also expected to read and understand scientific literature, taking from it the facts and concepts necessary for a real understanding of issues in science. You are required to pull many facts together from different sources to develop your own opinions about the moral and ethical problems facing modern society concerning technological advances. These are thinking skills that do not respond to simple memorization of facts and scientific vocabulary.

KEY IDEAS, PERFORMANCE INDICATORS, AND MAJOR UNDERSTANDINGS

Each commencement standard is subdivided into a number of Key Ideas. Key Ideas are broad, unifying statements about what you need to know. Within Standard 1, three Key Ideas are concerned with laboratory investigation and data analysis. Together, these unifying principles develop your ability to deal with data and understand how professional science is carried out in biology. Within Standard 4, seven Key Ideas present a set of concepts that are central to the science of biology. These Key Ideas develop your understanding of the essential characteristics of living things that allow them to be successful in diverse habitats.

Within each Key Idea, Performance Indicators are presented that indicate which skills you should be able to demonstrate through your mastery of the Key Idea. These Performance Indicators give guidance to both you and your teacher about what is expected of you as a student of biology.

Performance Indicators are further subdivided into Major Understandings. Major Understandings give specific concepts that you must master in order to achieve each Performance Indicator. It is from these Major Understandings that the Regents assessment material will be drawn.

LABORATORY COMPONENT

A meaningful laboratory experience is essential to the success of this or any other science course. You are expected to develop a good sense of how scientific inquiry is carried out by the professional scientist and how these same techniques can assist in the full understanding of concepts in science. The Regents requirement of 1,200 minutes of successful laboratory experience, coupled with satisfactory written reports of your findings, should be considered a minimum.

Students are required to complete four laboratory experiences required by the New York State Education Department and tested on the Regents Examination. See Barron's *Let's Review: Biology—The Living Environment, 4th edition* (Hunter, 2004) for a complete treatment of this new requirement.

Regents Examinations

Section 2 of the book consists of actual Biology Regents examinations and answers. These Regents examinations are based on the New York State Core Curriculum for the Living Environment.

Assessments: Format and Scoring

The format of the Regents assessment for The Living Environment is as follows, based on actual Regents examinations given to date.

Part A: 35 multiple-choice questions that test the student's knowledge of specific factual information. **All** questions must be answered on Part A. Maximum of **35** credits awarded.

Part B: Variable number of questions, representing a mixture of multiple-choice and constructed-response items. Questions may be based on the student's direct knowledge of biology, interpretation of experimental data, analysis of readings in science, and ability to deal with representations of biological phenomena. **All** questions must be answered on Part B. Maximum of **30** credits awarded.

Part C: Variable number of constructed-response questions. Questions may be based on the student's direct knowledge of biology, interpretation of experimental data, analysis of readings in science, and ability to deal with representations of biological phenomena. **All** questions must be answered on Part C. Maximum of **20** credits awarded.

Part D: Laboratory component prepared and administered as part of the Regents assessment. Maximum of **13** credits awarded (see below).

In June 2004, the Living Environment Regents Examination was changed to include a new Part D. This part of the examination aims to assess student knowledge of and skills on any of four required laboratory experiences supplied to schools by the New York State Education Department. The specific laboratory experiences required in any year will vary according to a preset schedule (see chart next page).

Part D is a new section of the Regents Examination. Questions on this section can be a combination of multiple-choice and constructed-response questions similar to those found in Parts A, B, and C of the Living Environment Regents Examination. The content of these questions will reflect the four specific laboratory experiences required for a particular year. You are strongly encouraged to include review of these laboratory experiences a part of your year-end Regents preparation activity.

The following chart summarizes the current laboratory requirement for New York State public schools:

LABORATORY REQUIREMENTS

Laboratory Title	Description
The Beaks of Finches	Explores the adaptive advantages of beaks with different physical characteristics
Relationships and Biodiversity	Explores the relationship between DNA structure and the biochemistry of inheritance
Making Connections	Explores the effects of physical activity on human metabolic activities
Diffusion Through a Membrane	Explores the nature of cross-membrane transport in living cells
Adaptations for Reproductive Success in Flowering Plants*	Not yet available
DNA Technology*	Not yet available
Environmental Conditions and Seed Germination*	Not yet available

*Not yet available as of this writing.

Studying questions from past Regents examinations is an invaluable aid in developing a mind set that will enable you to approach questions with understanding. Although exact questions are not repeated, question types are. If you practice questions that require interpretation, problem solving, and graph construction, you will do well on the entire exam. During the school year, the 30 required laboratory lessons teach you certain manipulative skills. Questions involving identification, measurement, and other laboratory procedures are based on the laboratory exercises. Review of past materials gives you insight as to the types of questions that you may be asked to answer. Study the questions in the Regents exams in this book diligently.

HOW TO STUDY

General Suggestions for Study

You've spent all year learning many different facts and concepts about biology—far more than you could ever hope to remember the "first time around." Your teacher has drilled you on these facts and concepts; you've done homework, taken quizzes and tests, performed laboratory experiments, and reviewed the material at intervals throughout the year.

Now it's time to put everything together. The Regents exam is only a few weeks away. If you and your teacher have planned properly, you will have finished all new information about 3 weeks before the exam. Now you have to make efficient use of the days and weeks ahead to review all that you've learned in order to score high on the Regents.

The task ahead probably seems impossible, but it doesn't have to be! You've actually retained much more of the year's material than you realize! The review process should be one that helps you to recall the many facts and concepts you've stored away in your memory. Your Barron's resources, including Barron's *Let's Review: The Living Environment,* will help you to review this material efficiently.

You also have to get yourself into the right frame of mind. It won't help to be nervous and stressed during the review process. The best way to avoid being stressed during any exam is to be well rested, prepared, and confident. We're here to help you prepare and to build your confidence. So let's get started on the road to a successful exam experience!

To begin, carefully read and follow the steps outlined below.

1. Start your review early; don't wait until the last minute. Allow at least 2 weeks to prepare for the Regents exam. Set aside an hour or two a day over the next few weeks for your review. Less than an hour a day is insufficient time for you to concentrate on the material meaningfully; more than 2 hours daily will yield diminishing returns on your investment of time.

2. Find a quiet, comfortable place to study. You should seat yourself at a well-lighted work surface, free of clutter, in a room without distractions of any kind. You may enjoy watching TV or listening to music curled up in a soft chair, but these and other diversions should be avoided when doing intense studying.

3. Make sure you have the tools you need, including this book, a pen and pencil, and some scratch paper for taking notes and doing calculations. Keep your class notebook at hand for looking up information between test-taking sessions. It will also help to have a good review text, such as

Barrron's *Let's Review: The Living Environment,* available for reviewing important concepts quickly and efficiently.

4. Concentrate on the material in the "Study Questions and Answers" section of this book. Read carefully and thoughtfully. Think about the questions that you review, and try to make sense out of them. Choose the answers carefully. (See the following section, "Using This Book for Study," for additional tips on question-answering techniques.)

5. Use available resources, including a dictionary and the glossary in this book, to look up the meanings of unfamiliar words in the practice questions. Remember that the same terms can appear on the Regents exam you will take, so take the opportunity to learn them now.

6. Remember: study requires time and effort. Your investment in study now will pay off when you take the Regents exam.

Using This Book for Study

This book is an invaluable tool if used properly. Read carefully and try to answer *all* the questions in Section 1 and on the practice exams. The more you study and practice, the more you will increase your knowledge of biology and the likelihood that you will earn a high grade on the Regents exam. To maximize your chances, use this book in the following way:

1. Answer all of the questions in the section entitled "Study Questions and Answers." Check your responses by using "Answers to Topic Questions," including "Wrong Choices Explained," following each question set. Record the number of correct responses on each topic in the "Self-Appraisal Guide" at the end of the section to identify your areas of strength and weakness. Use a good review text, such as Barron's *Let's Review: The Living Environment,* to study each area on which you did poorly. Finally, go back to the questions you missed on the first round and be sure that you fully understand what each question asks and why the correct answer is what it is.

2. When you have completed the questions in "Study Questions and Answers," go on to the examination section. Select the first complete examination and take it under test conditions.

3. Interpret the term *test conditions* as follows:
   • Be well rested; get a good night's sleep before attempting *any* exam.
   • Find a quiet, comfortable room in which to work.
   • Allow no distractions of any kind.
   • Select a well-lighted work surface free of clutter.
   • Have your copy of this book with you.
   • Bring to the room a pen, a pencil, some scratch paper, and a watch or alarm clock set for the 3-hour exam limit.

4. Take a deep breath, close your eyes for a moment, and RELAX! Tell yourself, "I know this stuff!" You have lots of time to take the Regents exam; use it to your advantage by reducing your stress level. Forget about your plans for later. For the present, your number 1 priority is to do your best, whether you're taking a practice exam in this book, or the real thing.

5. Read all test directions carefully. Note how many questions you must answer to complete each part of the exam. If test questions relate to a reading passage, diagram, chart, or graph, be sure you fully understand the given information before you attempt to answer the questions that relate to it.

6. When answering multiple-choice questions on the Regents exam, TAKE YOUR TIME! Pay careful attention to the "stem" of the question; read it over several times. These questions are painstakingly written by the test preparers, and every word is chosen to convey a specific meaning. If you read the question carelessly, you may answer a question that was never asked! Then read each of the four multiple-choice answers carefully, using a *pencil* to mark in the test booklet the answer you think is correct.

7. Remember that three of the multiple-choice answers are *incorrect;* these incorrect choices are called "distracters" because they seem like plausible answers to poorly prepared or careless students. To avoid being fooled by these distracters, you must think clearly, using everything you have learned about biology since the beginning of the year. This elimination process is just as important to your success on the Living Environment Regents exam as knowing the correct answer! If more than one answer seems to be correct, reread the question to find the words that will help you to distinguish between the correct answer and the distracters. When you have made your best judgment about the correct answer, circle the number in *pencil* in your test booklet.

8. Constructed-response questions appear in a number of different forms. You may be asked to select a term from a list, write the term on the answer sheet, and define the term. You may be asked to describe some biological phenomenon or state a biological fact using a complete sentence. You may be asked to read a value from a diagram of a measuring instrument and write that value in a blank on the answer paper. When answering this type of question, care should be exercised to follow directions precisely. If a complete sentence is called for, it must contain a subject and a verb, must be punctuated, and must be written understandably in addition to answering the scientific part of the question accurately. Values must be written clearly and accurately and include a unit of measure, if appropriate. Failure to follow the directions for a question may result in a loss of credit for that question.

9. A special type of constructed-response question is the essay or paragraph question. Typically, essay or paragraph questions provide an opportunity to earn multiple credits for answering the question correctly. As in the constructed-response questions described above, it is important that you follow the directions given if you hope to earn the maximum number of credits for the question. Typically, the question outlines exactly what must be included in your essay to gain full credit. Follow these directions step by step, double-checking to be certain that all question components are addressed in your answer. In addition, your essay or paragraph should follow the rules of good grammar and good communication so that it is readable and understandable. And, of course, it should contain correct information that answers all the parts of the question asked.

10. Graphs and charts are a special type of question that requires you to organize and represent data in graphical format. Typically for such questions, you are expected to place unorganized data in ascending order in a data chart or table. You may also be asked to plot organized data on a graph grid, connect the plotted points, and label the graph axes appropriately. Finally, questions regarding data trends and extrapolated projections may be asked, requiring you to analyze the data in the graph and draw inferences from it. As with all examination questions, always follow all directions for the question. Credit can be granted only for correctly following directions and accurately interpreting the data.

11. When you have completed the exam, relax for a moment. Check your time; have you used the entire 3 hours? Probably not. Resist the urge to quit. Go back to the beginning of the exam, and, in the time remaining, *retake the exam in its entirety*. Try to ignore the penciled notations you made the first time. If you come up with a different answer the second time through, read over the question with extreme care before deciding which response is correct. Once you have decided on the correct answer, mark your choice in ink in the answer booklet.

12. Score the exam using the Answer Key at the end of the exam. Review the "Answers Explained" section for each question to aid your understanding of the exam and the material. Remember that it's just as important to understand why the incorrect responses are incorrect as it is to understand why the correct responses are correct!

13. Finally, focus your between-exam study on your areas of weakness in order to improve your performance on the next practice exam. Complete all the practice exams in this book using the techniques outlined above.

Test-Taking Tips—A Summary

The following pages contain seven tips to help you achieve a good grade on the Living Environment Regents exam.

TIP 1
Be confident and prepared.

SUGGESTIONS
- Review previous tests.
- Use a clock or watch, and take previous exams at home under examination conditions (i.e., don't have the radio or television on).
- Get a review book. (The preferred book is Barron's *Let's Review: The Living Environment*.)
- Talk over the answers to questions on these tests with someone else, such as another student in your class or someone at home.
- Finish all your homework assignments.
- Look over classroom exams that your teacher gave during the term.
- Take class notes carefully.
- Practice good study habits.
- Know that there are answers for every question.
- Be aware that the people who made up the Regents exam want you to pass.
- Remember that thousands of students over the last few years have taken and passed a Biology Regents. You can pass too!
- Complete your study and review at least one day before taking the examination. Last-minute "cramming" may hurt, rather than enhance, your performance on the exam.

- Visit www.barronseduc.com or www.barronsregents.com for the latest information on the Regents exams.
- Talk over your answers to questions with someone else. Use Barron's web site to communicate directly with subject specialists.
- Be well rested when you enter the exam room. A good night's sleep is essential preparation for any examination.
- On the night prior to the exam day: lay out all the things you will need, such as clothing, pens, and admission cards.
- Bring with you two pens, two pencils, an eraser, and, if your school requires it, an identification card. Decide before you enter the room that you will remain for the entire 3-hour examination period, and either bring a wristwatch or sit where you can see a clock.
- Once you are in the exam room, arrange things, get comfortable, and attend to personal needs (the bathroom).
- Before beginning the exam, take a deep breath, close your eyes for a moment, and RELAX! Repeat this technique any time you feel yourself "tensing up" during the exam.
- Keep your eyes on your own paper; do not let them wander over to anyone else's paper.
- Be polite in making any reasonable demands of the exam room proctor, such as changing your seat or having window shades raised or lowered.

TIP 2

Read test instructions carefully.

SUGGESTIONS
- Be familiar with the format of the examination.
- Know how the test will be graded.
- Read all directions carefully. Be sure you fully understand supplemental information (reading passages, charts, diagrams, graphs) before you attempt to answer the questions that relate to it.
- Underline important words and phrases.
- Ask for assistance from the exam room proctor if you do not understand the directions.

TIP 3

Read each question carefully and read each choice before recording your answer.

SUGGESTIONS

• When answering the questions, TAKE YOUR TIME! Be sure to read the "stem" of the question and each of the four multiple-choice answers very carefully.
• If you are momentarily "stumped" by a question, put a check mark next to it and go on; come back to the question later if you have time.
• Remember that three of the multiple-choice answers (known as "distracters") are incorrect. If more than one answer seems to be correct, reread the question to find the words that will help you to distinguish between the correct answer and the distractors.
• When you have made your best judgment about the correct answer, circle the appropriate number in pencil on your answer sheet.

TIP 4

Budget your test time (3 hours).

SUGGESTIONS

• Bring a watch or clock to the test.
• The Regents examination is designed to be completed in 1½ to 2 hours.
• If you are absolutely uncertain of the answer to a question, mark your question booklet and move on to the next question.
• If you persist in trying to answer every difficult question *immediately,* you may find yourself rushing or unable to finish the remainder of the examination.
• When you have completed the exam, relax for a moment. Then go back to the beginning, and, in the time remaining, *retake the exam in its entirety.* Pay particular attention to questions you skipped the first time. Once you have decided on a correct response for multiple-choice questions, mark an "X" in ink through the penciled circle on the answer sheet.

• Plan to stay in the room for the entire three hours. If you finish early, read over your work — there may be some things that you omitted or that you may wish to add. You also may wish to refine your grammar, spelling, and penmanship.

TIP 5

Use your reasoning skills.

SUGGESTIONS
• Answer *all* questions.
• Relate (connect) the question to anything that you studied, wrote in your notebook, or heard your teacher say in class.
• Relate (connect) the question to any film, demonstration, or experiment you saw in class, any project you did, or to anything you may have learned from newspapers, magazines, or television.
• Look over the entire test to see whether one part of it can help you answer another part.

TIP 6

Don't be afraid to guess.

SUGGESTIONS
• In general, go with your first answer choice.
• Eliminate obvious incorrect choices.
• If still unsure of an answer, make an educated guess.
• There is no penalty for guessing; therefore, answer ALL questions. An omitted answer gets no credit.

TIP 7

Sign the Declaration.

SUGGESTIONS

- Be sure to sign the declaration on your answer sheet.
- Unless this declaration is signed, your paper cannot be scored.

New York State Regents Biology—The Living Environment: Core Curriculum and Assessment Map

The chart on pages 17–22 lists the Key Ideas and Performance Indicators from the New York State Regents Commencement Standard 1 and Standard 4 for *The Living Environment* (1999). For each Regents examination (assessment) listed across the top of the chart, item (question) numbers have been matched to the Performance Indicator(s) most closely associated with their content.

The information compiled in this chart represents a "map" showing how closely these assessments have mirrored the curriculum required by the state of New York for this course of study. It provides guidance to the teacher and student concerning the emphasis placed on each Performance Indicator by the New York State Education Department assessment development process on four recent Regents examinations.

By following this map, you will be sure to hit every major topic that is required by the state of New York for achievement of the Standards for *The Living Environment*. At the same time, you will be better able to focus your efforts on areas of needed study that will maximize your performance on the Regents examination without wasting valuable time on areas of the curriculum that are less likely to be represented on the examination.

Although some areas of the Core Curriculum have been less-well emphasized than others on Regents examinations to date, both teachers and students are cautioned that *all* parts of the Core Curriculum are required and are subject to testing. The Regents examination may be changed at any time and without notice to include items testing Performance Indicators that may not have been well emphasized in the past.

Students, you can find in Barron's *Let's Review: Biology—The Living Environment* (Hunter) an appropriate body of content, easy-to-read explanation, and practice exercises (with Answers Explained) developed especially for the New York State high school biology student. This resource will ease your way through *The Living Environment* course of study from start to finish and will prepare you well for the culminating Regents examination.

Teachers, you are encouraged to develop Standards-Based Learning Units (SBLUs) that will provide appropriate local content to illustrate these required Performance Indicators. Barron's *Let's Review: Biology—The Living Environment* (Hunter) provides an excellent source of material for the development of SBLUs that is consistent with the New York State Core Curriculum for this course. In addition, it includes many concepts that have traditionally been taught in the New York State Regents biology program since 1989 and relates this content to the new Regents Core Curriculum for *The Living Environment*.

LIVING ENVIRONMENT CURRICULUM MAP AND QUESTION INDEX

STANDARD 1, KEY IDEA 1: THE CENTRAL PURPOSE OF SCIENTIFIC INQUIRY IS TO DEVELOP EXPLANATIONS OF NATURAL PHENOMENA IN A CONTINUING AND CREATIVE PROCESS.

Performance Indicator	Description	August 2007	June 2008	August 2008	June 2009
1.1	The student should be able to elaborate on basic scientific and personal explanations of natural phenomena and develop extended visual models and mathematical formulations to represent one's thinking.				
1.2	The student should be able to hone ideas through reasoning, library research, and discussion with others, including experts.				
1.3	The student should be able to work toward reconciling competing explanations and clarify points of agreement and disagreement.	2	32		
1.4	The student should be able to coordinate explanations at different levels of scale, points of focus, and degrees of complexity and specificity, and recognize the need for such alternative representations of the natural world.	54			

STANDARD 1, KEY IDEA 2: BEYOND THE USE OF REASONING AND CONSENSUS, SCIENTIFIC INQUIRY INVOLVES THE TESTING OF PROPOSED EXPLANATIONS INVOLVING THE USE OF CONVENTIONAL TECHNIQUES AND PROCEDURES AND USUALLY REQUIRING CONSIDERABLE INGENUITY.

Performance Indicator	Description	August 2007	June 2008	August 2008	June 2009
2.1	The student should be able to devise ways of making observations to test proposed explanations.	35	34, 51, 52		
2.2	The student should be able to refine research ideas through library investigations, including electronic information retrieval and reviews of literature, and through peer feedback obtained from review and discussion.				
2.3	The student should be able to develop and present proposals including formal hypotheses to test explanations (i.e. predict what should be observed under specific conditions if the experiment is true).		63	61, 62	
2.4	The student should be able to carry out research for testing explanations, including selecting and developing techniques, acquiring and building apparatus, and recording observations as necessary.		31, 33	54, 63	53

STANDARD 1, KEY IDEA 3: THE OBSERVATIONS MADE WHILE TESTING PROPOSED EXPLANATIONS, WHEN ANALYZED USING CONVENTIONAL AND INVENTED METHODS, PROVIDE NEW INSIGHTS INTO NATURAL PHENOMENA.

Performance Indicator	Description	August 2007	June 2008	August 2008	June 2009
3.1	The student should be able to use various methods of representing and organizing observations (e.g. diagrams, tables, charts, graphs, equations, matrices) and insightfully interpret the organized data.	40, 51, 52, 53	39, 43, 44	47, 50, 51, 52	43, 44, 45, 46
3.2	The student should be able to apply statistical analysis techniques when appropriate to test if chance alone explains the results.				
3.3	The student should be able to assess correspondence between the predicted result contained in the hypothesis and the actual result, and reach a conclusion as to whether the explanation on which the prediction was based is supported.			31	
3.4	The student should be able to, based on the results of the test and through public discussion, revise the explanation and contemplate additional research.				
3.5	The student should be able to develop a written report for public scrutiny that describes the proposed explanation, including a literature review, the research carried out, its result, and suggestions for further research.				

STANDARD 4, KEY IDEA 1: LIVING THINGS ARE BOTH SIMILAR TO AND DIFFERENT FROM EACH OTHER AND FROM NONLIVING THINGS.

Performance Indicator	Description	August 2007	June 2008	August 2008	June 2009
1.1	The student should be able to explain how diversity of populations within ecosystems relates to the stability of ecosystems.	1, 43, 56, 57, 58	1, 2, 25, 42, 48	2, 3, 59, 60	1, 34, 47, 48, 49, 55
1.2	The student should be able to describe and explain the structures and functions of the human body at different organizational levels (e.g, systems, tissues, cells, organelles).	3, 5, 44, 46, 47	35, 53	8, 9, 33, 36, 40	2, 17
1.3	The student should be able to explain how a one-celled organism is able to function despite lacking the levels of organization present in more complex organisms.	7	4, 5, 9, 36	5, 33	3, 35

STANDARD 4, KEY IDEA 2: ORGANISMS INHERIT GENETIC INFORMATION IN A VARIETY OF WAYS THAT RESULT IN CONTINUITY OF STRUCTURE AND FUNCTION BETWEEN PARENTS AND OFFSPRING.

Performance Indicator	Description	August 2007	June 2008	August 2008	June 2009
2.1	The student should be able to explain how the structure and replication of genetic material result in offspring that resemble their parents.	8, 9, 11, 37, 38	3, 6, 7, 10, 11, 38, 50	6, 7, 12, 18, 19, 35	4, 5, 8, 15, 22, 28
2.2	The student should be able to describe and explain how the technology of genetic engineering allows humans to alter genetic makeup of organisms.	10, 12, 49, 50	12, 32	10, 13	10, 31

STANDARD 4, KEY IDEA 3: INDIVIDUAL ORGANISMS AND SPECIES CHANGE OVER TIME.

Performance Indicator	Description	August 2007	June 2008	August 2008	June 2009
3.1	The student should be able to explain the major patterns of evolution.	13, 14, 15, 18, 60	8, 13, 14, 16, 40, 41, 54, 65	11, 15, 16, 17, 48, 49, 55	11, 12, 16, 36, 37, 40, 59

STANDARD 4, KEY IDEA 4: THE CONTINUITY OF LIFE IS SUSTAINED THROUGH REPRODUCTION AND DEVELOPMENT.

Performance Indicator	Description	August 2007	June 2008	August 2008	June 2009
4.1	The student should be able to explain how organisms, including humans, reproduce their own kind.	6, 17, 19, 20, 39, 45	17, 18, 21, 22, 37	4, 21, 39, 43, 44, 45	6, 14, 51

STANDARD 4, KEY IDEA 5: ORGANISMS MAINTAIN A DYNAMIC EQUILIBRIUM THAT SUSTAINS LIFE.

Performance Indicator	Description	August 2007	June 2008	August 2008	June 2009
5.1	The student should be able to explain the basic biochemical processes in living organisms and their importance in maintaining dynamic equilibrium.	4, 33, 34	9, 15, 19	20, 23, 27, 37, 38, 41, 42	13, 42
5.2	The student should be able to explain disease as a failure of homeostasis.	21, 26, 55	20, 23, 59	65, 67	19, 20, 56, 57, 58
5.3	The student should be able to relate processes at the system level to the cellular level in order to explain dynamic equilibrium.	16, 22, 48	49, 56, 57, 64	22, 66	18, 21, 32, 33

STANDARD 4, KEY IDEA 6: PLANTS AND ANIMALS DEPEND ON EACH OTHER AND THEIR PHYSICAL ENVIRONMENT.

Performance Indicator	Description	August 2007	June 2008	August 2008	June 2009
6.1	The student should be able to explain factors that limit growth of individuals and populations.	28	24, 61	26, 32, 34	23, 24, 41
6.2	The student should be able to explain the importance of preserving diversity of species and habitats.	24, 25, 30, 41, 61	27, 55, 60, 62	1, 14, 30, 46	7, 9, 25, 26, 38, 39
6.3	The student should be able to explain how the living and nonliving environments change over time and respond to disturbances.	31, 32	28, 66	28	1

STANDARD 4, KEY IDEA 7: HUMAN DECISIONS AND ACTIVITIES HAVE A PROFOUND IMPACT ON THE PHYSICAL AND LIVING ENVIRONMENT.

Performance Indicator	Description	August 2007	June 2008	August 2008	June 2009
7.1	The student should be able to describe the range of interrelationships of humans with the living and nonliving environment.	42		58, 68	30
7.2	The student should be able to explain the impact of technological development and growth in the human population on the living and nonliving environment.	23, 59	29, 30, 45, 46, 47, 58, 67	24, 25, 56, 69, 71	29, 52, 55
7.3	The student should be able to explain how individual choices and societal actions can contribute to improving the environment.	27, 29, 62	26	29, 57, 64, 70	27, 52, 55

Tips for Teachers

CLASSROOM USE

All teachers will be able to use this book with their students as a companion to their regular textbooks and will find that their students gain considerable self-confidence and ability in test taking through its consistent use.

Teachers familiar with the Regents Biology Syllabus (1982) will notice a significant reduction in the amount of factual detail in the The Living Environment Core Curriculum (1999). This change is a reflection of a fundamental shift in philosophy about what skills and abilities students should have at the point of commencement at the upper-secondary level. It is assumed that science concepts have been taught and assessed at an age-appropriate level throughout their career, so that little additional detail needs to be presented at the upper-secondary level.

An excellent companion to this book (and any comprehensive biology text) is Barron's *Let's Review: Biology—The Living Environment* (Hunter). The factual material and organization of this book lend themselves well to the development of Standards-Based Learning Units (SBLUs). The level of detail is consistent with what students really have to know in order to do well on the New York State Regents Examination on the Living Environment.

APPLICATION-BASED CURRICULUM

The change in curriculum focus can be characterized as a switch from a fact-based curriculum to an application-based curriculum—one that is less concerned about content and more concerned about thinking. It is less about *how much* students know and more about *what they can do* with what they know. The latter, after all, is what real learning is all about; these are the abilities that will last a lifetime, not facts and scientific terminology.

This being said, it is acknowledged that students will have a difficult time expressing their views and making moral and ethical judgments about science if they lack a working knowledge of scientific principles and do not have at least a passing understanding of the terms used by biologists. For this reason, teachers and administrators will need to develop local curricula that complement the Core Curriculum. It is up to the teacher or administrator to decide what examples and factual knowledge best illustrate the concepts presented in the Core Curriculum, what concepts need to be reinforced and enhanced, what experiences will add measurably to students' understanding of science, and what examples of local interest should be included.

The teacher will immediately recognize the need to go beyond this level in the classroom, with examples, specific content, and laboratory experiences that complement and illuminate these Major Understandings. It is at this level that the locally developed curriculum is essential. Each school system is challenged to develop an articulated K-12 curriculum in mathematics, science, and technology that will position students to achieve a passing standard at the elementary and intermediate levels, such that success is maximized at the commencement level.

The addition of factual content must be accomplished without contradicting the central philosophy of the learning standards. If local curricula merely revert to the fact-filled syllabi of the past, then little will have been accomplished in the standards movement other than to add yet another layer of content and requirements on the heads of students. A balance must be struck between the desire to build students' ability to think and analyze and the desire to add to the content they are expected to master.

LABORATORY EXPERIENCE

A positive benefit of the reduction of factual detail in the Core Curriculum is that it should allow a more in-depth treatment of laboratory investigations to be planned and carried out than was possible under the previous syllabus. Laboratory experiences should be designed to address Standard 1 (inquiry techniques) but should also take into account Standards 2 (information systems), 6 (interconnectedness of content), and 7 (problem-solving approaches). They should also address the laboratory skills listed in Appendix A of the Core Curriculum.

In June 2004, the Living Environment Regents Examination was changed to include a new Part D. This part of the examination assesses student knowledge of and skills on any of four required laboratory experiences supplied to schools by the New York State Education Department. The specific laboratory experiences required in any year will vary according to a preset schedule (see chart on next page).

Part D is a new section of the Regents Examination. Questions on this section can be a combination of multiple-choice and constructed-response questions similar to those found in Parts A, B, and C of the Living Environment Regents Examination. The content of these questions reflect the four specific laboratory experiences required for a particular year. Teachers are strongly encouraged to include review of these laboratory experiences as part of their year-end Regents preparation activity.

The following chart summarizes the current laboratory requirement for New York State public schools:

Laboratory Title	Description
The Beaks of Finches	Explores the adaptive advantages of beaks with different physical characteristics
Relationships and Biodiversity	Explores the relationship between DNA structure and the biochemistry of inheritance
Making Connections	Explores the effects of physical activity on human metabolic activities
Diffusion Through a Membrane	Explores the nature of cross-membrane transport in living cells
Adaptations for Reproductive Success in Flowering Plants*	Not yet available
DNA Technology*	Not yet available
Environmental Conditions and Seed Germination*	Not yet available

*Not yet available as of this writing

Study Questions and Answers

QUESTIONS ON STANDARD 1—Scientific Inquiry: Students will use mathematical analysis, scientific inquiry, and engineering design, as appropriate, to pose questions, seek answers, and develop solutions.

Key Idea 1—Purpose of Scientific Inquiry: The central purpose of scientific inquiry is to develop explanations of natural phenomena in a continuing and creative process.

Performance Indicator	Description
1.1	The student should be able to elaborate on basic scientific and personal explanations of natural phenomena and develop extended visual models and mathematical formulations to represent one's thinking.
1.2	The student should be able to hone ideas through reasoning, library research, and discussion with others, including experts.
1.3	The student should be able to work toward reconciling competing explanations and clarify points of agreement and disagreement.
1.4	The student should be able to coordinate explanations at different levels of scale, points of focus, and degrees of complexity and specificity, and recognize the need for such alternative representations of the natural world.

Base your answers to questions 1 through 4 on the passage below and on your knowledge of biology.

To Tan or Not to Tan

Around 1870, scientists discovered that sunshine could kill bacteria. In 1903, Niels Finsen, an Icelandic researcher, won the Nobel Prize for his use of sunlight therapy against infectious diseases. Sunbathing then came into wide use as a treatment for tuberculosis, Hodgkin's disease (a form of cancer), and common wounds. The discovery of vitamin D, the "sunshine vitamin," reinforced the healthful image of the Sun. People learned that it was better to live in a sun-filled home than a dark dwelling. At that time, the relationship between skin cancer and exposure to the Sun was not known.

In the early twentieth century, many light-skinned people believed that a deep tan was a sign of good health. However, in the 1940s, the rate of skin cancer began to increase and reached significant proportions by the 1970s. At this time, scientists began to realize how damaging deep tans could really be.

Tanning occurs when ultraviolet radiation is absorbed by the skin, causing an increase in the activity of melanocytes, cells that produce the pigment melanin. As melanin is produced, it is absorbed by cells in the upper region of the skin, resulting in the formation of a tan. In reality, the skin is building up protection against damage caused by the ultraviolet radiation. It is interesting to note that people with naturally dark skin also produce additional melanin when their skin is exposed to sunlight.

Exposure to more sunlight means more damage to the cells of the skin. Research has shown that, although people usually do not get skin cancer as children, each time a child is exposed to the Sun without protection, the chance of that child getting skin cancer as an adult increases.

Knowledge connecting the Sun to skin cancer has greatly increased since the late 1800s. Currently, it is estimated that ultraviolet radiation is responsible for more than 90% of skin cancers. Yet, even with this knowledge, about 2 million Americans use tanning parlors that expose patrons to high doses of ultraviolet radiation. A recent survey showed that at least 10% of these people would continue to do so even if they knew for certain that it would give them skin cancer.

Many of the deaths due to this type of cancer can be prevented. The cure rate for skin cancer is almost 100% when it is treated early. Reducing exposure to harmful ultraviolet radiation helps to prevent it. During the past 15 years, scientists have tried to undo the tanning myth. If the word "healthy" is separated from the word "tan," maybe the occurrence of skin cancer will be reduced.

1. State *one* known benefit of daily exposure to the Sun. [1]

    _____

2. Explain what is meant by the phrase "the tanning myth." [1]

    _____

3. Which statement concerning tanning is correct?
    (1) Tanning causes a decrease in the ability of the skin to regulate body temperature.
    (2) Radiation from the Sun is the only radiation that causes tanning.
    (3) The production of melanin, which causes tanning, increases when skin cells are exposed to the Sun.
    (4) Melanocytes decrease their activity as exposure to the Sun increases, causing a protective coloration on the skin.   3_____

4. Which statement concerning ultraviolet radiation is *not* correct?
    (1) It may damage the skin.
    (2) It stimulates the skin to produce antibodies.
    (3) It is absorbed by the skin.
    (4) It may stimulate the skin to produce excess pigment.   4_____

5. Current knowledge concerning cells is a result of the investigations and observations of many scientists. The work of these scientists forms a well-accepted body of knowledge about cells. This body of knowledge is an example of a
    (1) hypothesis   (3) theory
    (2) controlled experiment   (4) research plan   5_____

6. In his theory of evolution, Lamarck suggested that organisms will develop and pass on to offspring variations that they need in order to survive in a particular environment. In a later theory of evolution, Darwin proposed that changing environmental conditions favor certain variations that promote the survival of organisms. Which statement is best illustrated by this information?
    (1) Scientific theories that have been changed are the only ones supported by scientists.
    (2) All scientific theories are subject to change and improvement.
    (3) Most scientific theories are the outcome of a single hypothesis.
    (4) Scientific theories are not subject to change.   6_____

Base your answers to questions 7 and 8 on the passage below and on your knowledge of biology.

The number in the parentheses ( ) at the end of a sentence is used to identify that sentence.

They Sure Do Look Like Dinosaurs

When making movies about dinosaurs, film producers often use ordinary lizards and enlarge their images many times (1). We all know, however, that although they look like dinosaurs and are related to dinosaurs, lizards are not actually dinosaurs (2).

Recently, some scientists have developed a hypothesis that challenges this view (3). These scientists believe that some dinosaurs were actually the same species as some modern lizard that had grown to unbelievable sizes (4). They think that such growth might be due to a special type of DNA called repetitive DNA, often referred to as "junk" DNA because scientists do not understand its functions (5). These scientists studied pumpkins that can reach sizes of nearly 1,000 pounds and found them to contain large amounts of repetitive DNA (6). Other pumpkins that grow to only a few ounces in weight have very little of this kind of DNA (7). In addition, cells that reproduce uncontrollably have almost always been found to contain large amounts of this DNA (8).

7. State *one* reason why scientists formerly thought of repetitive DNA as "junk." [1]

_____

_____

8. Write the number of the sentence that provides evidence supporting the hypothesis that increasing amounts of repetitive DNA are responsible for increased sizes of organisms. [1]

_____

ANSWERS EXPLAINED Questions 1–8

Standard 1, Key Idea 1, Performance Indicators 1–4

1. One response is required. Acceptable responses include:
* *Kills bacteria*
* *Produces vitamin D*
* *Treats diseases and/or wounds*

2. ·One response is required. Acceptable responses include:
*   *The "tanning myth" involves people believing that a tan is a sign of good health.*
*   *The "tanning myth" says that a good tan is good for people.*

3. **3** *The production of melanin, which causes tanning, increases when skin cells are exposed to the Sun* is the correct statement concerning tanning. Melanin is a dark pigment that is produced in specialized skin cells in response to ultraviolet radiation in sunlight or an artificial source. This information is found in the third paragraph of the passage.

WRONG CHOICES EXPLAINED:
(1) *Tanning causes a decrease in the ability of the skin to regulate body temperature* is not a correct statement concerning tanning. There is no known relationship between tanning and body temperature regulation.
(2) *Radiation from the Sun is the only radiation that causes tanning* is not a correct statement concerning tanning. Tanning can also occur when the skin is exposed to artificial sources of ultraviolet radiation. A reference is made in the fourth paragraph of the passage to "tanning parlors" where people can be exposed to artificial doses of ultraviolet radiation.
(4) *Melanocytes decrease their activity as exposure to the Sun increases, causing a protective coloration on the skin* is not a correct statement concerning tanning. Melanin is produced as a protective pigment that helps prevent deep penetration of ultraviolet radiation into the deep layers of the skin. When ultraviolet radiation is absorbed by melanocytes, their activity increases, not decreases.

4. **2** *It stimulates the skin to produce antibodies* is not a correct statement concerning ultraviolet radiation. There is no information in the passage relating to the production of antibodies as a result of absorption of ultraviolet radiation, and no known research indicates this type of relationship.

WRONG CHOICES EXPLAINED:
(1), (3), (4) *It may damage the skin, it is absorbed by the skin,* and *it may stimulate the skin to produce excess pigment* are all correct statements concerning ultraviolet radiation. Ultraviolet radiation is an invisible but extremely powerful form of electromagnetic radiation. It can penetrate unshielded living tissues and alter the genetic makeup of the cells it encounters. In humans, this radiation can cause the production of melanin from melanocytes; in extreme cases, it can stimulate the growth of skin cancer.

5. **3** The body of knowledge described in this question is an example of a *theory*. When scientists begin to study a phenomenon in nature, their first step is normally to investigate it through repeated observation and experimentation. As a result of the analysis of the large quantity of data gathered during this process, the scientists then formulate a theory ("well-accepted body of knowledge") that describes the phenomenon in a way that is consistent with the data.

WRONG CHOICES EXPLAINED:

(1) A *hypothesis* is not the body of knowledge described in this question. Scientists develop a hypothesis ("educated guess") around their preliminary observations concerning a natural phenomenon. The hypothesis may be proven accurate or inaccurate as a result of the experimentation used to test it. For this reason, a hypothesis cannot be considered a "well-accepted body of knowledge."

(2) A *controlled experiment* is not the body of knowledge described in this question. A controlled experiment is a scientific method used to test an experimental hypothesis. The data that results from a controlled experiment can be used to support the development of a "well-accepted body of knowledge," but it does not constitute that body of knowledge.

(4) A *research plan* is not the body of knowledge described in this question. A research plan may be a series of controlled experiments designed to test various aspects of a natural phenomenon. The data that results from the research plan can be used to support the development of a "well-accepted body of knowledge," but it does not constitute that body of knowledge.

6. **2** *All scientific theories are subject to change and improvement* is the statement best illustrated by the information given. Both Lamarck and Darwin developed their theories of evolution based on observations made and inferences drawn before there was a good understanding of the genetic basis of variation. Lamarck's earlier theory of "use and disuse" was disproven by later experiments of other scientists. Darwin's later theory of "natural selection," though much closer to the currently accepted scientific theory of evolution, has been modified and improved on by the work of later scientists who have had the benefit of modern-day research in genetics, paleontology, and other sciences.

WRONG CHOICES EXPLAINED:

(1) *Scientific theories that have been changed are the only ones supported by scientists* is not the statement best illustrated by the information given. Scientists generally support theories that have stood the test of good scientific research. A theory that has not changed, as long as it is still supported by such research, is generally supported by most scientists.

(3) *Most scientific theories are the outcome of a single hypothesis* is not the statement best illustrated by the information given. In fact, scientific theories are based on the results of many experiments that each contain their own independent hypotheses.

(4) *Scientific theories are not subject to change* is not the statement best illustrated by the information given. Scientists are constantly questioning and reevaluating scientific theories. It is likely that a vast majority of all scientific theories undergo at least some modification.

7.   One response is required that indicates a reason why scientists formerly thought of repetitive DNA as "junk." Acceptable responses include: [1]
  • *Scientists did not understand the function of repetitive DNA.*
  • *They didn't know what it did, and so they thought it was junk.*

8.   One credit is allowed for indicating that either sentence 6 or sentence 7 provides evidence supporting the hypothesis that increased amounts of repetitive DNA are responsible for increased sizes of organisms. These sentences give information about the results of scientific investigations that measured the amount of repetitive DNA in the cells of giant pumpkins and miniature pumpkins and found that giant pumpkins contain more of this kind of DNA than miniature pumpkins.

Key Idea 2—Methods of Scientific Inquiry: Beyond the use of reasoning and consensus, scientific inquiry involves the testing of proposed explanations involving the use of conventional techniques and procedures and usually requiring considerable ingenuity.

Performance Indicator	Description
2.1	The student should be able to devise ways of making observations to test proposed explanations.
2.2	The student should be able to refine research ideas through library investigations, including electronic information retrieval and reviews of literature, and through peer feedback obtained from review and discussion.
2.3	The student should be able to develop and present proposals including formal hypotheses to test explanations (i.e., predict what should be observed under specific conditions if the experiment is true).
2.4	The student should be able to carry out research for testing explanations, including selecting and developing techniques, acquiring and building apparatus, and recording observations as necessary.

Base your answers to questions 9 and 10 on the diagram below of the field of view of a light compound microscope and on your knowledge of microscopes.

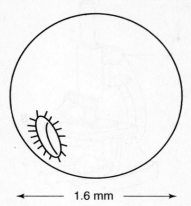

←——— 1.6 mm ———→

9. In order to center the organism in the field of view, the slide should be moved
(1) down and to the right
(2) down and to the left
(3) up and to the right
(4) up and to the left

9_____

10. The approximate length of the organism is
(1) 500 µm
(2) 1,600 µm
(3) 50 µm
(4) 1.6 µm

10_____

11. After viewing an organism under low power, a student switches to high power. The student should first
(1) adjust the mirror
(2) center the organism
(3) raise the objective and switch to high power
(4) close the diaphragm

11_____

12. Using one or more complete sentences, explain why a specimen viewed under the high-power objective of a microscope appears darker than when it is viewed under low power.

_____

_____

Base your answers to questions 13 and 14 on the diagram below of a compound light microscope.

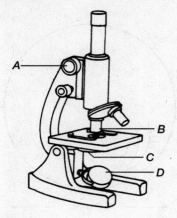

13. The letter *C* represents
(1) the mirror
(2) the diaphragm
(3) the eyepiece
(4) the high-power objective   13____

14. Select and name *one* of the labeled parts, and in one or more complete sentences describe its function.

_____

_____

15. The letter "p" as it normally appears in print is placed on the stage of a compound light microscope. Which best represents the image observed when a student looks through the microscope?
(1) p   (2) q   (3) b   (4) d   15____

16. To separate the parts of a cell by differences in density, a biologist would probably use
(1) a microdissection instrument
(2) an ultracentrifuge
(3) a compound light microscope
(4) an electron microscope   16____

17. The diagram below represents the field of view of a microscope. What is the approximate diameter, in micrometers, of the cell shown in the field?

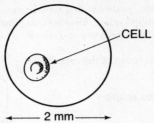

(1) 50 µm   (3) 1,000 µm
(2) 500 µm   (4) 2,000 µm   17 _____

18. Base your answer to the question on the diagram below.

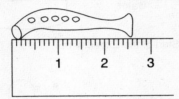

How many millimeters long is the organism resting on the metric ruler?

Questions 19 through 21 are based on the experiment described below.

A test tube was filled with a molasses solution, sealed with a membrane, and inverted into a beaker containing 200 mL of distilled water. A second test tube was filled with a starch solution, sealed with a membrane, and inverted into a beaker containing 200 mL of distilled water. After several hours, the water in each beaker was tested for the presence of molasses and starch.

The diagrams show the setup of the experiment.

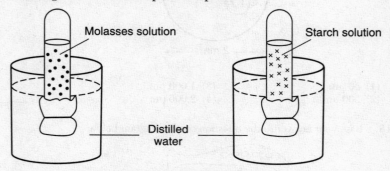

At the Start of the Experiment

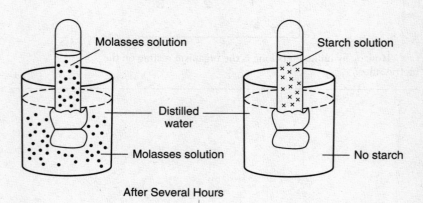

After Several Hours

Answer each question related to the experiment in one or more complete sentences.

19. What principle was being tested in the experiment?

20. What reagents were used in the experiment to test for the presence of molasses and starch?

_____

21. Draw one conclusion from this experiment.

_____

Questions 22 and 23 are based on the experiment described below.

An opaque disk was placed on several leaves of a geranium plant. The remaining leaves of the plant were untreated. After the plant had been exposed to sunlight, a leaf on which a disk had been placed was removed and tested as shown in parts *B* and *C* of the diagram below.

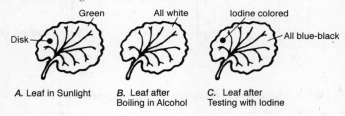

Green   All white   Iodine colored

Disk —   — All blue-black

A. Leaf in Sunlight   *B.* Leaf after   *C.* Leaf after
                           Boiling in Alcohol   Testing with Iodine

Answer each question related to the experiment in one or more complete sentences.

22. What conclusion can be drawn from the result of the experiment?

_____

23. What process was being investigated by the experiment?

_____

Questions 24 and 25 are based on the experiment described below.

A student added 15 mL of water to each of three test tubes, labeled A, B, and C. A 1-cc piece of raw potato was added to tube B. A 1-cc piece of cooked potato was added to tube C. Five drops of hydrogen peroxide (H_2O_2) were added to each test tube. The results are shown in the following diagram.

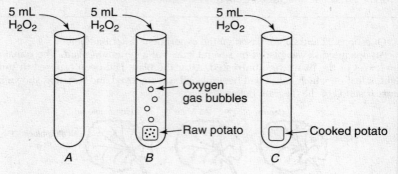

24. What conclusion can be drawn from the experiment?

_____

25. Which test tube is the control? Explain the reason for your choice.

_____

Base your answers to questions 26 through 28 on the diagram of the measuring device shown below.

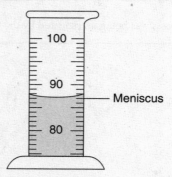

26. What is the name of this measuring device?

_____

27. In one complete sentence describe the procedure that you would follow to read the meniscus.

---

28. What must a student do to obtain a volume of 85 milliliters of liquid in this measuring device?
(1) Add 2.0 mL.   (3) Add 2.5 mL.
(2) Remove 2.0 mL.   (4) Remove 8.7 mL.   28 _____

ANSWERS EXPLAINED Questions 9–28

Standard 1, Key Idea 2, Performance Indicators 1–4

9. **2** Specimens viewed under the microscope appear upside-down, backward, and reversed.

WRONG CHOICES EXPLAINED:
(1), (3), (4) With any of these choices, the specimen would be moved out of the field of view.

10. **1** The field of view is given as 1.6 mm. 1 mm = 1000 μm. 1.6 mm × 1000 μm = 1600 μm. The diagram shows that three specimens would fit across the field of view. One-third of 1600 μm = 533 μm. Of the choices given, *500 μm* (choice 1) is closest to this value.

WRONG CHOICES EXPLAINED:
(2), (3), (4) Each of these choices is mathematically incorrect.

11. **2** The student should first *center the organism*. The field of view is smaller under high power; therefore, less of a specimen can be seen. If the organism is not centered, it may fall out of the field of view under high power.

WRONG CHOICES EXPLAINED:
(1) The *mirror is adjusted* for maximum light under low power. Because the diameter of the high-power objective is very small, it is impossible to adjust the light under high power.
(3) A compound light microscope is parfocal; that is, it is not necessary to *lift the high-power objective* to focus under high power. The specimen remains in focus when switching from low power to high power.
(4) *Closing the diaphragm* reduces the amount of light entering the objective. Therefore, the specimen would appear very dark and would be difficult to see.

12. *The diameter of the high-power objective is smaller than the diameter of the low-power objective. Less light enters through the high-power objective, and therefore the specimen appears darker.*

13. **2** The letter *C* represents *the diaphragm.*

WRONG CHOICES EXPLAINED:

(1) *The mirror* is represented by *D.*

(3), (4) *The eyepiece* and *the high-power objective* are not labeled on the diagram.

14. *Coarse adjustment (A)—used to focus a specimen under the low-power objective.*

or

Low-power objective (B)—along with the standard eyepiece, magnifies a specimen 100×.

or

Diaphragm (C)—regulates the amount of light entering the objectives.

or

Mirror (D)—provides a source of light that illuminates the specimen.

15. **4** The image of a specimen as seen under a microscope is upside-down (*d*). The right side is on the left side, and the top is on the bottom.

WRONG CHOICES EXPLAINED:

(1) In this choice (*p*) there is no change in the way the image of the letter appears.

(2) In this choice (*q*) the image of the letter is reversed in only one direction: The right and left sides are reversed.

(3) In this choice (*b*) the image of the letter is reversed in only one direction: The top and bottom are reversed.

16. **2** The *ultracentrifuge* is a machine that spins at a very high speed. A test tube of a liquid containing the parts of ruptured cells is placed in the machine. Each cell part has its own density (mass per unit volume). When the machine rotates, the cell parts fall to different levels in the test tube depending on their density.

WRONG CHOICES EXPLAINED:

(1) A *microdissection instrument* enables a biologist to remove a cell part from a single living cell. A micromanipulator is an example of such an instrument.

(3) A cell is transparent under a light microscope. Its structures cannot be seen unless the cell is stained. A *compound light microscope* can be used to view, but not to separate, cell parts.

(4)   *An electron microscope* uses beams of electrons to view freeze-dried specimens; it cannot be used to separate cell parts for study.

17.   **2**   Study the information given in the diagram. Notice that the diameter of the circle is 2 mm. Since 1 mm is equal to 1,000 µm, 2 mm are equal to 2,000 µm. In relation to the entire circle, how large is the cell? Is the cell one-half as large or one-fourth as large? Dividing the circle into four parts shows us that the diameter of the cell is about one-quarter the diameter of the circle. Dividing 4 into 2,000 results in *500 µm*.

WRONG CHOICES EXPLAINED:
(1) *50 µm* is too small. The cell is ten times larger than 50.
(3) *1,000 µm* is too large. The cell is not one-half the diameter.
(4) *2,000 µm* is the diameter of the circle. The cell is only one-fourth as large.

18.   The organism is *26 millimeters* long.

19.   The principle of *diffusion* was being tested in the experiment.

20.   *Benedict's solution* was used to test for the presence of molasses in the beaker. *Iodine* was used to test for the presence of starch in the beaker.

21.   *Molasses can diffuse through a membrane.*
or
      Starch cannot diffuse through a membrane.

22.   *No starch was produced in the area covered by the disk.*

23.   *The process of photosynthesis was being investigated.*

24.   *Raw potato contains an enzyme that breaks down hydrogen peroxide.*
or
      Cooking a potato destroys the enzyme that breaks down hydrogen peroxide.

25.   *Test tube A is the control.* A control is the part of the experiment that provides the basis of comparison for the variable being tested.

26.   The device is known as a *graduated cylinder*.

27.   *The meniscus should be read at eye level.*

28.   **2**   To obtain a volume of 85 mL, *2.0 mL* must be removed. The graduated cylinder contains 87 mL of liquid.

Key Idea 3—Analysis in Scientific Inquiry: The observations made while testing proposed explanations, when analyzed using conventional and invented methods, provide new insights into natural phenomena.

Performance Indicator	Description
3.1	The student should be able to use various methods of representing and organizing observations (e.g., diagrams, tables, charts, graphs, equations, matrices) and insightfully interpret the organized data.
3.2	The student should be able to apply statistical analysis techniques, when appropriate, to test if chance alone explains the results.
3.3	The student should be able to assess correspondence between the predicted result contained in the hypothesis and actual result, and reach a conclusion as to whether the explanation on which the prediction was based is supported.
3.4	The student should be able to, based on the results of the test and through public discussion, revise the explanation and contemplate additional research.
3.5	The student should be able to develop a written report for public scrutiny that describes the proposed explanation, including a literature review, the research carried out, its result, and suggestions for further research.

29. If curve *A* in the diagram represents a population of hawks in a community, what would most likely be represented by curve *B*?

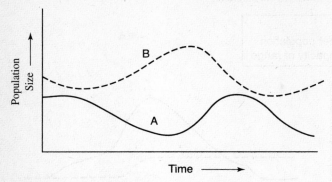

(1) the dominant trees in that community
(2) a population with which the hawks have a mutualistic relationship
(3) variations in the numbers of producers in that community
(4) a population on which the hawks prey   29 _____

Base your answers to questions 30 and 31 on this graph and on your knowledge of biology. The graph below depicts changes in the population growth rate of Kaibab deer.

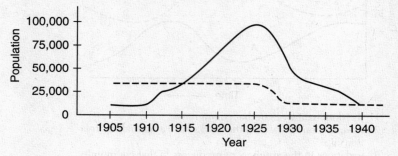

30. About how many deer could the range have supported in 1930 without some of them starving to death?
(1) 12,000   (3) 50,000
(2) 35,000   (4) 100,000   30____

31. In which year were the natural predators of the deer most likely being killed off faster than they could reproduce?
(1) 1905   (2) 1920   (3) 1930   (4) 1940   31____

32. Which process is illustrated by the diagram?

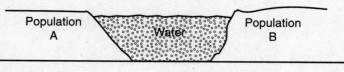

(1) migration   (3) speciation
(2) adaptive radiation   (4) isolation   32____

Base your answers to questions 33 through 36 on the information and data table below and on your knowledge of biology. The table shows the average systolic and diastolic blood pressure measured in millimeters of mercury (mm Hg) for humans between the ages of 2 and 14 years.

Data Table

Average Blood Pressure (mm Hg)		
Age	Systolic	Diastolic
2	100	60
6	101	64
10	110	72
14	119	76

DIRECTIONS (33–36): Using the information in the data table, construct a line graph on the grid provided, following the directions below.

33. Mark an appropriate scale on each labeled axis.

34. Plot the data for systolic blood pressure on your graph. Surround each point with a small triangle and connect the points.

35. Plot the data for diastolic blood pressure on your graph. Surround each point with a small circle and connect the points.

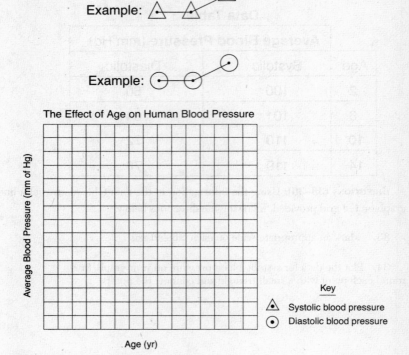

36. Using one or more complete sentences, state one conclusion that compares systolic blood pressure to diastolic blood pressure in humans between the ages of 2 and 14 years.

_____

_____

37. The graph below shows the results of an experiment.

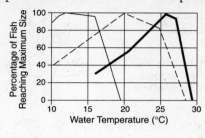

Key

```
——————  Brook trout
— — —   Northern pike
━━━━━   Largemouth bass
```

At 16°C, what percentage of the brook trout reached maximum size?

(1) 30%   (3) 75%

(2) 55%   (4) 95%   37_____

38. An experiment is represented in the diagram below.

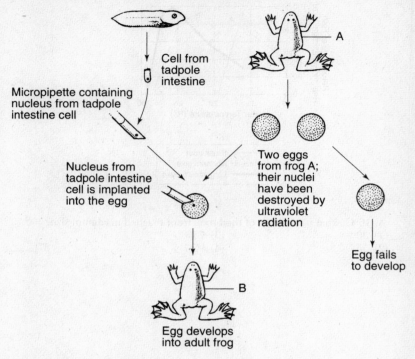

An inference that can be made from this experiment is that
(1) adult frog B will have the same genetic traits as the tadpole
(2) adult frog A can develop only from an egg and a sperm
(3) fertilization must occur in order for frog eggs to develop into adult frogs
(4) the nucleus of a body cell fails to function when transferred to other cell types

38_____

39. The charts below show the relationship of recommended weight to height in men and women age 25–29.

Height-Weight Charts

MEN Age 25–29 Weight (lb)				WOMEN Age 25–29 Weight (lb)			
Height Feet | Inches	Small Frame	Medium Frame	Large Frame	Height Feet | Inches	Small Frame	Medium Frame	Large Frame
5 2	128–134	131–141	138–150	4 10	102–111	109–121	118–131
5 3	130–136	133–143	140–153	4 11	103–113	111–123	120–134
5 4	132–138	135–145	142–156	5 0	104–115	113–126	122–137
5 5	134–140	137–148	144–160	5 1	106–118	115–129	125–140
5 6	136–142	139–151	146–164	5 2	108–121	118–132	128–143
5 7	138–145	142–154	149–168	5 3	111–124	121–135	131–147
5 8	140–148	145–157	152–172	5 4	114–127	124–138	134–151
5 9	142–151	148–160	155–176	5 5	117–130	127–141	137–155
5 10	144–154	151–163	158–180	5 6	120–133	130–144	140–159
5 11	146–157	154–166	161–184	5 7	123–136	133–147	143–163
6 0	149–160	157–170	164–188	5 8	126–139	136–150	146–167
6 1	152–164	160–174	168–192	5 9	129–142	139–153	149–170
6 2	155–168	164–178	172–197	5 10	132–145	142–156	152–173
6 3	158–172	167–182	176–202	5 11	135–148	145–159	155–176
6 4	162–176	171–187	181–207	6 0	138–151	148–162	158–179

The recommended weight for a 6'0" tall man with a small frame is closest to that of a

    (1) 5'10" man with a medium frame
    (2) 5'9" woman with a large frame
    (3) 6'0" man with a medium frame
    (4) 6'0" woman with a medium frame   39_____

Base your answers to questions 40 through 43 on the information below and on your knowledge of biology.

A group of biology students extracted the photosynthetic pigments from spinach leaves using the solvent acetone. A spectrophotometer was used to measure the percent absorption of six different wavelengths of light by the extracted pigments. The wavelengths of light were measured in units known as nanometers (nm). One nanometer is equal to one-billionth of a meter. The following data were collected:

> yellow light (585 nm)—25.8% absorption
> blue light (457 nm)—49.8% absorption
> orange light (616 nm)—32.1% absorption
> violet light (412 nm)—49.8% absorption
> red light (674 nm)—41.0% absorption
> green light (533 nm)—17.8% absorption

40. Complete all three columns in the data table below so that the wavelength of light either increases or decreases from the top to the bottom of the data table.

Color of Light	Wavelength of Light (nm)	Percent Absorption by Spinach Extract

DIRECTIONS (41–42): Using the information in the data table, construct a line graph on the grid provided, following the directions below.

41. Mark an appropriate scale on the axis labeled "Percent Absorption."

42. Plot the data from the data table on your graph. Surround each point with a small circle and connect the points.

Example: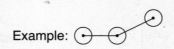

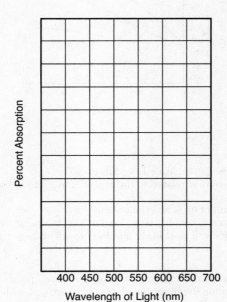

Percent Absorption

400 450 500 550 600 650 700

Wavelength of Light (nm)

43. Which statement is a valid conclusion that can be drawn from the data obtained in this investigation?
- (1) Photosynthetic pigments in spinach plants absorb blue light and violet light more efficiently than red light.
- (2) The data would be the same for all pigments in spinach plants.
- (3) Green light and yellow light are not absorbed by spinach plants.
- (4) All plants are efficient at absorbing violet light and red light. 43_____

44. The graph below represents the results of an investigation of the growth of three identical bacterial cultures incubated at different temperatures.

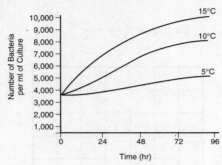

BIO/F126

Which inference can be made from this graph?
(1) Temperature is unrelated to the reproductive rate of bacteria.
(2) Bacteria cannot grow at a temperature of 5°C.
(3) Life activities in bacteria slow down at high temperatures.
(4) Refrigeration will most likely slow the growth of these bacteria.

44 _____

45. A study was conducted using two groups of ten plants of the same species. During the study, the plants were kept under identical environmental conditions. The plants in one group were given a growth solution every 3 days. The heights of the plants in both groups were recorded at the beginning of the study and at the end of a 3-week period. The data showed that the plants given the growth solution grew faster than those not given the solution.

When other researchers conduct this study to test the accuracy of the results, they should
(1) give growth solution to both groups
(2) make sure that the conditions are identical to those in the first study
(3) give an increased amount of light to both groups of plants
(4) double the amount of growth solution given to the first group

45 _____

46. Worker bees acting as scouts are able to communicate the distance of a food supply from the hive by performing a "waggle dance." The graph below shows the relationship between the distance of a food supply from the hive and the number of turns in the waggle dance every 15 seconds.

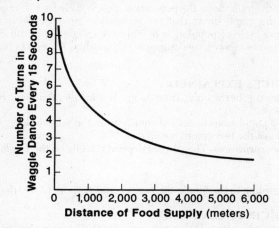

Using one or more complete sentences, state the relationship between the distance of the food supply from the hive and the number of turns a bee performs in the waggle dance every 15 seconds.

---

47. Based on experimental results, a biologist in a laboratory reports a new discovery. If the experimental results are valid, biologists in other laboratories should be able to perform
  (1) an experiment with a different variable and obtain the same results
  (2) the same experiment and obtain different results
  (3) the same experiment and obtain the same results
  (4) an experiment under different conditions and obtain the same results

47_____

ANSWERS EXPLAINED Questions 29–47

Standard 1, Key Idea 3, Performance Indicators 1–5

29. **4** The diagram shows the population growth cycle for two organisms. An examination of the graph shows that the population growth cycle of the hawks closely follows the cycle of population *B*. There is a slight lag in the cycles. This type of graph is used to show a predator-prey relationship. *The hawks prey on population B.*

WRONG CHOICES EXPLAINED:
    (1) Hawks are *not* herbivores. They do not live off the *dominant trees* in the community.
    (2) If the two populations benefited equally from the relationship (were *mutualistic*), the peaks of the two graphs would coincide.
    (3) Hawks are carnivores. They do not depend directly on the *producers* in the community.

30. **1** According to the graph, the range could support *12,000* deer in 1930.

WRONG CHOICES EXPLAINED:
    (2) The carrying capacity of the range was *35,000*. The *carrying capacity* is the maximum number of individuals that can be supported by the area. The number is usually constant unless severe environmental changes occur.
    (3) The actual number of deer occupying the range in 1930 was *50,000*.
    (4) A population of *100,000* deer was reached in 1925.

31. **2** In nature, a predator-prey relationship keeps the prey population in check. In *1920*, the population of deer increased. It was at this time that the predators were removed by human hunting.

WRONG CHOICES EXPLAINED:
    (1) The deer probably entered the region in *1905*. It takes time for an organism to adjust to a new environment.
    (3) In *1930*, the deer population, which had exceeded the carrying capacity, was declining. The decline was caused by starvation. The deer had consumed almost all of the available vegetation in the area.
    (4) By *1940*, the deer population reached the new carrying capacity of the range. The carrying capacity had been greatly reduced by overgrazing by the previously unchecked deer population.

32. **4** As the result of *isolation*, the members of populations A and B are separated and are prevented from interbreeding. Variations that occur in one area are not transmitted to the individuals in the other area. Consequently, over a long period of time, the genetic differences become accentuated, and the new variations are maintained.

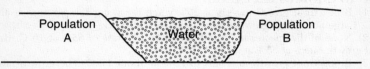

WRONG CHOICES EXPLAINED:

   (1) *Migration* is not possible when populations are separated by a geographical barrier such as water.

   (2) *Adaptive radiation* refers to a branching evolution and is not depicted in the diagram.

   (3) *Speciation* indicates formation of new species from a parent population. This is not indicated in the diagram.

33–35.

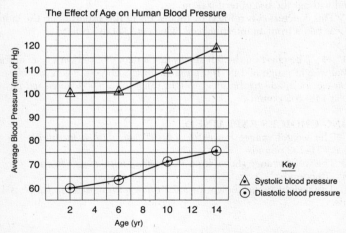

36.   *Systolic pressure is higher than diastolic pressure.* or *Both systolic pressure and diastolic pressure increase between the ages of 2 and 14.*

[*Note:* Other correct complete-sentence responses are acceptable.]

37.   **4**   Brook trout growth is represented by the lighter solid line on the graph. Tracing along the horizontal axis to 16°C and then tracing up to the point at which the brook trout line is encountered, we see that the growth rate is between 80% and 100% but closer to 100%; we estimate it is about 95%.

WRONG CHOICES EXPLAINED:

(1)   Brook trout growth rate is at *30%* when the water temperature is about 18°C. Largemouth bass growth rate is about 30% at 16°C.

(2)   None of the species indicated have a growth rate of 55% at 16°C. Brook trout growth rate is at 55% at about 17.5°C.

(3)   Brook trout growth rate falls to 75% at about 17°C. The growth rate of northern pike is at 75% at about 16°C.

38.   **1**   The diagram represents an experiment in cloning. Because the nucleus (which contains the genetic material) used in this experiment comes from a tadpole, the egg will produce new cells that have *genetic characteristics identical to those of the tadpole.*

WRONG CHOICES EXPLAINED:

(2)   It is unclear from this diagram how frog *A* came into being.

(3)   This conclusion is refuted by the results of this experiment; a new frog was created without the use of fertilization.

(4)   This conclusion is refuted by the results of this experiment; the tadpole nucleus was taken from an intestinal cell, which is a "body" cell.

39.   **4**   The chart on the left shows that a man 6'0" tall with a small frame has an ideal weight range of 149–160 pounds. Of the choices given, the closest comparison can be made to the ideal weight range of a 6'0" *woman with a medium frame* at 148–162 pounds.

WRONG CHOICES EXPLAINED:

(1)   The weight range shown for a *5'10" man with a medium frame* is 151–163 pounds.

(2)   The weight range shown for a *5'9" woman with a large frame* is 149–170 pounds.

(3)   The weight range shown for a 6'0" *man with a medium frame* is 157–170 pounds.

40.

Color of Light	Wavelength of Light (nm)	Percent Absorption by Spinach Extract
violet	412	49.8
blue	457	49.8
green	533	17.8
yellow	585	25.8
orange	616	32.1
red	674	41.0

41–42.

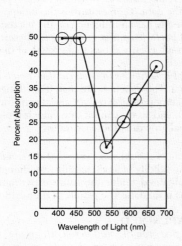

43.   **1**   The statement *photosynthetic pigments in spinach plants absorb blue light and violet light more efficiently than red light* is a valid conclusion that can be drawn from the data. The high point of the chart/graph data is clearly shown to be above the blue and violet wavelengths of light.

WRONG CHOICES EXPLAINED:

(2) The statement *the data would be the same for all pigments in spinach plants* is not supported by the results of this experiment. The chart/graph data show considerable variation in the experimental results as the wavelength of light varies.

(3) The statement *green light and yellow light are not absorbed by spinach plants* is not supported by the results of this experiment. Although the chart/graph data show a lower absorption rate at these wavelengths, there is still some absorption in this range.

(4) The statement *all plants are efficient at absorbing violet light and red light* is not supported by the results of this experiment. The experimental data are limited to the absorption of light by pigments found in one type of plant. These data cannot be extended to all plants unless all other types of plants are tested under the same experimental conditions and the results are found to be similar.

44. **4** Of those given, *refrigeration will most likely slow the growth of these bacteria* is the most reasonable inference that can be made from the graph data. The graph clearly shows a slower rate of growth (reproduction) at 5°C than at 10°C or 15°C.

WRONG CHOICES EXPLAINED:
(1) The inference that *temperature is unrelated to the reproductive rate of bacteria* is not supported by the data. Temperature is the independent (experimental) variable in this study. It clearly has an influence on the bacterial reproductive rate.

(2) The inference that *bacteria cannot grow at a temperature of 5°C* is not supported by the data. The graph clearly shows that growth at this temperature, while slow, occurs at a steady pace.

(3) The inference that *life activities in bacteria slow down at high temperatures* is not supported by the data. The data indicate that, if anything, bacterial activity increases with increasing temperature. No data are shown for bacterial growth at temperatures above 15°C, and so we cannot draw any inference about what happens to the rate of bacterial growth at these extremes.

45. **2** These researchers should *make sure that the conditions are identical to those in the first study.* The validity of any scientific experiment can be verified only if the same results are obtained under the same experimental conditions. Any change in these conditions invalidates the results of the verification study.

WRONG CHOICES EXPLAINED:
(1) If the researchers *give growth solution to both groups,* there will be no control group against which to compare the experimental group. The results of the verification study will be invalid because the experimental conditions will have been changed.

(3) If the researchers *give an increased amount of light to both groups of plants,* the original experimental method will be altered. The results of the verification study will be invalid because the experimental conditions will have been changed.

(4) If the researchers *double the amount of growth solution given to the first group,* the original experimental method will not be followed. The results of the verification study will be invalid because the experimental conditions will have been changed.

46. *The number of turns in the waggle dance decreases as the distance of the food supply from the hive increases.* Or *The closer to the hive the food source is located, the more turns there are in the waggle dance.* [*Note:* Any correct, complete-sentence answer is acceptable.]

47.   **3**   Other biologists in other laboratories should be able to perform *the same experiment and obtain the same results* if the experimental results are valid. Any experimental results obtained by one scientist must be validated through independent research by other scientists following the same procedures.

WRONG CHOICES EXPLAINED:
(1) If different scientists perform *an experiment with a different variable and obtain the same results,* the original experimental results will be invalidated.
(2), (4) If different scientists perform *the same experiment and obtain different results,* or *an experiment under different conditions and obtain the same results,* they will neither validate nor invalidate the results of the original experiment. All variables and conditions must be kept the same if the experimental results are to be properly tested.

QUESTIONS ON STANDARD 4—Biological Concepts: Students will understand and apply scientific concepts, principles, and theories pertaining to the physical setting and living environment and recognize the historical development of ideas in science.

Key Idea 1—Application of Scientific Principles: Living things are both similar to and different from each other and from nonliving things.

Performance Indicator	Description
1.1	The student should be able to explain how diversity of populations within ecosystems relates to the stability of ecosystems.
1.2	The student should be able to describe and explain the structures and functions of the human body at different organizational levels (e.g., systems, tissues, cells, organelles).
1.3	The student should be able to explain how a one-celled organism is able to function despite lacking the levels of organization present in more complex organisms.

48. In which life function is the potential energy of organic compounds converted to a form of stored energy that can be used by the cell?

    (1) transport   (3) excretion

    (2) respiration   (4) regulation   48_____

49. Which life activity is *not* required for the survival of an individual organism?

    (1) nutrition   (3) reproduction

    (2) respiration   (4) synthesis   49_____

50. Which function of human blood includes the other three?

    (1) transporting nutrients   (3) maintaining homeostasis

    (2) transporting oxygen   (4) collecting wastes   50_____

51. In the human body, the blood with the greatest concentration of oxygen is found in the

    (1) left atrium of the heart   (3) nephrons of the kidney

    (2) cerebrum of the brain   (4) lining of the intestine   51_____

52. Which type of vessel normally contains valves that prevent the backward flow of materials?

    (1) artery   (3) capillary

    (2) arteriole   (4) vein   52_____

DIRECTIONS (53–55): For each of questions 53 through 55, select the excretory structure, chosen from the list below, that best answers the question. Then record its number in the space provided at the right.

Excretory Structures

    (1) Alveolus

    (2) Nephron

    (3) Sweat gland

    (4) Liver

53. Which structure forms urine from water, urea, and salts?

_____

54. Which structure removes carbon dioxide and water from the blood?

_____

55. Which structure is involved in the breakdown of red blood cells?

_____

56. The bones of the lower arm are connected to the muscles of the upper arm by

    (1) ligaments   (3) cartilage

    (2) tendons   (4) skin   56_____

57. The diagram below shows the same type of molecules in area A and area B. With the passage of time, some molecules move from area A to area B.

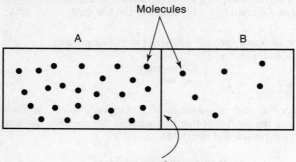

Molecules

A   B

Selectively Permeable Membrane

The movement is the result of the process of
(1) phagocytosis   (3) diffusion
(2) pinocytosis   (4) cyclosis   57_____

58. Which is the principal inorganic compound found in cytoplasm?
(1) lipid   (3) water
(2) carbohydrate   (4) nucleic acid   58_____

59. A specific organic compound contains only the elements carbon, hydrogen, and oxygen in a ratio of 1:2:1. This compound is most probably a
(1) nucleic acid   (3) protein
(2) carbohydrate   (4) lipid   59_____

60. Compared to ingested food molecules, end-product molecules of digestion are usually
(1) smaller and more soluble   (3) smaller and less soluble
(2) larger and more soluble   (4) larger and less soluble   60_____

61. The cellular function of the endoplasmic reticulum is to
(1) provide channels for the transport of materials
(2) convert urea to a form usable by the cell
(3) regulate all cell activities
(4) change light energy into chemical bond energy   61_____

62. In which organelles are polypeptide chains synthesized?
(1) nuclei   (3) ribosomes
(2) vacuoles   (4) cilia   62_____

63. Which organelle contains hereditary factors and controls most cell activities?
(1) nucleus   (3) vacuole
(2) cell membrane   (4) endoplasmic reticulum   63_____

64. Centrioles are cell structures involved primarily in
(1) cell division   (3) enzyme production
(2) storage of fats   (4) cellular respiration   64_____

65. Which cell structure contains respiratory enzymes?
(1) cell wall   (3) mitochondrion
(2) nucleolus   (4) vacuole   65_____

66. Which process is represented below?

enzymes
simple organic molecules → complex organic molecules + H_2O

(1) hydrolysis   (3) digestion
(2) synthesis   (4) respiration   66_____

67. Amino acids derived from the digestion of a piece of meat are transported to living cells of an animal. In the cell they are
(1) converted to cellulose
(2) used to attack invading bacteria
(3) synthesized into specific proteins
(4) incorporated into glycogen molecules   67_____

68. Which of the following variables has the *least* direct effect on the rate of a hydrolytic reaction regulated by enzymes?
(1) temperature   (3) carbon dioxide concen-
(2) pH   tration
                              (4) enzyme concentration   68_____

69. Which term refers to the chemical substance that aids in the transmission of the impulse through the area indicated by X?

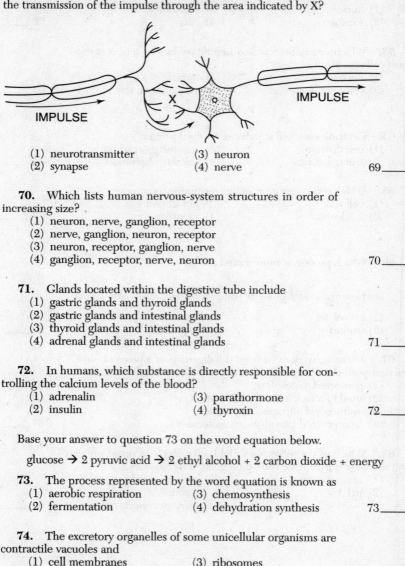

    (1) neurotransmitter   (3) neuron
    (2) synapse   (4) nerve   69_____

70. Which lists human nervous-system structures in order of increasing size?
    (1) neuron, nerve, ganglion, receptor
    (2) nerve, ganglion, neuron, receptor
    (3) neuron, receptor, ganglion, nerve
    (4) ganglion, receptor, nerve, neuron   70_____

71. Glands located within the digestive tube include
    (1) gastric glands and thyroid glands
    (2) gastric glands and intestinal glands
    (3) thyroid glands and intestinal glands
    (4) adrenal glands and intestinal glands   71_____

72. In humans, which substance is directly responsible for controlling the calcium levels of the blood?
    (1) adrenalin   (3) parathormone
    (2) insulin   (4) thyroxin   72_____

Base your answer to question 73 on the word equation below.

    glucose → 2 pyruvic acid → 2 ethyl alcohol + 2 carbon dioxide + energy

73. The process represented by the word equation is known as
    (1) aerobic respiration   (3) chemosynthesis
    (2) fermentation   (4) dehydration synthesis   73_____

74. The excretory organelles of some unicellular organisms are contractile vacuoles and
    (1) cell membranes   (3) ribosomes
    (2) cell walls   (4) centrioles   74_____

75. Which is a type of asexual reproduction that commonly occurs in many species of unicellular protists?
   (1) external fertilization   (3) binary fission
   (2) tissue regeneration   (4) vegetative propagation   75_____

ANSWERS EXPLAINED Questions 48–75

Standard 4, Key Idea 1, Performance Indicators 1–3

48. **2** *Respiration* is the life function by which ATP is made available to cells. Carbohydrate molecules are organic compounds. The breakdown of the carbohydrate molecules releases the energy stored in the bonds of the compounds. Potential energy is stored energy. The released potential energy is used to produce ATP.

WRONG CHOICES EXPLAINED:
(1) *Transport* is the life function by which materials are distributed throughout an organism.
(3) *Excretion* is the life function by which the wastes of metabolism are removed from an organism. Carbon dioxide, water, ammonia, and urea are metabolic wastes.
(4) *Regulation* is the life activity by which an organism responds to changes in its environment. The responses are controlled by the nervous system and the endocrine system.

49. **3** *Reproduction*, the life function through which a parent organism gives rise to offspring, is not necessary for the survival of the parent. Although reproduction is not required for the survival of an individual, it is necessary for the survival of a species. If a given species loses its potential for reproduction, it will become extinct.

WRONG CHOICES EXPLAINED:
(1) *Nutrition* is a collective term that refers to the biochemical processes by which cells extract nutrient molecules from food substances. The nutrients are used to build tissues, provide energy, and regulate the many biochemical activities that occur in cells. Without nutrition, cells die and, consequently, so do organisms. Each organism is dependent on adequate nutrition for survival.
(2) *Respiration* refers to the series of chemical changes that fuel molecules undergo to release chemical energy for cells. Respiration is necessary for the survival of the individual. Tissue cells cannot live without a means of obtaining chemical energy to power cellular activities such as active transport and metabolism. Of course, death of tissue cells means death of the individual.

(4) *Synthesis* occurs when small molecules are joined chemically to form large molecules. Enzymes, hormones, and body tissues are the results of syntheses, without which an individual organism cannot survive. Synthesis is a building-up process in which molecules vital to the life of the organism are produced.

50. **3** *Maintaining homeostasis* is the function of human blood that includes the other three. By transporting nutrients, oxygen, wastes, and other materials around the body, the blood helps to make essential materials available to every living body cell while removing potentially harmful materials from these tissues. Equal distribution of these materials helps to promote a steady state in the tissues essential to homeostatic balance.

WRONG CHOICES EXPLAINED:

(1), (2), (4) *Transporting nutrients, transporting oxygen,* and *collecting wastes* are all functions of the blood that are involved in maintaining homeostasis. Nutrients provide cells with dissolved food molecules. Oxygen is used by cells in the release of energy from these food molecules. Wastes such as urea and carbon dioxide are carried away from the cells for excretion into the environment.

51. **1** Blood that has just returned from the lungs has the greatest concentration of oxygen. The *left atrium of the heart* receives blood directly from the lungs.

WRONG CHOICES EXPLAINED:

(2) Brain tissue is one of the largest consumers of oxygen. The blood circulating in the *cerebellum* gives up most of its oxygen to the nerve cells.

(3) The largest concentration of metabolic wastes is found in the *nephrons*. The nephrons are filtering units in the kidney.

(4) The largest concentration of digested nutrients is found in the *lining of the intestine*. Absorption of nutrients occurs through the villi in the small intestine.

52. **4** *Veins* are blood vessels that carry blood to the heart. They contain valves that prevent the backflow of blood. The blood in veins is usually deoxygenated; the exception is the pulmonary vein in which the blood is rich in oxygen.

WRONG CHOICES EXPLAINED:

(1) *Arteries* are blood vessels that transport blood away from the heart. Arteries are rather thick-walled and pump blood in rhythm with the heart. They have no valves.

(2) Small arteries are *arterioles*. This type of blood vessel functions similarly to arteries. Arterioles lead into capillaries.

(3) *Capillaries* are the smallest blood vessels. They are one cell thick and permit diffusion of water, nutrients, gases, and other substances into and out of the bloodstream. Capillaries have no valves; they are the connecting vessels between arterioles and venules.

53. **2** The *nephron* is the unit of structure of the kidney. Each nephron has a glomerulus, Bowman's capsule, and kidney tubules. The kidney tubules filter out excess water, salts, and the wastes from protein metabolism. Urea and salts dissolved in water form urine.

54. **1** The *alveolus* is an air sac in the lung. It not only permits the diffusion of oxygen from the lungs into the bloodstream but also aids in the diffusion of carbon dioxide and water vapor out of the blood into the lungs.

55. **4** The *liver* is the largest gland in the body. One of its functions is to destroy old red blood cells and change the waste products into bile. The liver also synthesizes the anticoagulant known as heparin.

56. **2** *Tendons* are tough connective tissues made strong by fibers. Tendons connect muscles to bones. The movable joints function when muscles pull on tendons.

WRONG CHOICES EXPLAINED:

(1) *Ligaments* are strong connective tissues that contain elastic muscle fibers. Ligaments connect bone to bone.

(3) *Cartilage* is a supporting tissue that provides strength to body structures without rigidity. Cartilage supports structures such as the ears and nose and covers the ends of bones that form joints. The ground substance, or matrix, of cartilage is made of protein.

(4) *Skin* is composed of epithelial tissue. Skin serves as a body covering and has no function in the movement of bones or muscles.

57. **3** *Diffusion* is the process that results in the movement of molecules from a region of higher concentration (area A) to a region of lower concentration (area B). This net movement occurs until the concentrations of molecules have reached equilibrium between area A and area B.

WRONG CHOICES EXPLAINED:

(1), (2) *Phagocytosis* and *pinocytosis* are processes by which certain protists engulf their food and enclose it within a vacuole for digestion.

(4) *Cyclosis* refers to the streaming of cytoplasm in the cell, a simple form of intracellular transport.

58. **3** Inorganic compounds are compounds that do not contain carbon atoms. *Water,* the universal solvent, is the principal inorganic compound of cytoplasm. Water is the medium through which all chemical reactions take place in the cell.

WRONG CHOICES EXPLAINED:
(1), (2), (4) *Lipids, carbohydrates,* and *nucleic acids* are organic compounds. Organic compounds are carbon-containing compounds.

59. **2** Glucose is the building block of *carbohydrate* molecules. The ratio of carbon to hydrogen to oxygen is 1:2:1 in glucose and all reducing sugars. By dehydration synthesis, many glucose molecules form complex carbohydrates. However, the 1:2:1 ratio holds.

WRONG CHOICES EXPLAINED:
(1) A *nucleic acid* is composed of a phosphate group, a protein base, and a five-carbon sugar. DNA and RNA are nucleic acids. The CHO 1:2:1 ratio is not applicable.
(3) *Proteins* are built from amino acids, which, in addition to carbon, hydrogen, and oxygen, contain nitrogen. Some protein molecules also contain sulfur. The 1:2:1 ratio of elements does not apply to proteins because proteins are tissue builders whereas carbohydrates are fuel molecules.
(4) *Lipids* are fats and are composed of three fatty acid molecules and one glycerol molecule. The 1:2:1 ratio of carbon to hydrogen to oxygen does not apply to fats.

60. **1** The end products of digestion are usually *smaller and more soluble* than the ingested food molecules. Digestion makes available nutrient molecules that can diffuse across cell membranes and enter the cytoplasm of cells. Carbohydrates are broken down into glucose molecules. Fats are hydrolyzed into fatty acids and glycerol. Proteins are digested into their component amino acid molecules. Each of these end products of digestion is able to diffuse across cell membranes and enter into the biochemical activities of cells.

WRONG CHOICES EXPLAINED:
(2) Synthesis produces *larger molecules*. Larger molecules are more complex and are usually less soluble than smaller, simpler ones. Digestion results in smaller nutrient molecules.
(3) *Smaller molecules* are usually more soluble than larger ones. Digestion produces molecules that are more soluble than the complex nutrient molecules that were ingested.
(4) *Molecules derived from digestion* of ingested food are not larger than the molecules from which they came. Molecules produced by the digestion of complex carbohydrates, proteins, and fats are more soluble and are able to dissolve in water. Thus, these molecules can cross cell membranes.

61. **1** The endoplasmic reticulum is a network of membranes that extends throughout the cell. The membranes form channels that *provide for the movement of materials through the cell.*

WRONG CHOICES EXPLAINED:

(2) *Urea* is a metabolic waste. It is a poisonous nitrogen compound. Urea must be removed from the cells if an organism is to survive.

(3) The *nucleus* is the organelle in the cell that regulates all cellular activities.

(4) The *chloroplasts* are organelles in plant cells that contain the green pigment chlorophyll. Chloroplasts are necessary for the process of photosynthesis.

62. **3** Proteins are polypeptide chains. Proteins are synthesized in the *ribosomes*.

WRONG CHOICES EXPLAINED:

(1) The *nuclei* contain the genetic material carried in the chromosomes.

(2) *Vacuoles* are saclike organelles in the cytoplasm. Food vacuoles and contractile vacuoles are two common types of vacuoles.

(4) *Cilia* are microscopic hairs used for locomotion by some protozoans.

63. **1** The *nucleus* contains the hereditary factors. Nuclei of plant and animal cells house the chromosomes, which are composed of deoxyribonucleic acid. Molecules of DNA function as genes. Points on the chromosomes are genes. Genetic information is passed from parent to offspring by way of the genes. Chromosomes are part of the fine structure of the nucleus. DNA molecules contribute to the chemical structure. Genes are sites or points that dot the length of the chromosome. Genes, DNA molecules, and chromosomes function in passing along hereditary factors.

WRONG CHOICES EXPLAINED:

(2) The *cell membrane* encloses the contents of the cell and directs the flow of materials into and out of the cell. The cell membrane does not contribute to the passing of genetic material from one generation to the next. The function of the membrane is to control cellular transport.

(3) A *vacuole* is a fluid-filled space in the cytoplasm. Vacuoles help to regulate the internal pressure of the cell. The vacuoles in fat cells are filled with oil.

(4) The membranes that line the cytoplasmic canals within cells are known collectively as the *endoplasmic reticulum*. This cytoplasmic fine structure aids in the transport of molecules from the cell membrane to various sites within the cell. Neither the endoplasmic reticulum nor the vacuoles of the cell membrane contain hereditary structures.

64. **1** Centrioles are cell structures involved primarily in *cell division*. Centrioles are organelles that lie in the cytoplasm outside the nucleus; they are also found near the base of each flagellum and cilium. The centrioles of nonflagellated animal cells move to the spindle poles during cell division and seem to send out spindle fibers. The spindle fibers are attached to chromosomes and appear to pull the chromosomes from the center of the cell to the spindle poles.

WRONG CHOICES EXPLAINED:

(2) *Fats are stored* in cells. Fat in which the energy is channeled into heat production is stored in brown fat cells of hibernating mammals. At times, fat can be stored in arteries or accumulate around the heart. Fat cells are not involved in cell division.

(3) *Enzyme production* is controlled by the ribosomes that dot the membranes of the endoplasmic reticulum. Molecules of tRNA and mRNA regulate enzyme production.

(4) *Cellular respiration* is the process by which energy is released from glucose molecules. This process takes place in the mitochondria where oxygen is used as the final hydrogen carrier.

65.   **3**   Cellular respiration occurs in the mitochondria (plural of *mitochondrion*). Each step in the process of cellular respiration is regulated by enzymes. Respiratory enzymes are located in the mitochondria.

WRONG CHOICES EXPLAINED:

(1) The *cell wall* is composed of cellulose. Cell walls give shape and protection to plant cells.

(2) The *nucleolus* contains the materials needed for the synthesis of RNA. It is located in the nucleus of the cell.

(4) A *vacuole* is a rounded sac that serves as a storage place for food and waste products. Some vacuoles, such as contractile vacuoles, maintain a stable internal environment.

66.   **2**   *Synthesis* is the formation of complex molecules by combining simpler molecules. Water is removed from the simple molecules in this process.

WRONG CHOICES EXPLAINED:

(1) *Hydrolysis* is the addition of water to split complex molecules into simpler molecules. It is the opposite of synthesis.

(3) *Digestion* is another name for hydrolysis.

(4) *Respiration* is the process by which cells obtain energy. Glucose is converted to smaller molecules.

67.   **3**   Amino acids are the building blocks of proteins. The dehydration synthesis of amino acids *produces protein molecules*.

WRONG CHOICES EXPLAINED:

(1) *Cellulose* is a polysaccharide composed of hundreds of simple sugar molecules. The sugars were joined together by dehydration synthesis.

(2) Antibodies attack *invading bacteria*. Antibodies are protein molecules produced by special white blood cells.

(4) *Glycogen*, a polysaccharide, is a product of the dehydration synthesis of many glucose units.

68. **3** A *hydrolytic reaction* is a reaction in which a molecule is split. Enzymes are needed to speed up such a reaction. Any factor that affects the operation of the enzyme affects the speed at which the reaction takes place. The *concentration of carbon dioxide* has the least effect on enzyme activity.

WRONG CHOICES EXPLAINED:

(1) As the *temperature* is increased up to a point, the rate of the reaction increases. The increase in temperature increases the speed at which the enzyme and the substrate make contact with each other. The substrate is the molecule on which the enzyme acts. A very high temperature destroys the enzyme, and the reaction stops.

(2) Every enzyme works best at a particular *pH*. The enzymes in the stomach work in an acid environment, whereas the enzymes in the intestine work best in a basic medium.

(4) One molecule of an enzyme reacts with one molecule of a substrate. Increasing the *concentration of an enzyme* means that more substrate molecules will be acted on. The rate of the reaction will increase.

69. **1** A *neurotransmitter* is a chemical substance that is released by an impulse arriving at the terminal end of a neuron. The neurotransmitter diffuses across the synapse and stimulates the second nerve cell. Acetylcholine is an example of a neurotransmitter.

WRONG CHOICES EXPLAINED:

(2) A *synapse* is the space between the terminal end of one nerve cell and the dendrites of a second nerve cell. The area marked by an X in the diagram is a synapse.

(3) A *neuron* is a nerve cell that is specially adapted for the conduction of impulses.

(4) A *nerve* is made up of many neurons.

70. **1** A *neuron* is a single microscopic nerve cell. A *nerve* is composed of many nerve cells. A *ganglion* is a large mass of cell bodies of nerve cells; a ganglion functions as a coordinating center for impulses. A *receptor* is an organ specialized to receive environmental stimuli. The eye is an example of a receptor.

WRONG CHOICES EXPLAINED:

(2), (3), (4) In these three choices, either one or several structures are not arranged according to increasing size.

71. **2** *Gastric glands* are embedded in the walls of the stomach. They are duct glands that secrete gastric juice, a mixture of water, hydrochloric acid, rennin, and pepsin. Gastric juice begins the digestion of protein in the stomach. *Intestinal glands* are duct glands that line the walls of the small intestine. They secrete intestinal juice, a mixture of water, proteases, amylases, and lipases. Both types of glands lie within the digestive tube.

WRONG CHOICES EXPLAINED:

(1) Gastric glands are described above. *Thyroid glands* lie outside the digestive tract at the base of the neck, straddled across the larynx. The thyroid is a ductless gland that secretes the hormone known as thyroxin. Thyroid glands do not function in the biochemical process of digestion.

(3) *Thyroid glands* and intestinal glands are described above. Thyroxin controls the metabolism of cells. The explanation above shows why this choice is wrong.

(4) *Adrenal glands* are dual endocrine glands that lie on top of each kidney. They are not within the digestive tract. The adrenal medulla, the inner gland, secretes the hormone adrenaline, also known as epinephrine. This hormone enables the body to function in emergencies. The adrenal cortex secretes about six active hormones, including cortisone, the antiarthritis hormone.

72.   **3**   *Parathormone* is the hormone secreted by the parathyroid glands. The parathyroids are buried in the thyroids. Parathormone controls the level of calcium in the blood. Lack of blood calcium causes muscles to go into tetany. Tetany, or cramping, of the heart muscle causes death.

WRONG CHOICES EXPLAINED:

(1) *Adrenalin* is the hormone of the adrenal medulla, a ductless gland called the "gland of combat." Adrenaline stimulates the heart to beat faster, increases the rate of breathing, and controls the constriction and dilation of the arteriole walls.

(2) *Insulin* is secreted by the beta cells of the islets of Langerhans, which lie in the pancreas. Insulin controls sugar metabolism; specifically, it makes cell walls permeable to glucose and encourages the phosphorylation of fructose.

(4) *Thyroxin* is released by the thyroid gland. The rate of cellular metabolism is controlled by thyroxin. Iodine is used in the synthesis of thyroxin. People whose thyroid glands fail to develop become cretins; they are mentally retarded and physically undersized.

73.   **2**   Another name for anaerobic respiration is *fermentation*. In the process of fermentation, glucose is converted to energy, alcohol, and carbon dioxide.

WRONG CHOICES EXPLAINED:

(1) *Aerobic respiration* is another name for cellular respiration. This process requires oxygen. The following is an equation for aerobic respiration.

$$\text{glucose} + \text{oxygen} \rightarrow \text{pyruvic acid} \rightarrow \text{carbon dioxide} + \text{water} + \text{energy}$$

(3) *Chemosynthesis* is the synthesis of carbohydrates from inorganic compounds without the use of sunlight as a source of energy. Chemosynthesis is a form of autotrophic nutrition. It is carried out only by certain species of bacteria such as nitrifying bacteria.

(4) *Dehydration synthesis* is the method by which simple molecules are converted to complex molecules.

74. **1** *Cell membranes* and contractile vacuoles are excretory organelles of some unicellular organisms. The cell membrane is a selectively permeable membrane. It permits the diffusion of carbon dioxide and ammonia, two metabolic waste gases.

WRONG CHOICES EXPLAINED:

(2) *Cell walls* are composed of nonliving materials. Many canals penetrate through these walls, allowing the unrestricted passage of molecules.

(3) Proteins are synthesized in *ribosomes*.

(4) *Centrioles* are rodlike particles found in the centrosome. They function during the processes of mitosis and meiosis. Centrioles are found only in animal cells.

75. **3** A unicellular protist (e.g., an ameba) is composed of a single cell. When this cell divides by mitosis, the process is known as *binary fission*.

WRONG CHOICES EXPLAINED:

(1) *External fertilization* is an element of sexual reproduction in many aquatic multicellular species. Both the sexual nature of this process and the fact that it is carried out by multicellular animals eliminate this as a correct choice.

(2) *Tissue regeneration* implies a process that occurs in multicellular organisms.

(4) *Vegetative propagation* is a form of asexual reproduction common to certain species of multicellular plants; it cannot be carried out by unicellular protists.

Key Idea 2—Genetic Continuity: Organisms inherit genetic information in a variety of ways that result in continuity of structure and function between parents and offspring.

Performance Indicator	Description
2.1	The student should be able to explain how the structure and replication of genetic material result in offspring that resemble their parents.
2.2	The student should be able to describe and explain how the technology of genetic engineering allows humans to alter genetic makeup of organisms.

76. Corn plants grown in the dark will be white and usually much taller than genetically identical corn plants grown in light, which will be green and shorter. The most probable explanation for this is that the
  (1) corn plants grown in the dark were all mutants for color and height
  (2) expression of a gene may be dependent on the environment
  (3) plants grown in the dark will always be genetically albino
  (4) phenotype of a plant is independent of its genotype   76_____

77. In order for a substance to act as a carrier of hereditary information, it must be
  (1) easily destroyed by enzyme action
  (2) exactly the same in all organisms
  (3) present only in the nuclei of cells
  (4) copied during the process of mitosis   77_____

78. During synapsis in meiosis, portions of one chromosome may be exchanged for corresponding portions of its homologous chromosome. This process is known as
  (1) nondisjunction   (3) crossing-over
  (2) polyploidy   (4) hybridization   78_____

79. A DNA nucleotide is composed of three parts. These three parts may be
  (1) phosphate, adenine, and thymine
  (2) phosphate, deoxyribose, and thymine
  (3) phosphate, glucose, and cytosine
  (4) adenine, thymine, and cytosine   79_____

80. A double-stranded DNA molecule replicates as it unwinds and "unzips" along weak
  (1) hydrogen bonds   (3) phosphate groups
  (2) carbon bonds   (4) ribose groups   80_____

Base your answers to questions 81 through 84 on your knowledge of biology and the diagrams below. The diagram on the left represents a portion of a double-stranded DNA molecule. The diagrams at the right represent specific combinations of nitrogenous bases found in compounds transporting specific amino acids.

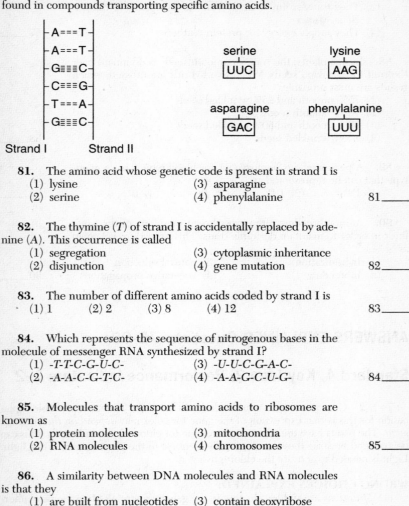

Strand I   Strand II

81. The amino acid whose genetic code is present in strand I is
(1) lysine   (3) asparagine
(2) serine   (4) phenylalanine   81_____

82. The thymine (*T*) of strand I is accidentally replaced by adenine (*A*). This occurrence is called
(1) segregation   (3) cytoplasmic inheritance
(2) disjunction   (4) gene mutation   82_____

83. The number of different amino acids coded by strand I is
(1) 1   (2) 2   (3) 8   (4) 12   83_____

84. Which represents the sequence of nitrogenous bases in the molecule of messenger RNA synthesized by strand I?
(1) -T-T-C-G-U-C-   (3) -U-U-C-G-A-C-
(2) -A-A-C-G-T-C-   (4) -A-A-G-C-U-G-   84_____

85. Molecules that transport amino acids to ribosomes are known as
(1) protein molecules   (3) mitochondria
(2) RNA molecules   (4) chromosomes   85_____

86. A similarity between DNA molecules and RNA molecules is that they
(1) are built from nucleotides   (3) contain deoxyribose
(2) are double-stranded   sugar
                                   (4) contain uracil   86_____

87. What is the function of DNA molecules in the synthesis of proteins?
(1) They catalyze the formation of peptide bonds.
(2) They determine the sequence of amino acids in a protein.
(3) They transfer amino acids from the cytoplasm to the nucleus.
(4) They supply energy for protein synthesis.   87____

88. In pea plants, the trait for smooth seeds is dominant over the trait for wrinkled seeds. When two hybrids are crossed, which results are most probable?
(1) 75% smooth and 25% wrinkled seeds
(2) 100% smooth seeds
(3) 50% smooth and 50% wrinkled seeds
(4) 100% wrinkled seeds   88____

89. A person who is homozygous for blood type A has a genotype that can be represented as
(1) I^aI^b (2) I^aI^a (3) I^ai (4) ii 89____

90. Animal breeders often cross breed members of the same litter in order to maintain desirable traits. This procedure is known as
(1) hybridization   (3) natural selection
(2) inbreeding   (4) vegetative propagation   90____

ANSWERS EXPLAINED Questions 76–90

Standard 4, Key Idea 2, Performance Indicators 1–2

76. **2** Corn plants grown in the dark will be white. The most probable explanation for this is that expression of the gene for color may *depend on the environment*. The plants have the genetic information for chlorophyll production. This can be assumed because they are genetically identical to the plants grown in the light. Light is needed to activate the chlorophyll gene.

WRONG CHOICES EXPLAINED:
(1) *Mutations* are sudden changes in the genetic material. Mutations are inherited. Because the plants grown in the dark were genetically identical to those grown in the light, neither group lacked the genetic information for chlorophyll production.

(3) *Albinism* is a condition resulting from the absence of a normal gene for color. Both groups of plants had the normal gene for color.

(4) The *phenotype* is the physical appearance of the organism. The phenotype depends on the genotype.

77. **4** The hereditary information is contained in the chromosomes. During mitosis, the *chromosomes duplicate*. The duplication of chromosomes ensures the equal distribution of identical genetic material to the new cells.

WRONG CHOICES EXPLAINED:

(1) If the hereditary information is *destroyed*, the cells cannot function. The chromosomes contain the information necessary for carrying out all cellular activities.

(2) *No two organisms are exactly alike.* No two organisms have the same hereditary material. Identical twins are the only exception to these statements.

(3) Plasmagenes are *genes located outside the nucleus*. Drug resistance in some bacteria is transmitted through plasmagenes.

78. **3** *Crossing-over* is the exchange of chromosomal material between homologous pairs of chromosomes. This process occurs during synapsis in meiosis.

WRONG CHOICES EXPLAINED:

(1) *Nondisjunction* is the failure of homologous chromosomes to separate from each other during meiosis. Cells with extra chromosomes and cells with too few chromosomes result from nondisjunction.

(2) *Polyploidy* is a condition in which the cells have extra sets of chromosomes beyond the normal 2n number.

(4) *Hybridization* is the crossing of two organisms that are distinctly different from each other. The purpose is to bring together new combinations of genes. Usually the individual with the new gene combinations is more sturdy than either parent. A tangelo is a cross between a tangerine and a grapefruit.

79. **2** A DNA nucleotide is composed of a *deoxyribose sugar molecule, a phosphoric acid molecule, and a nitrogen base*.

WRONG CHOICES EXPLAINED:

(1) *Adenine and thymine* are bases.

(3) *Glucose* is not the sugar molecule in DNA.

(4) *Adenine, thymine, and cytosine* are bases.

80. **1** The two strands of DNA are held together by *hydrogen bonds*. The hydrogen bonds form weak links between the base pairs of each strand.

WRONG CHOICES EXPLAINED:

(2) The sugars and bases of the nucleotides of DNA are organic compounds. Each individual compound is made up of *carbon bonds*.

(3) The nucleotides in each strand are joined together by *phosphate groups*.

(4) There are no *ribose groups* in DNA.

81. **1** *Lysine* is the amino acid whose genetic code is present on DNA strand I. Strand I of the DNA molecule contains two triplet codons (a triplet codon is a three-base sequence): AAG-CTG. Each triplet codon represents a specific amino acid. The amino acids are carried by tRNA molecules. The tRNA codon matches the DNA codon (except that U replaces T). There are two possible tRNA codons that could match the DNA strand I sequence; they are AAG and CUG. Of these, only the AAG tRNA appears in the diagram. The AAG tRNA carries the amino acid known as lysine.

DNA strand I	mRNA	tRNA	
A	U	A	
A	U	A	→ serine
G	C	G	
C	G	C	
T	A	U	→ unnamed amino
G	C	G	acid

WRONG CHOICES EXPLAINED:

(2) The DNA triplet code for *serine* is TTC. This codon does not appear on strand I of the DNA molecule in the diagram.

(3) The DNA triplet code for *asparagine* is GAC. This codon does not appear on strand I of the DNA molecule in the diagram.

(4) The DNA triplet code for *phenylalenine* is TTT. This codon does not appear on strand I of the DNA molecule in the diagram.

82. **4** A gene controls the production of a protein. The substitution of one base for another changes the triplet code. One amino acid will be substituted for another. The result is a *mutation*. The replacement of glutamic acid by valine in a hemoglobin molecule causes sickle cell anemia.

WRONG CHOICES EXPLAINED:

(1) *Segregation* is the separation of alleles from each other during the formation of gametes.

(2) *Disjunction* is the separation of homologous chromosomes during the process of meiosis.

(3) *Cytoplasmic inheritance* is the inheritance of genes located in the cytoplasm, not in the nucleus. The cytoplasmic genes are called plasmagenes.

83. **2** *Two* different amino acids are coded by strand I. Strand I has two triplet codes, six bases.

WRONG CHOICES EXPLAINED:
(1) Only three bases would have to be shown in the diagram to code for *1* amino acid.
(3) 24 bases are needed for *8* amino acids.
(4) 48 bases are needed for *12* amino acids.

84. **3** *UUCGAC* is the correct sequence. Base pairing is an important concept in DNA duplication and RNA synthesis. Adenine pairs with thymine; cytosine pairs with guanine. There is no thymine in RNA. Uracil takes its place.

WRONG CHOICES EXPLAINED:
(1) *TTCGUC* is not correct. Because thymine is present in the base sequences, the molecule cannot be RNA.
(2) *AACGTC* is not correct. The base sequences are not complementary to either strand I or strand II.
(4) *AAGCUG* is not correct. The base sequences are complementary to strand II in the diagram not strand I.

85. **2** Amino acids are transported to the ribosomes by *RNA molecules* known as transfer RNA, tRNA.

WRONG CHOICES EXPLAINED:
(1) *Protein molecules* are synthesized in the ribosomes. The code for the synthesis is contained in mRNA.
(3) Cellular respiration takes place in the *mitochondria*.
(4) *Chromosomes* are structures found in the nucleus. They are composed of DNA and protein. The genes are located on the chromosomes.

86. **1** Both DNA and RNA *are built from nucleotides*.

WRONG CHOICES EXPLAINED:
(2) Only DNA is *double-stranded*. RNA is single-stranded.
(3) *Deoxyribose* is the sugar in the DNA nucleotides. Ribose is the sugar in the RNA nucleotides.
(4) The base thymine is replaced by *uracil* in RNA nucleotides.

87. **2** The *sequence of amino acids* in a protein is determined by the triplet codes in DNA. The codes are carried to the ribosomes when mRNA is synthesized. A DNA strand is the template in mRNA synthesis.

WRONG CHOICES EXPLAINED:
(1) Enzymes *catalyze the formation of peptide bonds*. A peptide bond is a C–N bond formed by the dehydration synthesis of amino acids.

(3) *Amino acids are transferred* from the cytoplasm to the ribosomes, not to the nucleus. Transfer RNA is the carrier molecule.

(4) ATP molecules *supply the energy for protein synthesis*.

88. **1** A *hybrid* is an individual that has two different alleles for a particular trait. The hybrid may be represented by the symbols *Ss*. The hybrid is smooth; the smooth trait is dominant over the wrinkled trait. When two hybrids are crossed, *75% of the offspring will have the smooth trait and 25% of the offspring will have the wrinkled trait.*

	S	s
S	SS	Ss
s	Ss	ss

WRONG CHOICES EXPLAINED:

(2), (3), (4) These choices are incorrect based on the information provided by the Punnett square.

89. **2** The term *homozygous* means *pure for the trait*. A homozygous individual has two identical alleles for a gene. There are three alleles for blood type: I^a, I^b, and i. The allele I^a produces a protein for blood type A; the allele I^b produces a protein for blood type B; the allele i does not produce either protein. The type of blood is determined by the combination of alleles. The genotype refers to the allelic combination. Because the person in the question is homozygous for type A blood, his genotype is I^aI^a.

WRONG CHOICES EXPLAINED:

(1) The I^a and I^b alleles are both dominant over the i allele. When both I^a and I^b alleles occur in the same person, the person has type AB blood.

(3) The type of blood represented by the genotype I^ai is type A.

(4) The genotype of a person with type O blood is ii.

90. **2** The mating of members of the same litter to maintain desirable traits is known as *inbreeding*. Because mating pairs come from the same litter, they are genetically similar to each other. Inbreeding is used to maintain pure breeds.

WRONG CHOICES EXPLAINED:

(1) *Hybridization,* or outbreeding, is the mating of organisms with contrasting traits. It is the opposite of inbreeding.

(3) Factors in the environment select the organisms that are best adapted to survive in the environment. This principle is known as *natural selection*. It is an essential feature in the theory of evolution.

(4) *Vegetative propagation* is asexual reproduction in plants.

Key Idea 3—Organic Evolution: Individual organisms and species change over time.

Performance Indicator	Description
3.1	The student should be able to explain the major patterns of evolution.

91. In modern classification, protozoa and algae are known as molds, and bacteria are known as
    (1) bryophytes   (3) protists
    (2) plants   (4) animals   91_____

92. Most modern biologists agree that an ideal classification system should reflect
    (1) nutritional similarities among organisms
    (2) habitat requirements of like groups
    (3) distinctions between organisms based on size
    (4) evolutionary relationships among species   92_____

93. Which term includes the other three?
    (1) genus   (3) kingdom
    (2) species   (4) phylum   93_____

94. In one modern classification system, organisms are grouped into three
    (1) kingdoms   (3) genera
    (2) phyla   (4) species   94_____

95. Which is one basic assumption of the heterotroph hypothesis?
    (1) More complex organisms appeared before less complex organisms.
    (2) Living organisms did not appear until there was oxygen in the atmosphere.
    (3) Large autotrophic organisms appeared before small photosynthesizing organisms.
    (4) Autotrophic activity added molecular oxygen to the environment.   95_____

96. According to the heterotroph hypothesis, scientists believe that life arose in
    (1) a desert environment   (3) a vacuum
    (2) a forest environment   (4) an ocean environment   96_____

97. From an evolutionary standpoint, the greatest advantage of sexual reproduction is the
    (1) variety of organisms produced
    (2) appearance of similar traits generation after generation
    (3) continuity within a species
    (4) small number of offspring produced   97_____

98. According to modern theories of evolution, which of the following factors would be *least* effective in bringing about species changes?
    (1) geographic isolation   (3) genetic recombination
    (2) changing environments   (4) asexual reproduction   98_____

99. A factor that tends to cause species to change is a
    (1) stable environment   (3) recombination of genes
    (2) lack of migration   (4) decrease in mutations   99_____

100. If a fossil mammoth were discovered frozen in ice, its cells could be analyzed to determine whether its proteins were similar to those of the modern elephant. This type of investigation is known as comparative
    (1) anatomy   (3) biochemistry
    (2) embryology   (4) ecology   100_____

101. If members of the same species have been geographically isolated from each other for an extended period of time, which will they most likely exhibit?
    (1) mutations identical to each other
    (2) random recombination occurring in the same manner
    (3) evolution of traits of high adaptive value for their particular environments
    (4) evolution into two new species which will have no problem interbreeding   101_____

102. Skeletal similarities between two animals of different species are probably due to the fact that both species
    (1) live in the same environment
    (2) perform the same functions
    (3) are genetically related to a common ancestor
    (4) have survived until the present time   102_____

103. The best means of discovering if there is a close evolutionary relationship between animals is to compare

    (1) blood proteins   (3) foods consumed

    (2) use of forelimbs   (4) habitats occupied   103 _____

104. In the process of evolution, the effect of the environment is to

    (1) prevent the occurrence of mutations

    (2) act as a selective force on variations in species

    (3) provide conditions favorable for the formation of fossils

    (4) provide stable conditions favorable to the survival of all species   104 _____

105. In a stable population in which the gene frequencies have been constant for a long time, the rate of evolution

    (1) increases   (3) remains the same

    (2) decreases   (4) increases, then decreases   105 _____

106. Certain strains of bacteria that were susceptible to penicillin in the past have now become resistant. The probable explanation for this is that

    (1) the mutation rate must have increased naturally

    (2) the strains have become resistant because they needed to do so for survival

    (3) a mutation was retained and passed on to succeeding generations because it had high survival value

    (4) the principal forces influencing the pattern of survival in a population are isolation and mating   106 _____

107. The frequency of traits that presently offer high adaptive value to a population may *decrease* markedly in future generations if

    (1) conditions remain stable   (3) all organisms with these

    (2) the environment changes   traits survive

                            (4) mating remains random   107 _____

108. Since the publication of Darwin's theory, evolutionists have developed the concept that

    (1) a species produces more offspring than can possibly survive

    (2) the individuals that survive are those best fitted to the environment

    (3) through time, favorable variations are retained in a species

    (4) mutations are partially responsible for the variations within a species   108 _____

109. One factor that Darwin was unable to explain satisfactorily in his theory of evolution was
    (1) natural selection   (3) survival of the fittest
    (2) overproduction   (4) the source of variations   109 _____

ANSWERS EXPLAINED Questions 91–109

Standard 4, Key Idea 3, Performance Indicator 1

91. **3** The *protists* include all unicellular organisms and organisms that have both plant and animal features within one cell. Protozoa and algae are protists.

WRONG CHOICES EXPLAINED:
(1) *Bryophytes* are multicellular green plants that do not have vascular tissue. Mosses are examples of bryophytes.
(2) Multicellular photosynthetic organisms make up the *plant* kingdom, which includes both vascular and nonvascular plants.
(4) The *animal* kingdom is composed of multicellular organisms that cannot manufacture their own food. The organisms within this kingdom lack cell walls, and most are capable of some type of locomotion.

92. **4** A classification system should reflect *evolutionary relationships among species*. Evolutionary relationships are determined on the basis of the similarities in the anatomy, embryology, and biochemistry among organisms.

WRONG CHOICES EXPLAINED:
(1) All animals from protozoans to humans utilize the same nutrients in a similar manner. The process of photosynthesis is the same in tree cells and unicellular algae. *Nutritional similarities among organisms* are *not* useful in a system of classification.
(2) Both the whale and the fish live in an ocean environment. However, the whale is a mammal. Other than *sharing the same habitat*, the whale has no fish characteristics.
(3) Algae and protozoans are both microscopic organisms. However, algae have plant characteristics and protozoans have animal characteristics. *Size cannot be used* as the basis for a system of classification.

93. **3** According to the classification system, the largest grouping of organisms is the *kingdom*. Following this, the other groups, in order, are phylum, class, order, family, genus, species. Depending on its chief characteristics, an organism is placed in one of five kingdoms, Monera, Protist, Fungi, Animal, or Plant.

94. **1** In one modern classification system, organisms are grouped into three *kingdoms:* Animal, Plant, and Protist. Organisms that are not typical plants or animals are classified as protists. Examples are protozoa, slime molds, and bacteria.

WRONG CHOICES EXPLAINED:

(2) A *phylum* (plural phyla) is a large grouping that consists of classes, orders, families, genera, and species.

(3) A *genus* (plural genera) is a classification group composed of species. Members of a genus are more closely related than groups belonging to a given phylum.

(4) The *species* is the unit of classification. All members of a species are so closely related that they can mate and produce viable offspring.

95. **4** A heterotroph is an organism that must get its food from a source outside its own body cells; it cannot synthesize its food from inorganic materials. The heterotroph hypothesis proposes that the first living things on Earth were heterotrophs that obtained their food from the organic materials in the primitive seas. Autotrophs are organisms, such as green plants, that can synthesize their own food. At some stage in Earth's history, *autotrophic activity* used up the carbon dioxide in the air and, as a consequence of photosynthesis, *added molecular oxygen to the atmosphere.*

WRONG CHOICES EXPLAINED:

(1) Coacervates and then relatively *simple cells developed before more complex organisms.*

(2) Living heterotrophs *appeared before molecular oxygen was added to the atmosphere.*

(3) *Small cells that carried on photosynthesis appeared before the more complex vascular and seed plants.* The course of evolution is from the simple to the complex.

96. **4** According to the heterotroph hypothesis, life on earth evolved through a sequence of stages. The gases of the primitive atmosphere, such as methane, ammonia, and hydrogen, were washed down by heavy rains into the *early oceans.* They were acted on by ultraviolet radiation, cosmic rays, the earth's heat, and radioactivity. The bonding together of the molecules resulted in the formation of larger organic molecules.

WRONG CHOICES EXPLAINED:

(1) A *desert* environment could not support "first" life. The intense heat and the rapid evaporation of water are conditions that do not allow for the movement or maintenance of molecules in a fluid medium.

(2) A *forest* environment does not provide the pools of warm water on a continuous basis necessary for aggregate molecules to form.

(3) A *vacuum*, a place without air, cannot support life.

97.   **1**   Sexual reproduction helps to maximize the number of different allelic combinations that occur in offspring, leading to a greater *variety of organisms produced* within the species' population as a whole. Individuals displaying favorable traits in a changing environment are more likely to survive and to pass these traits on to their offspring, a fact that helps to promote evolutionary change.

WRONG CHOICES EXPLAINED:

(2), (3)   *Appearance of similar traits generation after generation* and *continuity within a species* describe conditions that promote stability and uniformity within species, both of which are maximized during asexual reproduction.

(4)   Evolutionarily speaking, the larger the number of offspring produced during reproduction, the more successful a particular species variety tends to be in competing with other varieties of the same species. *Small number of offspring produced,* therefore, is not an evolutionary advantage of sexual or any other type of reproduction.

98.   **4**   Variations among organisms are necessary for speciation. *Asexual reproduction* is least effective in bringing about changes in species. There are no variations among organisms that are reproduced asexually. These organisms are genetically like their parents.

WRONG CHOICES EXPLAINED:

(1)   *Geographic isolation* increases the chance that a group of organisms will develop a new gene pool, which will give rise to a new species.

(2)   *Changes in the environment* cause shifts in the gene pool. The genes that ensure the survival of organisms increase in the pool. Thus, the environment changes the characteristics of the original population, and a new species is formed.

(3)   *Genetic recombination* occurs during meiosis and fertilization. The shuffling of genes results in the appearance of new characteristics in a population.

99.   **3**   The *recombination of genes* is one factor that tends to cause species to change. Mutations are changes in genes. When like mutations combine in the fertilized egg, the new characteristic will be expressed in the offspring. If this mutation adds to the survival value of the organism, the gene change will be passed on to progeny because individuals having this mutation will live to reproduce.

WRONG CHOICES EXPLAINED:

(1)   A *stable environment* will probably not cause species to change. Beneficial mutations become effective in changing environments. For example, the mutations that produced white fur in the polar bear were beneficial. At one time, the polar regions were tropical. A change to a glacial environment was accompanied by a change or changes in the animal species that inhabited the region. Bears with a dark coat color became immediate targets for natural enemies.

(2)   *Migration* aids the recombination of genes because organisms have greater opportunities for interbreeding.

(4)   A *decrease in mutations* does not aid speciation but slows it.

100. **3** *Biochemistry* is the study of the chemistry of living organisms. Proteins are compounds found only in living organisms.

WRONG CHOICES EXPLAINED:

(1) *Anatomy* is the study of the structure of organisms.

(2) *Embryology* is the study of the development of embryos.

(4) *Ecology* is the study of the relationship of living organisms to each other and to their environment.

101. **3** Geographic isolation involves the separation of organisms by natural barriers. Each group of isolated individuals develops its own gene pool because each group lives under different environmental conditions. In the case of members of the same species who have been geographically isolated from each other, the *selection for individuals with special survival traits* is different in each environment.

WRONG CHOICES EXPLAINED:

(1) *Mutations* are the raw materials for evolution. Although mutations might have been the same in each group, they do not have the same adaptive value in each group.

(2) *Random recombination* occurs in the same manner in each group. However, the recombination process operates on two distinctly different gene pools.

(4) Usually, *different species cannot mate*. If their mating happens to be successful, their offspring will not be fertile.

102. **3** Morphology is the study of the structure and form of living things. When the arm of a human and the wing of a bird are studied, they are seen to have similar bone structure. This indicates that both organisms descended from a *common ancestor*. They have both undergone many changes since then and are now very different from each other. However, they still retain some of the same genes and therefore show a similarity in many parts of their bodies, including the arrangement of the bones in their forelimbs.

103. **1** A close evolutionary relationship between animals can be shown by a study of their *blood proteins*. The precipitin test is used to show such a relationship. A rabbit can be sensitized to human blood by being injected with human serum. When the sensitized rabbit serum is mixed with human serum, a white precipitate forms. If the sensitized rabbit serum is mixed with serum from a chicken, there is no reaction. However, a precipitate does form when the sensitized rabbit serum is mixed with the serum of a chimpanzee. In a like manner, the serum of a dog and a wolf show precipitation with serum sensitized to dog serum.

WRONG CHOICES EXPLAINED:

(2) The *forelimbs* of many unrelated animals may be used for the same purpose.

(3) Many unrelated animals *consume similar food*.

(4) Many unrelated animals *live in the same habitat*.

104. **2** There are many variations among the organisms of a species. Some variations allow an organism to survive best in a particular environment. Such factors as climate, food supply, and type of predators determine which organisms are *best adapted to that environment.*

WRONG CHOICES EXPLAINED:
(1) *Mutations* occur naturally and randomly. Mutations increase the variations among organisms.
(3) *Fossils* are the remains of organisms that lived in the past. They present evidence that evolution has occurred.
(4) There are many factors in an environment that influence the survival of different species. A *stable environment* preserves the species that have already adapted to that environment.

105. **3** Stable gene pools are a hallmark of nonevolving populations. The rate of evolution in such populations neither increases nor decreases, but *remain the same.*

WRONG CHOICES EXPLAINED:
(1), (2), and (4) all refer to changing rates of evolution that do not occur.

106. **3** Bacteria resistant to penicillin developed as a result of mutation. Organisms that did not receive the mutated gene were killed by the antibiotic. Those in which gene mutation occurred survived and *passed the mutation on to succeeding generations.*

WRONG CHOICES EXPLAINED:
(1) *The mutation rate did not increase.* The survivors had the mutated gene that allowed the bacteria to resist the effects of penicillin. These resistant strains reproduced, creating populations that replaced the nonresistant strains.
(2) *Need does not determine mutation.* Mutations are chance occurrences.
(4) Survival of a species depends on the ability of its members to obtain food, carry out respiration, and *reproduce successfully. Isolation* does not increase species survival.

107. **2** As long as environmental conditions remain stable, the alleles controlling traits that promote individual survival in that environment tend to be maintained at a high level in the population's gene pool. When *the environment changes,* however, the factors that promoted survival may no longer be present. Selection pressure may then operate to reduce the frequency of the once-prevailing alleles in favor of alleles controlling other, contrasting traits that increase individuals' chances for survival in the new environment.

WRONG CHOICES EXPLAINED:

(1) As long as *conditions remain stable,* selection pressures on individuals displaying favorable traits remain low, promoting the maintenance of a high frequency of alleles controlling these traits.

(3) If *all organisms with these traits survive,* their genes will be passed on to future generations at a high rate. This will help to maintain a high frequency of alleles controlling these traits in the population.

(4) If *mating remains random,* the probability that alleles will pair in unrestricted combinations will remain high, helping to ensure that the laws of probability will operate freely and that gene frequencies for existing traits will remain stable.

108.   **4**   Darwin proposed his theory of evolution in 1856. His theory did not explain how variation arose in organisms. In 1901, Hugo De Vries discovered the existence of mutations. *Mutations accounted for the rise of variations in organisms.*

WRONG CHOICES EXPLAINED:

(1) One of the principles of Darwin's theory of evolution stated that a species produced *more offspring than could possibly survive.*

(2) Another principle stated that the *individuals that survived were those best suited to the environment.*

(3) *The variations favored by the environment are retained within a species.* It is the environment that determines which variations are favorable.

109.   **4**   Darwin's theory of evolution was completed and published before Mendel completed his study of inheritance in the garden pea. Darwin could not explain *how variations occurred* or how they were passed on from parent to offspring.

WRONG CHOICES EXPLAINED:

(1) Darwin's theory of *natural selection* was divided into five distinct principles that formulated his concept of evolution. These ideas were set forth in his book *The Origin of the Species by Natural Selection.*

(2) *Overproduction* was one of the principles of Darwin's theory. He explained that for a species to continue in existence, it must overproduce in order to maintain the species number. For example, one female codfish lays about 9 million eggs. Not all of these eggs are fertilized and not all codfish lay about 9 million eggs. Not all of these eggs are fertilized and not all codfish fry reach adulthood. If the 9 million eggs per female were fertilized and if these zygotes developed into adult fish, the seas would be overrun with codfish. However, if the number of gametes produced by codfish were greatly reduced, the species would die out. This overproduction of gametes is necessary to maintain codfish survival.

(3) Another of Darwin's principles was *survival of the fittest.* No two organisms are alike; each has variations. These variations may either help or hinder the organism in its struggle for existence. An organism with variations that help it reach food faster is more fit and has a better potential for survival than a slower, less fit member of the species.

Key Idea 4—Reproductive Continuity: The continuity of life is sustained through reproduction and development.

Performance Indicator	Description
4.1	The student should be able to explain how organisms, including humans, reproduce their own kind.

110. Which occurs in a plant cell but *not* in an animal cell during mitotic cell division?

    (1) formation of spindle fibers   (3) formation of a cell plate
    (2) chromosome duplication   (4) cytoplasmic division   110_____

111. A plant cell with 12 chromosomes undergoes normal mitosis. What is the total number of chromosomes in each of the resulting daughter cells?

    (1) 24   (2) 12   (3) 6   (4) 4   111_____

112. Asexual reproduction *differs* from sexual reproduction in that, in asexual reproduction,

    (1) new organisms are usually genetically identical to the parent
    (2) the reproductive cycle involves the production of gametes
    (3) nuclei of sex cells fuse to form a zygote
    (4) offspring show much genetic variation   112_____

113. In most multicellular animals, meiotic cell division occurs in specialized organs known as

    (1) gonads   (3) kidneys
    (2) gametes   (4) cytoplasmic organelles   113_____

114. Which is an important adaptation for reproduction among land animals?

    (1) fertilization of gametes outside the body of the female
    (2) fertilization of gametes within the body of the female
    (3) production of sperm cells with thick cell walls
    (4) production of sperm cells with thin cell walls   114_____

115. In humans, a single primary sex cell may produce four gametes. These gametes are known as

    (1) diploid egg cells   (3) polar bodies
    (2) monoploid egg cells   (4) sperm cells   115_____

116. In sexual reproduction, the 2n chromosome number is restored as a direct result of
    (1) fertilization   (3) cleavage
    (2) gamete formation   (4) meiosis   116_____

117. In human females, the main function of the follicle-stimulating hormone (FSH) secreted by the pituitary gland is to
    (1) stimulate the adrenal glands to produce cortisone
    (2) stimulate activity in the ovaries
    (3) control the metabolism of calcium
    (4) regulate the rate of oxidation in the body   117_____

118. If the first stage of an uninterrupted human menstrual cycle is the follicle stage, the last stage includes the
    (1) formation of sperm   (3) buildup of the uterine
        cells in the testis   lining
    (2) release of a mature egg (4) shedding of the uterine
                                     lining   118_____

119. Which statement best describes internal fertilization?
    (1) It does not require motile gametes.
    (2) It helps to make terrestrial life possible.
    (3) It requires the presence of many eggs.
    (4) It normally occurs in the male.   119_____

120. What are the normal chromosome numbers of a sperm, egg, and zygote, respectively?
    (1) monoploid, monoploid, and monoploid
    (2) monoploid, diploid, and diploid
    (3) diploid, diploid, and diploid
    (4) monoploid, monoploid, and diploid   120_____

121. When compared with the number of gametes produced from a single primary sex cell during oogenesis, the number of gametes produced from a single human primary sex cell during spermatogenesis is usually
    (1) four times as great   (3) half as great
    (2) twice as great   (4) the same   121_____

122. In human males, sperm cells are suspended in a fluid medium. The main advantage gained from this adaptation is that the fluid

(1) removes polar bodies from the surface of the sperm
(2) activates the egg nucleus so that it begins to divide
(3) acts as a transport medium for sperm
(4) provides currents that propel the egg down the oviduct   122____

Base your answers to questions 123 and 124 on the diagrams and the information below.

A   **B**   **C**   **D**
BEAN   CHIMPANZEE   CHICKEN   AMEBA

123. Which organisms were produced as a result of fertilization?

(1) A, B, and C, only   (3) C and D, only
(2) B and C, only   (4) B, C, and D, only   123____

124. Structures that function in the storage of food to be used by growing embryonic cells are indicated by

(1) 1 and 3   (3) 2 and 4
(2) 2 and 3   (4) 3 and 4   124____

Base your answers to questions 125 through 127 on your knowledge of biology and the information below.

A biologist cut a flap of ectoderm from the top of a developing embryo. He did not remove the piece of ectoderm but just folded it back. Then he cut out the mesoderm underneath and completely removed it. He folded the flap of ectoderm back in place. The ectoderm healed; however, a complete nervous system did not develop.

125. This experiment was most likely performed immediately after

(1) cleavage   (3) fertilization
(2) gestation   (4) gastrulation   125____

126. This experiment interfered with the process of
(1) differentiation   (3) cleavage
(2) zygote formation   (4) ovulation   126_____

127. This experiment demonstrates that the
(1) ectoderm is solely responsible for development of the nervous system
(2) nervous system is destroyed during surgical operations
(3) mesoderm influences the development of the nervous system
(4) digestive enzymes have a major role in the development of embryonic layers   127_____

128. In a developing embryo, the mesoderm layer normally gives rise to
(1) epidermal tissue   (3) digestive tract lining
(2) skeletal tissue   (4) respiratory tract lining   128_____

129. What is the function of the placenta in a mammal?
(1) It surrounds the embryo and protects it from shock.
(2) It allows mixing of the maternal and fetal blood.
(3) It permits the passage of nutrients and oxygen from the mother to the fetus.
(4) It replaces the heart of the fetus until the fetus is born.   129_____

ANSWERS EXPLAINED Questions 110–129

Standard 4, Key Idea 4, Performance Indicator 1

110. **3** Mitosis is the process by which two identical nuclei are formed. Mitotic cell division is usually followed by cytoplasmic division. A plant cell has a rigid cell wall. Division of the cytoplasm begins with the appearance of a *cell plate* between the two nuclei. The cell plate is composed of membrane fragments from the endoplasmic reticulum.

WRONG CHOICES EXPLAINED:
(1) *Spindle fibers* are elastic-like protein fibers. Chromosome movement is controlled by spindle fibers.
(2) In order for two nuclei to be identical, they must have the same number and kind of chromosomes. The *chromosomes duplicate* before the nucleus divides. The mitotic process is the same in both plant and animal cells.

(4) *Cytoplasmic division*, or cytokinesis, usually follows nuclear division in both plant and animal cells.

111. **2** Chromosomes are structures in the nucleus. During mitosis, two cells with identical chromosomes are formed. Because the cell had 12 chromosomes, the daughter cells must also have *12* chromosomes.

WRONG CHOICES EXPLAINED:

(1) A cell with *24* chromosomes has twice the diploid number. The condition in which there are extra sets of chromosomes is known as polyploidy.

(3) A cell with *6* chromosomes has one-half the diploid number. Monoploid cells arise through meiosis.

(4) A cell with *4* chromosomes can only arise through a complete breakdown of the mitotic or meiotic process.

112. **1** In asexual reproduction, new organisms are produced by a single parent. Asexual reproduction involves the mitotic process. The *genetic material of the offspring is identical to that of the parent*.

WRONG CHOICES EXPLAINED:

(2) *Gametes* are produced by sexually reproducing organisms.

(3) The fusion of sex cells (gametes) to *form a zygote* is characteristic of sexually reproducing organisms.

(4) *Genetic variation* among offspring is characteristic of sexually reproducing organisms. The process of meiosis through synapsis and segregation ensures new combinations of genetic material.

113. **1** *Gonads* are sex glands. In these glands, gametes, or sex cells, are produced from primary sex cells that undergo meiosis, also known as reduction division. Male gonads are called testes, and female gonads, ovaries.

WRONG CHOICES EXPLAINED:

(2) *Gametes* are sex cells and not organs. Sex cells are specialized for fertilization.

(3) *Kidneys* are organs of excretion and are specialized for filtering metabolic wastes out of the blood. The nephron is the unit of structure and function in the kidney. Meiosis does not take place in kidney cells.

(4) *Cytoplasmic organelles* such as mitochondria, ribosomes, lysosomes, and endoplasmic reticula are not organs.

114. **2** A gamete is a reproductive cell that must fuse with another gamete to produce a new individual. Sperm cells and egg cells are gametes. Fertilization is the fusion of an egg cell and a sperm cell. In land animals, *fertilization occurs within the body of the female* and is known as internal fertilization.

WRONG CHOICES EXPLAINED:

(1) External fertilization, the *union of gametes outside the female's body*, occurs in animals that live in a watery environment. Fish and amphibians reproduce by external fertilization.

(3), (4) The question refers to reproduction in animals. Animal cells, including gametes, do not have *cell walls*.

115. **4** Primary sex cells give rise to gametes. Gametes are formed by the process of meiosis. In meiosis, a diploid cell divides twice to form four monoploid cells. In humans the four gametes, which are identical in size, are known as *sperm cells*.

WRONG CHOICES EXPLAINED:

(1) Chromosomes occur in pairs. The diploid number of chromosomes is the full number of chromosomes of all the pairs. Meiosis is cell division in which the nucleus receives one member of each pair of chromosomes. The nucleus of an *egg cell* thus contains half the diploid chromosome number, or the monoploid number.

(2) In formation of the egg cell, the cytoplasm does not divide equally. One large monoploid cell, the egg cell, and three very small cells (polar bodies) are produced from one primary sex cell.

(3) The three small monoploid cells accompanying the egg cell are known as *polar bodies*. Polar bodies degenerate and do *not* function in fertilization.

116. **1** The diploid chromosome number is represented as *2n,* and the monoploid number as *n*. When two gametes in the *n* condition combine, a *2n* cell is produced. *Fertilization* is the union of two gametes.

WRONG CHOICES EXPLAINED:

(2) *Gamete formation* reduces the chromosome number from *2n* to *n*.

(3) *Cleavage* is mitotic cell division without growth. It is the process by which a fertilized egg cell becomes a multicellular embryo.

(4) *Meiosis,* or reduction division, reduces the chromosome number of diploid cells.

117. **2** Follicles contain immature egg cells. The follicles are found in the ovary. FSH *stimulates ovarian* follicle development.

WRONG CHOICES EXPLAINED:

(1) ACTH stimulates the *adrenal glands* to produce cortisone. ACTH is secreted by the pituitary gland.

(3) *Calcium metabolism* is controlled by parathormone. The parathyroid gland secretes parathormone.

(4) Thyroxin secreted by the thyroid gland is the major regulator of the rate of *oxidation*. The hormones from the adrenal glands and the pancreas also play a role in the oxidation of glucose.

118.   **4**   The menstrual cycle is a series of changes that occur within the female reproductive system. The events of the cycle prepare the uterus to receive an embryo. The lining of the uterus is built up. If the cycle is not interrupted, the egg is not fertilized and no embryo is formed. In the last stage of the cycle, the *lining of the uterus disintegrates and is shed*.

WRONG CHOICES EXPLAINED:
(1) *Sperm cells* are produced by the male.
(2) Ovulation (*release of a mature egg*) occurs midway through the menstrual cycle.
(3) Once a month the *uterus* is prepared to receive an embryo. What happens to the lining of the uterus depends upon presence or absence of an embryo. If an embryo is present, the uterus continues to develop and the menstrual cycle is interrupted.

119.   **2**   *It helps to make terrestrial life possible* is the correct response. Internal fertilization, as its name implies, occurs within the body of the parent (usually female). The conditions in the female reproductive tract provide an ideal environment for the survival and pairing of gametes, helping to ensure that fertilization occurs successfully. This method of reproduction is especially helpful in the survival of terrestrial animal species, who live where harsh conditions (such as drying, heat, and cold) can easily damage or kill gametes released into the environment for external fertilization.

WRONG CHOICES EXPLAINED:
(1) *Motile gametes* (such as human sperm cells) are common in species employing both external and internal fertilization. Motility (ability to move) enables the sperm cells to swim toward the egg cell in either environment.
(3) Because of the dangers posed to fragile gametes in any environment, the *presence of many eggs* is characteristic of species employing external fertilization. Species using internal fertilization produce relatively few eggs in the reproductive process.
(4) In most species internal *fertilization occurs within the body of the female*, not that of the *male*.

120.   **4**   *Monoploid, monoploid, and diploid* is the correct combination. Sperm cells and egg cells are monoploid (*n*) gametes formed during the process of meiotic cell division. A zygote results from the fusion of two monoploid nuclei in fertilization and so must be diploid (*2n*) in chromosome number.

WRONG CHOICES EXPLAINED:
(1), (2), (3) Each of these distracters contains an incorrect combination of choices (see above).

121. **1** In the process of oogenesis, a single primary sex cell gives rise to a single monoploid egg cell and three nonfunctional monoploid polar bodies. The process of spermatogenesis yields four functional monoploid sperm cells for each primary sex cell. Therefore, a comparison of these two processes leads to the conclusion that, per primary sex cell, spermatogenesis yields *four times* as many gametes as oogenesis does.

WRONG CHOICES EXPLAINED:
  (2), (3), (4) Each of these distracters contains a mathematical comparison that is not consistent with the explanation above.

122. **3** The fluid surrounding human sperm cells *acts as a transport medium for sperm*. This fluid is known as semen. Its primary function is to provide a protective watery medium for sperm cells as they enter the female reproductive tract.

WRONG CHOICES EXPLAINED:
  (1) *Removes polar bodies from the surface of the sperm* is a "nonsense" distracter. Polar bodies are not associated with sperm production.
  (2) *Activates the egg nucleus so that it begins to divide* is not a function of semen. The egg is stimulated to divide by the act of fertilization. Semen is not directly involved in this process.
  (4) *Provides currents that propel the egg down the oviduct* is not a function of semen. Cilia that line the oviduct are responsible for establishing fluid currents that both carry the egg downward toward the uterus and carry sperm upward toward the ovary. Semen is not directly involved in this process.

123. **1** Fertilization is one of the processes in sexual reproduction. Sexual reproduction is the method of reproduction in the bean plant, chimpanzee, and chicken (*A, B, and C, only*). The ameba reproduces asexually by binary fission.

WRONG CHOICES EXPLAINED:
  (2) The bean plant (*B*) was omitted in this choice.
  (3) The ameba (*D*) is an incorrect answer. Both the bean (*A*) and the chimpanzee (*B*) were omitted in this choice.
  (4) The bean plant (*A*) was omitted and the ameba (*D*) is an incorrect answer.

124. **3** The structures that function in the storage of food for the embryonic cells are labeled *2* and *4*. Structure *2* is the cotyledon of the seed. The yolk *sac* is structure *4*.

WRONG CHOICES EXPLAINED:
  (1) Structure *3* refers to the wall of the uterus. The embryo of a chimpanzee is nourished through the placenta, not the uterine wall. Structure *1* is the leaf of the embryo bean plant.
  (2) Although structure *2* is a correct answer, structure *3* is incorrect.
  (4) Although structure *4* is a correct answer, structure *3* is incorrect.

125. **4** The experiment was performed after *gastrulation*. Gastrulation is a stage in embryonic development that gives rise to three germ layers of cells. The three germ layers are the ectoderm, mesoderm, and endoderm.

WRONG CHOICES EXPLAINED:
(1) *Cleavage* is a stage of embryonic development in which the zygote undergoes rapid mitotic divisions. The final result is a ball of cells.
(2) *Gestation* is a prebirth period. It is the time a developing embryo spends in the uterus.
(3) The union of a sperm cell nucleus with an egg cell nucleus is called *fertilization*. The result of the process is a zygote.

126. **1** The experiment interfered with the development of a nervous system. The development of special tissues and organisms is known as *differentiation*.

WRONG CHOICES EXPLAINED:
(2) *Zygote formation* must occur before an embryo can develop.
(3) The process of *cleavage* provides the embryo with hundreds of undifferentiated cells.
(4) *Ovulation* is the release of an egg from the ovary.

127. **3** The experiment demonstrates that the development of the *nervous system is influenced by the presence of the mesoderm*. The nervous system does not develop when the mesoderm is removed.

WRONG CHOICES EXPLAINED:
(1) If the *ectoderm was solely responsible for the development of the nervous system*, the nervous system would have developed after the mesoderm was removed.
(2) There was no *nervous system* present when the surgery was performed.
(4) The experiment was not concerned with the reasons for the *development of the embryonic layers*.

128. **2** Each of the three germ layers of the embryo is responsible for the development of the systems of the body. The *skeletal system* develops from the mesoderm. The muscle system, circulatory system, and excretory system also evolve from the mesoderm.

WRONG CHOICES EXPLAINED:
(1) *Epidermal cells* form the outer covering or skin of the body. The skin and nervous system develop from the ectoderm.
(3), (4) The linings of the *digestive* and *respiratory tracts* develop from the endoderm.

129.   **3**   The placenta is an area of spongy tissue in the uterus. It is very rich in blood vessels. The placenta functions as a respiratory and excretory organ of the fetus. *Vital materials are exchanged between the capillaries of the fetus and the capillaries of the mother.*

WRONG CHOICES EXPLAINED:

(1) The amnion is a fluid-filled sac surrounding the embryo. The fluid bathes the cells of the fetus and *protects it against shock.*

(2) The circulatory systems of the mother and the fetus are separate from each other. *Blood does not flow from one system into the other.*

(4) The *embryo develops its own heart.* The placenta provides an area for the diffusion of materials into and out of the fetus.

Key Idea 5—Dynamic Equilibrium and Homeostasis: Organisms maintain a dynamic equilibrium that sustains life.

Performance Indicator	Description
5.1	The student should be able to explain the basic bio-chemical processes in living organisms and their importance in maintaining dynamic equilibrium.
5.2	The student should be able to explain disease as a failure of homeostasis.
5.3	The student should be able to relate processes at the system level to the cellular level in order to explain dynamic equilibrium.

130.   Which energy conversion occurs in the process of photosynthesis?

(1) Light energy is converted to nuclear energy.
(2) Chemical bond energy is converted to nuclear energy.
(3) Light energy is converted to chemical bond energy.
(4) Mechanical energy is converted to light energy   130_____

131.   An environmental change that would most likely increase the rate of photosynthesis in a bean plant would be an increase in the
  (1) intensity of green light
  (2) concentration of nitrogen in the air
  (3) concentration of oxygen in the air
  (4) concentration of carbon dioxide in the air   131 _____

132.   During photosynthesis, molecules of oxygen are released as a result of the "splitting" of water molecules. This is a direct result of the
  (1) dark reaction   (3) formation of PGAL
  (2) light reaction   (4) formation of CO_2 132 _____

133.   An organism that makes its own food without the direct need for any light energy is known as a
  (1) chemosynthetic heterotroph   (3) photosynthetic het-
  (2) chemosynthetic autotroph   erotroph
                                  (4) photosynthetic autotroph   133 _____

134.   While looking through a microscope at a section of a leaf from a freshwater plant, a student observed some cells in which chloroplasts were moving around with the cytoplasm. This type of movement is known as
  (1) pinocytosis   (3) osmosis
  (2) synapsis   (4) cyclosis   134 _____

135.   By what process does carbon dioxide pass through the stomates into the leaf ?
  (1) diffusion   (3) respiration
  (2) osmosis   (4) pinocytosis   135 _____

136.   Two end products of aerobic respiration are
  (1) oxygen and alcohol   (3) carbon dioxide and water
  (2) oxygen and water   (4) carbon dioxide and oxy-
                                   gen   136 _____

137.   Homeostatic regulation of the body is made possible through the coordination of all body systems. This coordination is achieved mainly by
  (1) respiratory and   (3) nervous and endocrine
       reproductive systems   systems
  (2) skeletal and excretory (4) circulatory and digestive
       systems   systems   137 _____

138. Phenylketonuria (PKU) is an inherited condition characterized by mental retardation. The symptoms of the disorder result from an inability to synthesize a single type of
(1) enzyme   (3) blood cell
(2) nutrient   (4) brain cell   138_____

139. All the children of a hemophiliac male and a normal female are normal with respect to blood clotting. However, some of their grandsons are hemophiliacs. This is an example of the pattern of hereditary known as
(1) sex determination   (3) incomplete dominance
(2) sex linkage   (4) multiple alleles   139_____

140. What is the total number of chromosomes in a typical body cell of a person with Down's syndrome?
(1) 22   (2) 23   (3) 44   (4) 47   140_____

ANSWERS EXPLAINED Questions 130–140

Standard 4, Key Idea 5, Performance Indicators 1–3

130. **3** In the process of photosynthesis, carbon dioxide and water molecules are converted to glucose. Light is the energy source for the reaction. *Light energy is transformed into the chemical bond energy* of the glucose molecules.

WRONG CHOICES EXPLAINED:
(1), (2), (4) The Law of Conservation of Energy states that energy cannot be created or destroyed, but it can be changed from one form to another. All the choices refer to this law, but only choice (3) occurs in living organisms.

131. **4** Experiments have shown that the rate of photosynthesis depends on the availability of carbon dioxide. The greater the *concentration of carbon dioxide*, the greater the rate of photosynthesis.

WRONG CHOICES EXPLAINED:
(1) Chlorophyll is a light-absorbing pigment found in chloroplasts. However, it does not absorb much of the green wavelength of light. An increase in the *intensity of green light* has no effect on the photosynthetic rate.
(2) Carbohydrates are the products of photosynthesis. They do not contain atoms of nitrogen. The nitrogen needed by plants comes from nitrates in the soil, not from *the concentration of nitrogen in the air*.

(3) Oxygen is released during photosynthesis. Therefore, *the concentration of oxygen in the air* has no direct effect on photosynthesis.

132. **2**   Sunlight provides the energy needed to split water molecules into hydrogen and oxygen. Because light is required, this part of photosynthesis is known as the *light reaction.*

WRONG CHOICES EXPLAINED:
   (1) The *dark reaction* does not use light energy. In this reaction, the hydrogen released from the light reaction is combined with carbon dioxide.
   (3) *PGAL*, phosphoglyceric aldehyde, is the first stable compound formed during the dark reaction. This compound is later converted to glucose.
   (4) *Carbon dioxide is not formed* but is used during photosynthesis.

133. **2**   An autotroph manufactures its own food. A *chemosynthetic autotroph* produces its own food without the use of light energy. It obtains its energy from certain chemical reactions that take place in the cell.

WRONG CHOICES EXPLAINED:
   (1) A *chemosynthetic organism* cannot be a heterotroph. Heterotrophs do not have the ability to manufacture their own food. All animals are heterotrophs.
   (3) *Photosynthetic organisms* manufacture their own food. They cannot be heterotrophs.
   (4) A *photosynthetic autotroph* utilizes light energy. All green plants are photosynthetic autotrophs.

134. **4**   The movement of chloroplasts in the plant cell was due to the movement of cytoplasm in the cell. *Cyclosis* is the streaming of cytoplasm in a cell.

WRONG CHOICES EXPLAINED:
   (1) *Pinocytosis* is the formation of a pocket by an infolding of the cell membrane. Large molecules are brought into the cell by this process.
   (2) *Synapsis* is the pairing of homologous chromosomes during meiosis.
   (3) *Osmosis* is the movement of water across a selectively permeable membrane. Osmosis is the diffusion of water.

135. **1**   Carbon dioxide passes through the stomates of a leaf by *diffusion.* Diffusion is passive transport. Molecules move along a concentration gradient from an area of high density to an area of lower density.

WRONG CHOICES EXPLAINED:
   (2) *Osmosis* is the diffusion of water.
   (3) *Respiration* is an energy-releasing process.
   (4) *Pinocytosis* is active transport. Cells use energy to draw in large molecules by the infolding of their cell membranes.

136. **3** Respiration is a process by which cells release energy from glucose molecules. Aerobic respiration requires the presence of oxygen. In the process of aerobic respiration, glucose is oxidized to *carbon dioxide and water*. Both compounds are end products, the results of a chemical reaction.

WRONG CHOICES EXPLAINED:

(1), (2), (4) All three choices are incorrect because oxygen is consumed in aerobic respiration. Oxygen is not the end product of the reaction.

137. **3** Homeostasis refers to the steady state of control of the cell and, in turn, the entire body. The biochemical processes that take place in body cells occur in even and regular sequences. The cells, tissues, and organs in all body systems must function cooperatively so that the organism can carry out its life functions effectively. The coordination of all these biochemical activities is made possible by the work of the *nervous and endocrine systems*. The nervous system carries impulses from sense organs to the brain or spinal cord and then to effector organs such as muscles or glands. The endocrine system secretes hormones that control the functions of certain glands, tissues, and organs. Together the nervous and endocrine systems maintain the homeostasis of the body.

WRONG CHOICES EXPLAINED:

(1) The *respiratory system* is specialized for the intake and distribution of oxygen. It also expels waste gases. The *reproductive system* is specialized for the developing of embryos. Both systems are controlled by the nervous and endocrine systems. In mammals, hormones control gestation and birth.

(2) The *skeletal system* gives support to the body. Hormones control the growth of long bones. Nerve cell fibers help the muscles to function. The *excretory system* coordinates waste removal from the body. Hormones control water loss and reabsorption by the kidney tubules.

(4) Blood and lymph circulate by way of the *circulatory system*. Hydrolysis of food takes place in the *digestive system*. Both systems depend on the nervous and endocrine systems for the coordination of mechanical and biochemical activities.

138. **1** According to the one gene–one enzyme theory, a single gene is responsible for the production of a single enzyme. Because PKU is an inherited defect, the gene for an *enzyme* is absent in the victim.

WRONG CHOICES EXPLAINED:

(2) *Nutrients* are *not* synthesized by animals. Nutrients must be taken in from organic sources.

(3) *Blood cells* are *not* affected nor involved in the PKU disorder.

(4) The development of *brain cells* is affected by the PKU condition. Because of the lack of an enzyme, phenylalanine is converted to phenylpyruvic acid, which accumulates in brain tissue. The result is mental retardation.

139. **2** A trait that appears more often in one sex than in the other sex is said to be *sex-linked*. The gene for the trait is located on a sex chromosome. Hemophilia and color blindness are examples of sex-linked traits.

WRONG CHOICES EXPLAINED:

(1) *Sex determination* is controlled by a pair of sex chromosomes. There are two kinds of sex chromosomes, an X and a Y chromosome. A female has two X chromosomes (XX). A male has one X chromosome and one Y chromosome (XY).

(3) *Incomplete dominance* is a type of inheritance in which neither allele in a hybrid is dominant. The hybrid shows a trait completely different from either parent. The inheritance of color in a Japanese four-o'clock flower is an example of incomplete dominance, or blending.

(4) Alleles are different forms of the same gene. The term *multiple alleles* implies that a gene has more than two forms. Inheritance of human blood type involves multiple alleles.

140. **4** The normal number of chromosomes in human body cells is 46. Down's syndrome results from the presence of an extra chromosome. The extra or 47th chromosome is due to meiotic nondisjunction.

WRONG CHOICES EXPLAINED:

(1), (3) The numbers 22 and 44 do not apply to any known normal human chromosome number.

(2) The number 23 is the monoploid number of chromosomes in human gametes.

Key Idea 6—Interdependence of Living Things: Plants and animals depend on each other and their physical environment.

Performance Indicator	Description
6.1	The student should be able to explain factors that limit growth of individuals and populations.
6.2	The student should be able to explain the importance of preserving diversity of species and habitats.
6.3	The student should be able to explain how the living and nonliving environments change over time and respond to disturbances.

141.   Animals *cannot* synthesize nutrients from inorganic raw materials. Therefore, animals obtain their nutrients by
   (1) combining carbon dioxide with water
   (2) consuming preformed organic compounds
   (3) hydrolyzing large quantities of simple sugars
   (4) oxidizing inorganic molecules for energy   141_____

142.   Which organisms carry out heterotrophic nutrition?
   (1) ferns   (3) fungi
   (2) grasses   (4) mosses   142_____

143.   Which activity is an example of intracellular digestion?
   (1) a grasshopper chewing blades of grass
   (2) a maple tree converting starch to sugar in its roots
   (3) an earthworm digesting proteins in its intestine
   (4) a fungus digesting dead leaves   143_____

144.   A hydra ingests a daphnia, digests it, and later egests some materials. All these events are most closely associated with the life process known as
   (1) transport   (3) growth
   (2) synthesis   (4) nutrition   144_____

145.   Some bacteria are classified as saprophytes because they are organisms that
   (1) feed on other living things
   (2) feed on dead organic matter
   (3) manufacture food by photosynthesis
   (4) contain vascular bundles   145_____

146.   Aerobic organisms are dependent on autotrophs. One reason for this dependency is that most autotrophs provide the aerobic organisms with
   (1) oxygen   (3) nitrogen gas
   (2) carbon dioxide   (4) hydrogen   146_____

147.   Of the following, the greatest amount of the Earth's food production is thought to occur in
   (1) coastal ocean waters   (3) taiga forests
   (2) desert biomes   (4) tundra biomes   147_____

148.   Most of the minerals within an ecosystem are recycled and returned to the environment by the direct activities of organisms known as

(1) producers   (3) decomposers
(2) secondary consumers   (4) primary consumers   148_____

149.   Which type of organism is *not* shown in the following representation of a food chain?

grass → mouse → snake → hawk

(1) herbivore   (3) producer
(2) decomposer   (4) carnivore   149_____

150.   In the food chain shown below, which organism represents a primary consumer?

(1) grasshopper   (3) frog
(2) grass   (4) snake   150_____

151.   A lake contains minnows, mosquito larvae, sunfish, algae, and pike. Which of these organisms would probably be present in the largest number?

(1) minnows   (3) sunfish
(2) larvae   (4) algae   151_____

152.   An abiotic factor that affects the ability of pioneer organisms such as lichens to survive is the

(1) type of climax vegetation   (3) type of substratum
(2) species of algae   (4) species of bacteria   152_____

153.   In order to avoid predators, the clown fish hides unharmed in the stinging tentacles of the sea anemone. The clown fish attracts food to the sea anemone. This is an example of a type of relationship known as

(1) mutualism   (3) predator–prey
(2) commensalism   (4) parasitism   153_____

154. Which world biome has the greatest number of organisms?

(1) tundra

(2) tropical forest

(3) temperate deciduous forest

(4) marine   154 _____

155. In a particular area, living organisms and the nonliving environment function together as

(1) a population

(2) a community

(3) an ecosystem

(4) a species   155 _____

ANSWERS EXPLAINED Questions 141–155

Standard 4, Key Idea 6, Performance Indicators 1–3

141. **2** The portions of food that are usable to an animal are known as nutrients. Carbohydrates, lipids, and proteins are the organic nutrients needed by all organisms. Animals cannot synthesize their own nutrients. They must eat other organisms that contain the *preformed organic nutrients*.

WRONG CHOICES EXPLAINED:

(1) *Carbon dioxide and water* are inorganic compounds. These compounds are converted to nutrients by plants only. The process is called photosynthesis.

(3) Hydrolysis is the breakdown of compounds to simpler molecules through the action of enzymes in the presence of water. Glucose is a *simple sugar*. The hydrolyzing of glucose results in the release of energy in a cell.

(4) The use of *inorganic molecules* for the production of energy occurs only in certain species of bacteria. Chemosynthesis does not occur in members of the animal kingdom.

142. **3** *Fungi* are nongreen plants. They are heterotrophs, which means that they cannot manufacture their own food. Organisms that carry out heterotrophic nutrition must take in preformed organic molecules.

WRONG CHOICES EXPLAINED:

(1), (2), (4) *Ferns*, *grasses*, and *mosses* are green plants or bryophytes. Green plants are *autotrophs*, which means that they manufacture their own food.

143. **2** Digestion that occurs within a cell is known as intracellular digestion. Plants do not have special digestive systems. Digestion, or the *conversion of starch to sugar*, occurs within the individual cells of a plant, including those of the root.

WRONG CHOICES EXPLAINED:

(1) *Grasshoppers* have a digestive system. Digestion is extracellular and takes place in a digestive tube outside the body cells. Chewing a blade of grass is an example of mechanical digestion taking place in the mouth.

(3) *Earthworms* also have a digestive system. Proteins are digested outside the body cells in a portion of the digestive system known as the intestine.

(4) *Fungi* demonstrate a special form of extracellular digestion. Digestive enzymes are secreted into the external environment. The nutrients from the digested food diffuse into the cells.

144.　**4**　*Nutrition* is the life process most closely associated with a hydra ingesting, digesting, and egesting a daphnia. Ingestion is the process by which food materials are taken into the body of an organism such as a hydra. Digestion is the process by which the complex food molecules within the daphnia are hydrolyzed to soluble end products. Egestion is the process by which the undigestible materials of the daphnia's body are expelled from the body of the hydra.

WRONG CHOICES EXPLAINED:

(1) *Transport* is the life process by which soluble foods and other materials are circulated through the body of an organism such that they reach all parts of the organism's body.

(2) *Synthesis* is a process by which complex materials are constructed from simpler chemical components. The processes described in the question represent hydrolysis, the opposite of synthesis.

(3) *Growth* is a process that involves an increase in cell number and cell size, leading to an increase in the size of the organism. The end products of digestion can be used to supply raw materials for such growth.

145.　**2**　*Saprophytes* are organisms that *feed on dead organic matter*. Fungi and the bacteria of decay are examples of saprophytes.

WRONG CHOICES EXPLAINED:

(1) Heterotrophs *live on or off other living organisms*. A dog flea is an example of a heterotroph also known as a parasite.

(3) Autotrophs *manufacture their own food by photosynthesis*. Algae, mosses, and grasses are examples of autotrophs.

(4) Higher plants *contain vascular bundles*. These plants, called *tracheophytes*, include ferns, conifers, and flowering plants.

146.　**1**　Aerobic organisms need *oxygen* for cellular respiration. Some autotrophs are photosynthetic organisms. Oxygen is released by photosynthesis. Aerobic organisms depend on the autotrophs to release oxygen into the environment.

WRONG CHOICES EXPLAINED:

(2) *Carbon dioxide* is a waste product from the cellular respiration of aerobic organisms.

(3) *Nitrogen gas* makes up 78% of the atmosphere. However, the nitrogen cannot be used in the gaseous form by aerobic organisms and most autotrophs.

(4) *Hydrogen* does *not* exist as a gas on our planet. It is combined with other elements. The hydrogen needed by organisms comes mostly from water and organic compounds.

147. **1** The area of greatest food production is in the region where the greatest rate of photosynthesis occurs. The area must be rich in minerals, water, gases, and light. The *coastal ocean waters* meet these requirements.

WRONG CHOICES EXPLAINED:

(2) There is very little precipitation in the *desert*. Water is the factor that limits plant growth.

(3) The water in the *taiga* is frozen part of the year.

(4) The *tundra* is a frozen plain. Water is frozen almost all year long on the tundra.

148. **3** Bacteria of decay are *decomposers* that release minerals from decaying plant and animal bodies and return them to the environment.

WRONG CHOICES EXPLAINED:

(1) *Producers* are autotrophs, that is, green plants that synthesize food by photosynthesis from carbon dioxide and water in the presence of sunlight.

(2) *Secondary consumers* are animals that eat other animals. For example, a frog feeds on flies. Thus, a frog is a secondary consumer.

(4) *Primary consumers* are organisms that feed on plants only. Herbivores are primary consumers.

149. **2** *Decomposers* are not shown in the food chain. They are organisms that live on dead things. Fungi and bacteria are decomposers.

WRONG CHOICES EXPLAINED:

(1) An *herbivore* is a primary consumer; that is, it eats vegetation. The mouse is the herbivore in the food chain.

(3) A *producer* is a green plant. It depends on sunlight to synthesize its own food. The grass is the producer in the food chain.

(4) A *carnivore* is an animal that eats the flesh of other animals. Both the snake and the hawk are carnivores.

150. **1** The *grasshopper* is a primary consumer because it feeds on vegetation.

WRONG CHOICES EXPLAINED:

(2) *Grass* is an autotroph, that is, a producer or self-feeder. Grass is a green plant that can make its own food.

(3) A *frog* is a secondary consumer; it eats insects that are plant eaters.

(4) A *snake* is a secondary consumer; it eats animals that are primary consumers.

151.   **4**   The organisms in the question make up a food chain. The number of organisms at each level of the food chain decreases as one moves down the chain. The pyramid of energy shown below represents this fact. The producers, which form the base of the pyramid, are the most numerous. *Algae* are the producers in this food chain.

WRONG CHOICES EXPLAINED:

(1) The *minnows* are carnivores. They are the secondary consumers.

(2) The *larvae* are herbivores. They are primary consumers.

(3) The *sunfish* are tertiary consumers and occupy the top level of the pyramid. The organisms at the top of the pyramid are the least numerous.

152.   **3**   A *substratum* is the surface on which organisms grow. Lichens grow on rocks. Rocks are nonliving. The nonliving parts of the environment make up the abiotic factors.

WRONG CHOICES EXPLAINED:

(1) *Climax vegetation* is the type of vegetation that occupies an area in its final stage of succession. Plants make up the biotic, or living, environment.

(2) *Algae* are living organisms. They are part of the biotic (living) environment.

(4) *Bacteria* also make up the biotic environment.

153.   **1**   A relationship between two organisms in which both benefit from the association is known as *mutualism.* The clown fish is protected by the sea anemone. The sea anemone is sessile. The clown fish draws food to the sea anemone.

WRONG CHOICES EXPLAINED:

(2) *Commensalism* is a relationship between two organisms in which one organism is benefited by the association. The second organism is neither harmed nor benefited by the association. Barnacles attached to a whale are an example of a commensal relationship.

(3) A predator is a carnivore that hunts, kills, and eats its prey. The prey is the hunted organism. A *predator-prey* relationship is important in controlling the population of both.

(4) *Parasitism* is a relationship between two organisms in which one organism is benefited by the association whereas the second organism is harmed by the association. The parasitized organism is called the host.

154. **4** A biome is a large area dominated by one major type of vegetation and one type of climate. The *marine* biome has the greatest number of organisms.

WRONG CHOICES EXPLAINED:

(1) The *tundra* is a region where the ground is frozen all year long. Mosses and lichens are the dominant vegetation.

(2) The *tropical rain forest* is dominated by broadleaf plants. The region is always warm, and the rainfall is abundant and continuous.

(3) The *temperate deciduous forest* is made up of trees that shed their leaves once a year. The winters are cold, and the summers are warm. The rainfall is distributed throughout the year.

155. **3** *An ecosystem* is an area in which communities of living organisms interact with the nonliving environment.

WRONG CHOICES EXPLAINED:

(1) A *population* is all the organisms of a particular species living in a given area.

(2) A *community* is made up of populations of different species that interact with each other.

(4) A *species* is a group of organisms whose members are able to interbreed with each other. The offspring resulting from the matings are fertile and can reproduce.

Key Idea 7—Human Impact on the Environment: Human decisions and activities have a profound impact on the physical and living environment.

Performance Indicator	Description
7.1	The student should be able to describe the range of interrelationships of humans with the living and nonliving environment.
7.2	The student should be able to explain the impact of technological development and growth in the human population on the living and nonliving environment.
7.3	The student should be able to explain how individual choices and societal actions can contribute to improving the environment.

156. Human impact on the environment is most often more dramatic than the impact of most other living things because humans have a greater
    (1) need for water   (3) ability to adapt to change
    (2) need for food   (4) ability to alter the
                                      environment   156_____

157. Which human activity would have the most direct impact on the oxygen–carbon dioxide cycle?
    (1) reducing the rate of ecological succession
    (2) decreasing the use of water
    (3) destroying large forest areas
    (4) enforcing laws that prevent the use of leaded gasoline   157_____

158. Fertilizers used to improve lawns and gardens may interfere with the equilibrium of an ecosystem because they
    (1) cause mutations in all plants
    (2) cannot be absorbed by roots
    (3) can be carried into local water supplies
    (4) cause atmospheric pollution   158_____

159.   The tall wetland plant purple loosestrife was brought from Europe to the United States in the early 1800s as a garden plant. The plant's growth is now so widespread across the United States that it is crowding out a number of native plants. This situation is an example of

    (1) the results of the use of pesticides

    (2) the recycling of nutrients

    (3) the flow of energy present in all ecosystems

    (4) an unintended effect of adding a species to an ecosystem   159_____

160.   Choose *one* ecological problem from the list below.

Ecological Problems

Global warming

Destruction of the ozone shield

Loss of biodiversity

Discuss the ecological problem you chose. In your answer be sure to state:

- the problem you selected and *one* human action that may have caused the problem [1]
- *one* way in which the problem can negatively affect humans [1]
- *one* positive action that could be taken to reduce the problem [1]

_____

_____

_____

Base your answers to questions 161 through 163 on the information below and on your knowledge of biology.

The planning board of a community held a public hearing in response to complaints by residents concerning a waste-recycling plant. The residents claimed that the waste-hauling trucks were polluting air, land, and water and that the garbage has brought an increase in rats, mice, and pathogenic bacteria to the area. The residents insisted that the waste-recycling plant be closed permanently.

Other residents recognized the health risks but felt that the benefits of waste recycling outweighed the health issues.

161.   Identify two specific health problems that could result from living near the waste-recycling plant [2]

_____

162. Identify two specific health problems that could result from living near the waste-recycling plant. [2]

163. State one ecological benefit of recycling wastes. [1]

164. Which organism is a near-extinct species?
(1) Japanese beetle   (3) blue whale
(2) dodo bird   (4) passenger pigeon   164_____

165. Which human activity has probably contributed most to the acidification of lakes in the Adirondack region?
(1) passing environmental protection laws
(2) establishing reforestation projects in lumbered areas
(3) burning fossil fuels that produce air pollutants containing sulfur and nitrogen
(4) using pesticides for the control of insects that feed on trees   165_____

166. Compared to a natural forest, the wheat field of a farmer lacks
(1) heterotrophs   (3) autotrophs
(2) significant biodiversity   (4) stored energy   166_____

167. Which factor is not considered by ecologists when they evaluate the impact of human activities on an ecosystem?
(1) amount of energy released from the Sun
(2) quality of the atmosphere
(3) degree of biodiversity
(4) location of power plants   167_____

168. A new type of fuel gives off excessive amounts of smoke. Before this type of fuel is widely used, an ecologist would most likely want to know
(1) what effect the smoke will have on the environment
(2) how much it will cost to produce the fuel
(3) how long it will take to produce the fuel
(4) if the fuel will be widely accepted by consumers   168_____

169.   Which of the following is the most ecologically promising method of insect control?
- (1) interference with insect reproductive processes
- (2) stronger insecticides designed to kill higher percentages of insects
- (3) physical barriers to insect pests
- (4) draining marshes and other insect habitats   169_____

170.   Which is an example of biological control of a pest species?
- (1) DDT was used to destroy the red mite.
- (2) Most of the predators of a deer population were destroyed by humans.
- (3) Gypsy moth larvae (tree defoliators) are destroyed by beetle predators that were cultured and released.
- (4) Drugs were used in the control of certain pathogenic bacteria.   170_____

171.   To ensure environmental quality for the future, each individual should
- (1) acquire and apply knowledge of ecological principles
- (2) continue to take part in deforestation
- (3) use Earth's finite resources
- (4) add and take away organisms from ecosystems   171_____

172.   Ladybugs were introduced as predators into an agricultural area of the United States to reduce the number of aphids (pests that feed on grain crops). Describe the positive and negative effects of this method of pest control. Your response must include at least:

- two advantages of this method of pest control [2]

- two possible dangers of using this method of pest control [2]

_____

_____

_____

_____

_____

173. Some people claim that certain carnivores should be destroyed because they kill beneficial animals. Explain why these carnivores should be protected. Your answer must include information concerning:

- prey population growth [1]

- extinction [1]

- the importance of carnivores in an ecosystem [1]

_____

_____

_____

_____

ANSWERS EXPLAINED Questions 156–173

Standard 4, Key Idea 7, Performance Indicators 1–3

156.   **4**   The fact that humans have a greater *ability to alter the environment* means that human impact on the environment is often more dramatic than that of most other living things. In addition to our ability to make physical changes in the environment, humans have the unique ability to alter the environment chemically, introducing many materials that are not found in nature and that cannot be converted to useful products by nature.

WRONG CHOICES EXPLAINED:
(1), (2) On an individual basis, humans' *need for water* and *need for food* are not significantly greater than those of other living things. However, the fact is that our large population places incredible demands on the environment to supply these basic resources. As a result, our tendency to destroy natural habitats to create additional water and agricultural resources is a significant factor affecting the natural world.

(3) On an individual basis, humans' *ability to adapt to change* is not significantly greater than that of other living things. However, as a species, we have created artificial environments to protect ourselves from harsh environmental conditions. To the extent that these artificial environments are dependent on energy and other natural resources, their construction and maintenance have resulted in significant alterations of the natural world.

157.   **3**   *Destroying large forest areas* is the human activity that would have the most direct impact on the oxygen-carbon cycle. Reducing the number of trees over a large area would decrease the forest's ability to absorb carbon dioxide and water and convert them to atmospheric oxygen and glucose. The millions of leaves in a forest are capable of releasing many tons of oxygen gas to the atmosphere. The massive bodies of forest trees can likewise store tons of carbon in the form of complex carbohydrates such as cellulose.

WRONG CHOICES EXPLAINED:
(1) *Reducing the rate of ecological succession* is not the human activity that would have the most direct impact on the oxygen-carbon cycle. Ecological succession is a process by which one plant-animal community is replaced over time by other plant-animal communities until a stable climax community is established. Reducing its rate would only have the effect of prolonging each successive community longer than might otherwise be expected but would not directly alter the cycling of carbon and oxygen.

(2) *Decreasing the use of water* is not the human activity that would have the most direct impact on the oxygen-carbon cycle. Water is a precious resource in many parts of the world. Reducing water use so as to conserve it would represent a positive impact of human activity on the environment but would not directly alter the cycling of carbon and oxygen.

(4) *Enforcing laws that prevent the use of leaded gasoline* is not the human activity that would have the most direct impact on the oxygen-carbon cycle. Lead is a dangerous heavy metal pollutant released when leaded gasoline is burned. Enforcing laws that limit its use would represent a positive impact of human activity on the environment, but would not directly alter the cycling of carbon and oxygen.

158.   **3**   Fertilizers used to improve lawns and gardens may interfere with the equilibrium of an ecosystem because they *can be carried into local water supplies*. Once dissolved fertilizers enter streams, ponds, wetlands, or lakes, they provide an abundant nutrient source for the growth of algae. As masses of algae die off in the water environment, their decomposition can rob the water of oxygen needed for the survival of fish and other water-dwelling populations, causing their elimination from the habitat. When these species disappear, other species that depend on them for food must migrate or starve. Because the changes caused by the entry of fertilizers into water environments are so significant, it can be said that ecosystem equilibrium is destroyed.

WRONG CHOICES EXPLAINED:
(1) It is not true that fertilizers *cause mutations in all plants*. Some compounds with chemical structures similar to that of fertilizers are known to stimulate rapid gene mutation in plant cells that may lead to the death of the plant. However, the class of chemical compounds known as fertilizers do not have this effect on all plants.

(2) It is not true that fertilizers *cannot be absorbed by all plants*. When dissolved in water, fertilizers can easily enter plants by being absorbed via simple diffusion into root hairs.

(4) It is not normally true that fertilizers *cause atmospheric pollution*. Most fertilizers are relatively stable chemical compounds that are solids at normal temperatures. For this reason fertilizers are not normally responsible for atmospheric pollution unless they are applied in a gaseous form (such as ammonia) or become airborne (when attached to dry soil particles).

159. **4** The situation described in the question is an example of *an unintended effect of adding a species to an ecosystem*. Although purple loosestrife has adapted well to North American habitats, its rapid growth in wetland environments has stressed or eliminated populations of cattail, pickerelweed, and other native plant species. The introduction of nonnative purple loosestrife to the North American continent has had an unintended negative effect on these native species and on the balance of nature established over many centuries.

WRONG CHOICES EXPLAINED:
(1), (2), (3) The situation described in the question is not an example of *the results of the use of pesticides*, *the recycling of nutrients*, or *the flow of energy present in all ecosystems*. The introduction of a nonnative plant (a living thing) is not the same as the introduction of a chemical pesticide, the recycling of nutrients, or the flow of energy (nonliving things).

160. A three-part response is required that must include the following points:
• One human activity that may have caused the ecological problem selected from the list [1]
• One way the problem may negatively affect humans [1]
• One positive action that could be taken to reduce the problem [1]

Note: No credit is awarded for discussing an ecological problem not on the list.

Acceptable responses include: [3]
• *Global warming is a worldwide ecological problem that may be caused by the release of carbon dioxide and other gases in automobile exhaust. [1] This problem may negatively affect humans if the warming conditions disrupt weather patterns and lead to droughts, floods, or other natural disasters. [1] One positive action that could be taken to help the problem would be to find an energy source for automobiles that would not release carbon dioxide into the atmosphere. [1]*
• *An ecological problem affecting humans is destruction of the ozone layer, which is caused by the use of chemicals known as CFCs as propellants in aerosol sprays. [1] This is a problem for*

humans because the ozone layer protects us from ultraviolet radiation from the sun; without this protection we would have an increased chance of getting skin cancer. [1] A way to help solve this problem would be to ban the use of CFCs in aerosols. [1]

- *Loss of biodiversity is an ecological problem that negatively impacts humans. This problem is caused whenever humans destroy a natural habitat and convert it to other uses. [1] The overall health of our environment depends on the diversity of species that fill different roles in nature. When species diversity and environmental health are reduced, our health is threatened as well. [1] This problem can be reversed only if we use education to learn that protecting natural species is just as important as protecting our own. [1]*

161. Two responses are required. Acceptable responses include:

- *Asthma*
- *Respiratory infections*
- *Allergic reactions*
- *Cancer*
- *Bacterial infections*
- *Viral infections*
- *Disease linked to a pathogen*
- *Poisoning linked to toxic contamination of groundwater*

162. One response is required. Acceptable responses include:

- *Particles in the air*
- *Presence of viruses or bacteria on trucks*
- *Chemicals in air or water*
- *Carcinogens*
- *Mold and fungus spores*

163. One response is required. Acceptable responses include:

- *Conservation of natural resources*
- *Protection of finite resources*
- *Energy conservation*
- *Reduction in pollution*
- *Landfills last longer*
- *Preservation of open space resources*

164. **3** The blue whale is near extinction because of uncontrolled hunting by humans.

WRONG CHOICES EXPLAINED:

(1) The Japanese beetle is a plant pest that was accidentally introduced into the United States. Its population is kept in check by the praying mantis, its predator.

(2) The dodo bird became extinct because of hunting by humans.

(4) The passenger pigeon became extinct in the 1900s due to hunting by humans.

165.   **3**   *Burning fossil fuels that produce air pollutants containing sulfur and nitrogen* is the human activity that has probably contributed the most to the acidification of lakes in the Adirondack region. These pollutants combine with water in the atmosphere to form sulfuric and nitric acids. These acids then enter lakes in rainfall and runoff, adding to the acidic quality of the lake water and killing many susceptible species.

WRONG CHOICES EXPLAINED:

(1) *Passing environmental protection laws* is not an activity that results in the acidification of lakes. In fact, it is a positive human activity that can help to limit the production and release of such gases into the atmosphere.

(2) *Establishing reforestation projects in lumbered areas* is not an activity that results in the acidification of lakes. In fact, it is a positive human activity that can help to replace trees lost because of the acidification of soils by acid rain

(4) *Using pesticides for the control of insects that feed on trees* is not an activity that results in the acidification of lakes. It is a negative human activity carried out to protect commercial crops from destruction and does not normally result in the production of sulfur and nitrogen gases.

166.   **2**   *Significant biodiversity* is the factor lacking in a wheat field as compared to a natural forest. *Biodiversity* is a term relating to the variety of life forms in an environment. Natural environments, including forests, are typically made up of thousands of species that interact to provide a balanced, ecologically responsive community. By contrast, farm fields are often limited to a small number of different species, and predominantly a single species. Communities lacking in biodiversity are unstable and prone to collapse when environmental conditions change.

WRONG CHOICES EXPLAINED:

(1) *Heterotrophs* are not lacking in a farm field compared to a forest. Heterotrophs are found within a wheat field, although their number and variety are normally limited to those that use wheat or its by-products as food.

(3) *Autotrophs* are not lacking in a farm field compared to a forest. Wheat is a type of autotroph, as are the various weed species that may be interspersed among the wheat plants in the field.

(4) *Stored energy* is not lacking in a farm field compared to a forest. As the wheat grows in the field, it absorbs the Sun's energy and stores it as the chemical bond energy of carbohydrates and other organic compounds.

167. **1** The *amount of energy released from the Sun* is not normally considered by an ecologist when evaluating the impact of human activities on an ecosystem. The amount of solar energy emitted by the Sun is generally constant and out of our direct control. Because it is not a variable that can be directly affected by human activities, it is usually not a consideration in decisions of this kind.

WRONG CHOICES EXPLAINED:

(2) The *quality of the atmosphere* is often a factor considered by ecologists in evaluating the impact of human activities on an ecosystem. Many human activities introduce chemical contaminants into the atmosphere. These chemicals may have a negative impact on the health and survival of humans and other species.

(3) The *degree of biodiversity* is often a factor considered by ecologists in evaluating the impact of human activities on an ecosystem. Human activities often put pressure on natural species, eliminating those unable to migrate or adapt. As biodiversity in an area declines, so does environmental stability. This situation threatens the health and survival of humans and other species.

(4) The *location of power plants* is often a factor considered by ecologists in evaluating the impact of human activities on an ecosystem. Fossil fuel plants can pollute the atmosphere and consume valuable petroleum products. Nuclear plants can release radiation and heat into the environment. Hydroelectric, solar, wind, and geothermal plants can destroy natural habitats because of space considerations. Each of these consequences can affect the health and survival of humans and other species.

168. **1** An ecologist would want to know *what effect the smoke will have on the environment* before a new type of fuel is widely used. By understanding this effect, the ecologist can make more informed judgments about whether the smoke will harm the environment and human health.

WRONG CHOICES EXPLAINED:

(2), (3), (4) An ecologist is less likely to want to know *how much it will cost to produce the fuel, how long it will take to produce the fuel,* and *if the fuel will be widely accepted by consumers.* Although these are important questions for the manufacturer, they do not provide critical information for the ecologist, whose main concern is the protection of environmental quality for humans and other organisms.

169. **1** Interference with insect reproductive processes is known as biological control. It is the most promising method of controlling insects because it is the least ecologically damaging.

WRONG CHOICES EXPLAINED:

(2) The use of insecticides is a chemical control of insects. Insecticides kill both harmful and helpful insects. The chemicals accumulate in the bodies of birds, fish, and mammals and interfere with their normal life activities.

(3) It is impossible to set up physical barriers for insects because they are motile and are also carried from place to place by animals and humans.

(4) Draining marshes and other insect habitats has helped to control many insects such as mosquitoes. However, this method interferes with the life cycles of useful organisms living in the area.

170. **3** Insecticides are chemical pest controls. Biological controls are other insect species that feed on or in some way prey on an insect pest species. The example given here is control of the gypsy moth larvae by a certain species of beetle.

WRONG CHOICES EXPLAINED:

(1) DDT is an insecticide and represents chemical control. DDT is no longer used because it destroyed the insect food of birds and other wildlife.

(2) Humans upset the balance of nature (the balance of natural communities) by killing off deer predators. The deer population then increased so dramatically that deer starved to death because there was not enough food to support them.

(4) The use of drugs to cure disease is an example of chemical control of pathogens.

171. **1** Each individual should *acquire and apply knowledge of ecological principles* in order to ensure environmental quality for the future. By understanding how environmental principles operate, we can make more informed judgments about activities that may harm the environment and human health.

WRONG CHOICES EXPLAINED:

(2) If each individual were to *continue to take part in deforestation*, environmental quality would be degraded. Because forests are a natural part of the environment, eliminating them disturbs the balance of nature and can have significant negative consequences for environmental quality.

(3) If each individual were to *use Earth's finite resources*, environmental quality would be threatened. As these resources are used up, fewer remain for future generations. In addition, processing these resources consumes energy, produces pollutants, and adds to the solid waste problem.

(4) If each individual were to *add and take away organisms from ecosystems*, environmental quality would be diminished. Each natural community has established itself based on the particular niches filled by each type of organism. Adding to or taking away from this community upsets the balance of nature and would likely cause negative consequences.

172.   Write one or more paragraphs describing positive and negative effects of this method of pest control. Include the following points:

- Two advantages of this method of pest control [2]
  - ➤ Chemicals are not added to the environment.
  - ➤ Biological controls are more specific than chemical controls.
  - ➤ Ladybugs are less likely to kill beneficial organisms.
  - ➤ Desirable garden plants are protected from aphid attacks.
  - ➤ Birds and other unintended victims of pesticide use are spared.
  - ➤ Human health is protected against the toxic effects of pesticides.
- Two possible dangers of using this method of pest control [2]
  - ➤ The control insects may eat the food of other organisms.
  - ➤ The population of natural predators of the aphids may be eliminated or greatly reduced.
  - ➤ The control organism may become overpopulated.
  - ➤ The control organisms may themselves become pests.

Sample paragraph: *The method of pest control described is known as "biological control." This method of insect control has some distinct advantages over chemical controls: First, biological controls don't release toxic chemicals into the air and water, a fact that helps to protect wildlife and humans from being unintended victims of chemical pesticides. Second, biological controls are usually specific, which means that beneficial insects such as ladybugs and preying mantises aren't harmed. [2] There are also some things we should be careful of in the use of biological controls: First, we should know a lot about the control organism to be sure that it doesn't crowd out our native beneficial organisms. Second, we should remember that the control organism could become a pest, too, if it gets too numerous in the environment. [2]*

173.   Write one or more paragraphs explaining why carnivores should be protected. Include the following points:

- Information concerning prey population growth [1]
  - ➢ If predators are destroyed, the prey population will increase.
  - ➢ If unchecked by predation or disease, a natural population will tend to increase in number geometrically.
- Information concerning extinction [1]
  - ➢ If too many carnivores of a particular species are killed, the species may become extinct.
  - ➢ Extinction is a definite possibility when any species has too few members alive to carry out effective breeding.
  - ➢ Complete elimination of any species from its natural range can destabilize the ecosystem.
- Information concerning the importance of carnivores in an environment [1]
  - ➢ By feeding on herbivores, carnivores help keep certain species of plants from being eliminated because of overgrazing in a particular area.
  - ➢ Without predators to limit its number, a prey population could exceed the capacity of its range, resulting in widespread starvation and death of the prey population.
  - ➢ Carnivorous animals are part of the natural scheme that promotes ecological equilibrium.

Sample paragraph: *Carnivores are important in an ecosystem because by reducing the number of prey organisms, the food organisms of the prey are kept from being eliminated from the environment. [1] If the predators were destroyed, the prey population would increase [1], perhaps to the point of consuming so many of the plants that the prey feed on that these plants would become extinct. [1]*

STANDARD/KEY IDEA	QUESTION NUMBERS	NUMBER OF CORRECT RESPONSES	NUMBER OF INCORRECT RESPONSES
1.1 Purpose of Scientific Inquiry	1–8		
1.2 Methods of Scientific Inquiry	9–28		
1.3 Analysis in Scientific Inquiry	29–47		
4.1 Application of Scientific Principles	48–75		
4.2 Genetic Continuity	76–90		
4.3 Organic Evolution	91–109		
4.4 Reproductive Continuity	110–129		
4.5 Dynamic Equilibrium and Homeostasis	130–140		
4.6 Interdependence of Living Things	141–155		
4.7 Human Impact on the Environment	156–173		

Glossary

PROMINENT SCIENTISTS

Crick, Francis A 20th-century British scientist who, with James Watson, developed the first workable model of DNA structure and function.

Darwin, Charles A 19th-century British naturalist whose theory of organic evolution by natural selection forms the basis for the modern scientific theory of evolution.

Fox, Sidney A 20th-century American scientist whose experiments showed that Stanley Miller's simple chemical precursors could be joined to form more complex biochemicals.

Hardy, G. H. A 20th-century British mathematician who, with W. Weinberg, developed the Hardy-Weinberg principle of gene frequencies.

Lamarck, Jean An 18th-century French scientist who devised an early theory of organic evolution based on the concept of "use and disuse."

Linnaeus, Carl An 18th-century Dutch scientist who developed the first scientific system of classification, based on similarity of structure.

Mendel, Gregor A 19th-century Austrian monk and teacher who was the first to describe many of the fundamental concepts of genetic inheritance through his work with garden peas.

Miller, Stanley A 20th-century American scientist whose experiments showed that the simple chemical precursors of life could be produced in the laboratory.

Morgan, Thomas Hunt A 20th-century American geneticist whose pioneering work with Drosophila led to the discovery of several genetic principles, including sex linkage.

Watson, James A 20th-century American scientist who, with Francis Crick, developed the first workable model of DNA structure and function.

Weinberg, W. A 20th-century German physician who, with G. H. Hardy, developed the Hardy-Weinberg principle of gene frequencies.

Weismann, August A 19th-century German biologist who tested Lamarck's theory of use and disuse and found it to be unsupportable by scientific methods.

BIOLOGICAL TERMS

abiotic factor Any of several nonliving, physical conditions that affect the survival of an organism in its environment.

absorption The process by which water and dissolved solids, liquids, and gases are taken in by the cell through the cell membrane.

accessory organ In human beings, any organ that has a digestive function but is not part of the food tube. (See **liver; gallbladder; pancreas.**)

acid A chemical that releases hydrogen ion (H+) in solution with water.

acid precipitation A phenomenon in which there is thought to be an interaction between atmospheric moisture and the oxides of sulfur and nitrogen that results in rainfall with low pH values.

active immunity The immunity that develops when the body's immune system is stimulated by a disease organism or a vaccination.

active site The specific area of an enzyme molecule that links to the substrate molecule and catalyzes its metabolism.

active transport A process by which materials are absorbed or released by cells against the concentration gradient (from low to high concentration) with the expenditure of cell energy.

adaptation Any structural, biochemical, or behavioral characteristic of an organism that helps it to survive potentially harsh environmental conditions.

addition A type of chromosome mutation in which a section of a chromosome is transferred to a homologous chromosome.

adenine A nitrogenous base found in DNA and RNA molecules.

adenosine triphosphate (ATP) An organic compound that stores respiratory energy in the form of chemical-bond energy for transport from one part of the cell to another.

adrenal cortex A portion of the adrenal gland that secretes steroid hormones which regulate various aspects of blood composition.

adrenal gland An endocrine gland that produces several hormones, including **adrenaline.** (See **adrenal cortex; adrenal medulla.**)

adrenal medulla A portion of the adrenal gland that secretes the hormone adrenaline, which regulates various aspects of the body's metabolic rate.

adrenaline A hormone of the adrenal medulla that regulates general metabolic rate, the rates of heartbeat and breathing, and the conversion of glycogen to glucose.

aerobic phase of respiration The reactions of aerobic respiration in which two pyruvic acid molecules are converted to six molecules of water and six molecules of carbon dioxide.

aerobic respiration A type of respiration in which energy is released from organic molecules with the aid of oxygen.

aging A stage of postnatal development that involves differentiation, maturation, and eventual deterioration of the body's tissues.

air pollution The addition, due to technological oversight, of some unwanted factor (e.g., chemical oxides, hydrocarbons, particulates) to our air resources.

albinism A condition, controlled by a single mutant gene, in which the skin lacks the ability to produce skin pigments.

alcoholic fermentation A type of anaerobic respiration in which glucose is converted to ethyl alcohol and carbon dioxide.

allantois A membrane that serves as a reservoir for wastes and as a respiratory surface for the embryos of many animal species.

allele One of a pair of genes that exist at the same location on a pair of homologous chromosomes and exert parallel control over the same genetic trait.

allergy A reaction of the body's immune system to the chemical composition of various substances.

alveolus One of many "air sacs" within the lung that function to absorb atmospheric gases and pass them on to the bloodstream.

amino acid An organic compound that is the component unit of proteins.

amino group A chemical group having the formula —NH_2 that is found as a part of all amino acid molecules.

ammonia A type of nitrogenous waste with high solubility and high toxicity.

amniocentesis A technique for the detection of genetic disorders in human beings in which a small amount of amniotic fluid is removed and the chromosome content of its cells analyzed. (See **karyotyping.**)

amnion A membrane that surrounds the embryo in many animal species and contains a fluid to protect the developing embryo from mechanical shock.

amniotic fluid The fluid within the amnion membrane that bathes the developing embryo.

amylase An enzyme specific for the hydrolysis of starch.

anaerobic phase of respiration The reactions of aerobic respiration in which glucose is converted to two pyruvic acid molecules.

anaerobic respiration A type of respiration in which energy is released from organic molecules without the aid of oxygen.

anal pore The egestive organ of the paramecium.

anemia A disorder of the human transport system in which the ability of the blood to carry oxygen is impaired, usually because of reduced numbers of red blood cells.

angina pectoris A disorder of the human transport system in which chest pain signals potential damage to the heart muscle due to narrowing of the opening of the coronary artery.

Animal One of the five biological kingdoms; it includes multicellular organisms whose cells are not bounded by cell walls and which are incapable of photosynthesis (e.g., human being).

Annelida A phylum of the Animal Kingdom whose members (annelids) include the segmented worms (e.g., earthworm).

antenna A receptor organ found in many arthropods (e.g., grasshopper), which is specialized for detecting chemical stimuli.

anther The portion of the stamen that produces pollen.

antibody A chemical substance, produced in response to the presence of a specific antigen, which neutralizes that antigen in the immune response.

antigen A chemical substance, usually a protein, that is recognized by the immune system as a foreign "invader" and is neutralized by a specific antibody.

anus The organ of egestion of the digestive tract.

aorta The principal artery carrying blood from the heart to the body tissues.

aortic arches A specialized part of the earthworm's transport system that serves as a pumping mechanism for the blood fluid.

apical meristem A plant growth region located at the tip of the root or tip of the stem.

appendicitis A disorder of the human digestive tract in which the appendix becomes inflamed as a result of bacterial infection.

aquatic biome An ecological biome composed of many different water environments.

artery A thick-walled blood vessel that carries blood away from the heart under pressure.

arthritis A disorder of the human locomotor system in which skeletal joints become inflamed, swollen, and painful.

Arthropoda A phylum of the Animal Kingdom whose members (arthropods) have bodies with chitinous exoskeletons and jointed appendages (e.g., grasshopper).

artificial selection A technique of plant/animal breeding in which individual organisms displaying desirable characteristics are chosen for breeding purposes.

asexual reproduction A type of reproduction in which new organisms are formed from a single parent organism.

asthma A disorder of the human respiratory system in which the respiratory tube becomes constricted by swelling brought on by some irritant.

atrium In human beings, one of the two thin-walled upper chambers of the heart that receive blood.

autonomic nervous system A subdivision of the peripheral nervous system consisting of nerves associated with automatic functions (e.g., heartbeat, breathing).

autosome One of several chromosomes present in the cell that carry genes controlling "body" traits not associated with primary and secondary sex characteristics.

autotroph An organism capable of carrying on autotrophic nutrition. Self feeder.

autotrophic nutrition A type of nutrition in which organisms manufacture their own organic foods from inorganic raw materials.

auxin A biochemical substance, plant hormone, produced by plants that regulates growth patterns.

axon An elongated portion of a neuron that conducts nerve impulses, usually away from the cell body of the neuron.

base A chemical that releases hydroxyl ion (OH^-) in solution with water.

bicarbonate ion The chemical formed in the blood plasma when carbon dioxide is absorbed from body tissues.

bile In human beings, a secretion of the liver that is stored in the gallbladder and that emulsifies fats.

binary fission A type of cell division in which mitosis is followed by equal cytoplasmic division.

binomial nomenclature A system of naming, used in biological classification, that consists of the genus and species names (e.g., *Homo sapiens*).

biocide use The use of pesticides that eliminate one undesirable organism but that have, due to technological oversight, unanticipated effects on beneficial species as well.

biological controls The use of natural enemies of various agricultural pests for pest control, thereby eliminating the need for biocide use—a positive aspect of human involvement with the environment.

biomass The total mass of living material present at the various trophic levels in a food chain.

biome A major geographical grouping of similar ecosystems, usually named for the climax flora in the region (e.g., Northeast Deciduous Forest).

biosphere The portion of the earth in which living things exist, including all land and water environments.

biotic factor Any of several conditions associated with life and living things that affect the survival of living things in the environment.

birth In placental mammals, a stage of embryonic development in which the baby passes through the vaginal canal to the outside of the mother's body.

blastula In certain animals, a stage of embryonic development in which the embryo resembles a hollow ball of undifferentiated cells.

blood The complex fluid tissue that functions to transport nutrients and respiratory gases to all parts of the body.

blood typing An application of the study of immunity in which the blood of a person is characterized by its antigen composition.

bone A tissue that provides mechanical support and protection for bodily organs, and levers for the body's locomotive activities.

Bowman's capsule A cup-shaped portion of the nephron responsible for the filtration of soluble blood components.

brain An organ of the central nervous system that is responsible for regulating conscious and much unconscious activity in the body.

breathing A mechanical process by which air is forced into the lungs by means of muscular contraction of the diaphragm and rib muscles.

bronchiole One of several subdivisions of the bronchi that penetrate the lung interior and terminate in alveoli.

bronchitis A disorder of the human respiratory system in which the bronchi become inflamed.

bronchus One of the two major subdivisions of the breathing tube; the bronchi are ringed with cartilage and conduct air from the trachea to the lung interior.

Bryophyta A phylum of the Plant Kingdom that consists of organisms lacking vascular tissues (e.g., moss).

budding A type of asexual reproduction in which mitosis is followed by unequal cytoplasmic division.

bulb A type of vegetative propagation in which a plant bulb produces new bulbs that may be established as independent organisms with identical characteristics.

cambium The lateral meristem tissue in woody plants responsible for annual growth in stem diameter.

cancer Any of a number of conditions characterized by rapid, abnormal, and uncontrolled division of affected cells.

capillary A very small, thin-walled blood vessel that connects an artery to a vein and through which all absorption into the blood fluid occurs.

carbohydrate An organic compound composed of carbon, hydrogen, and oxygen in a 1:2:1 ratio (e.g., $C_6H_{12}O_6$).

carbon-14 A radioactive isotope of carbon used to trace the movement of carbon in various biochemical reactions, and also used in the "carbon dating" of fossils.

132   Glossary

carbon-fixation reactions A set of biochemical reactions in photosynthesis in which hydrogen atoms are combined with carbon and oxygen atoms to form PGAL and glucose.

carbon-hydrogen-oxygen cycle A process by which these three elements are made available for use by other organisms through the chemical reactions of respiration and photosynthesis.

carboxyl group A chemical group having the formula—COOH and found as part of all amino acid and fatty acid molecules.

cardiac muscle A type of muscle tissue in the heart and arteries that is associated with the rhythmic nature of the pulse and heartbeat.

cardiovascular disease In human beings, any disease of the circulatory organs.

carnivore A heterotrophic organism that consumes animal tissue as its primary source of nutrition. (See **secondary consumer.**)

carrier An individual who, though not expressing a particular recessive trait, carries this gene as part of his/her heterozygous genotype.

carrier protein A specialized molecule embedded in the cell membrane that aids the movement of materials across the membrane.

cartilage A flexible connective tissue found in many flexible parts of the body (e.g., knee); common in the embryonic stages of development.

catalyst Any substance that speeds up or slows down the rate of a chemical reaction. (See **enzyme.**)

cell plate A structure that forms during cytoplasmic division in plant cells and serves to separate the cytoplasm into two roughly equal parts.

cell theory A scientific theory that states, "All cells arise from previously existing cells" and "Cells are the unit of structure and function of living things."

cell wall A cell organelle that surrounds and gives structural support to plant cells; cell walls are composed of cellulose.

central nervous system The portion of the vertebrate nervous system that consists of the brain and the spinal cord.

centriole A cell organelle found in animal cells that functions in the process of cell division.

centromere The area of attachment of two chromatids in a double-stranded chromosome.

cerebellum The portion of the human brain responsible for the coordination of muscular activity.

cerebral hemorrhage A disorder of the human regulatory system in which a broken blood vessel in the brain may result in severe dysfunction or death.

cerebral palsy A disorder of the human regulatory system in which the motor and speech centers of the brain are impaired.

cerebrum The portion of the human brain responsible for thought, reasoning, sense interpretation, learning, and other conscious activities.

cervix A structure that bounds the lower end of the uterus and through which sperm must pass in order to fertilize the egg.

chemical digestion The process by which nutrient molecules are converted by chemical means into a form usable by the cells.

chemosynthesis A type of autotrophic nutrition in which certain bacteria use the energy of chemical oxidation to convert inorganic raw materials to organic food molecules.

chitin A polysaccharide substance that forms the exoskeleton of the grasshopper and other arthropods.

chlorophyll A green pigment in plant cells that absorbs sunlight and makes possible certain aspects of the photosynthetic process.

chloroplast A cell organelle found in plant cells that contains chlorophyll and functions in photosynthesis.

Chordata A phylum of the Animal Kingdom whose members (chordates) have internal skeletons made of cartilage and/or bone (e.g., human being).

chorion A membrane that surrounds all other embryonic membranes in many animal species, protecting them from mechanical damage.

chromatid One strand of a double-stranded chromosome.

chromosome mutation An alteration in the structure of a chromosome involving many genes. (See **nondisjunction; translocation; addition; deletion.**)

cilia Small, hairlike structures in paramecia and other unicellular organisms that aid in nutrition and locomotion.

classification A technique by which scientists sort, group, and name organisms for easier study.

cleavage A series of rapid mitotic divisions that increase cell number in a developing embryo without corresponding increase in cell size.

climax community A stable, self-perpetuating community that results from an ecological succession.

cloning A technique of genetic investigation in which undifferentiated cells of an organism are used to produce new organisms with the same set of traits as the original cells.

closed transport system A type of circulatory system in which the transport fluid is always enclosed within blood vessels (e.g., earthworm, human).

clot A structure that forms as a result of enzyme-controlled reactions following the rupturing of a blood vessel and serves as a plug to prevent blood loss.

codominance A type of intermediate inheritance that results from the simultaneous expression of two dominant alleles with contrasting effects.

codon See **triplet codon.**

Coelenterata A phylum of the Animal Kingdom whose members (coelenterates) have bodies that resemble a sack (e.g., hydra, jellyfish).

coenzyme A chemical substance or chemical subunit that functions to aid the action of a particular enzyme. (See **vitamin.**)

cohesion A force binding water molecules together that aids in the upward conduction of materials in the xylem.

commensalism A type of symbiosis in which one organism in the relationship benefits and the other is neither helped nor harmed.

common ancestry A concept central to the science of evolution which postulates that all organisms share a common ancestry whose closeness varies with the degree of shared similarity.

community A level of biological organization that includes all of the species populations inhabiting a particular geographic area.

comparative anatomy The study of similarities in the anatomical structures of organisms, and their use as an indicator of common ancestry and as evidence of organic evolution.

comparative biochemistry The study of similarities in the biochemical makeups of organisms, and their use as an indicator of common ancestry and as evidence of organic evolution.

comparative cytology The study of similarities in the cell structures of organisms, and their use as an indicator of common ancestry and as evidence of organic evolution.

comparative embryology The study of similarities in the patterns of embryological development of organisms, and their use as an indicator of common ancestry and as evidence of organic evolution.

competition A condition that arises when different species in the same habitat attempt to use the same limited resources.

complete protein A protein that contains all eight essential amino acids.

compound A substance composed of two or more different kinds of atom (e.g., water: H_2O).

compound light microscope A tool of biological study capable of producing a magnified image of a biological specimen by using a focused beam of light.

conditioned behavior A type of response that is learned, but that becomes automatic with repetition.

conservation of resources The development and application of practices to protect valuable and irreplaceable soil and mineral resources—a positive aspect of human involvement with the environment.

constipation A disorder of the human digestive tract in which fecal matter solidifies and becomes difficult to egest.

consumer Any heterotrophic animal organism (e.g., human being).

coronary artery An artery that branches off the aorta to feed the heart muscle.

coronary thrombosis A disorder of the human transport system in which the heart muscle becomes damaged as a result of blockage of the coronary artery.

corpus luteum A structure resulting from the hormone-controlled transformation of the ovarian follicle that produces the hormone progesterone.

corpus luteum stage A stage of the menstrual cycle in which the cells of the follicle are transformed into the corpus luteum under the influence of the hormone LH.

cotyledon A portion of the plant embryo that serves as a source of nutrition for the young plant before photosynthesis begins.

cover-cropping A proper agricultural practice in which a temporary planting (cover crop) is used to limit soil erosion between seasonal plantings of main crops.

crop A portion of the digestive tract of certain animals that stores food temporarily before digestion.

cross-pollination A type of pollination in which pollen from one flower pollinates flowers of a different plant of the same species.

crossing-over A pattern of inheritance in which linked genes may be separated during synapsis in the first meiotic division, when sections of homologous chromosomes may be exchanged.

cuticle A waxy coating that covers the upper epidermis of most leaves and acts to help the leaf retain water.

cutting A technique of plant propagation in which vegetative parts of the parent plant are cut and rooted to establish new plant organisms with identical characteristics.

cyclosis The circulation of the cell fluid (cytoplasm) within the cell interior.

cyton The "cell body" of the neuron, which generates the nerve impulse.

cytoplasm The watery fluid that provides a medium for the suspension of organelles within the cell.

cytoplasmic division The separation of daughter nuclei into two new daughter cells.

cytosine A nitrogenous base found in both DNA and RNA molecules.

daughter cell A cell that results from mitotic cell division.

daughter nucleus One of two nuclei that form as a result of mitosis.

deamination A process by which amino acids are broken down into their component parts for conversion into urea.

death The irreversible cessation of bodily functions and cellular activities.

deciduous A term relating to broadleaf trees which shed their leaves in the fall.

decomposer Any saprophytic organism that derives its energy from the decay of plant and animal tissues (e.g., bacteria of decay, fungus); the final stage of a food chain.

decomposition bacteria In the nitrogen cycle, bacteria that break down plant and animal protein and produce ammonia as a by-product.

dehydration synthesis A chemical process in which two organic molecules may be joined after removing the atoms needed to form a molecule of water as a by-product.

deletion A type of chromosome mutation in which a section of a chromosome is separated and lost.

dendrite A cytoplasmic extension of a neuron that serves to detect an environmental stimulus and carry an impulse to the cell body of the neuron.

denitrifying bacteria In the nitrogen cycle, bacteria that convert excess nitrate salts into gaseous nitrogen.

deoxygenated blood Blood that has released its transported oxygen to the body tissues.

deoxyribonucleic acid (DNA) A nucleic acid molecule known to be the chemically active agent of the gene; the fundamental hereditary material of living organisms.

deoxyribose A five-carbon sugar that is a component part of the nucleotide unit in DNA only.

desert A terrestrial biome characterized by sparse rainfall, extreme temperature variation, and a climax flora that includes cactus.

diabetes A disorder of the human regulatory system in which insufficient insulin production leads to elevated blood sugar concentrations.

diarrhea A disorder of the human digestive tract in which the large intestine fails to absorb water from the waste matter, resulting in watery feces.

diastole The lower pressure registered during blood pressure testing. (See **systole.**)

differentiation The process by which embryonic cells become specialized to perform the various tasks of particular tissues throughout the body.

diffusion A form of passive transport by which soluble substances are absorbed or released by cells.

digestion The process by which complex foods are broken down by mechanical or chemical means for use by the body.

dipeptide A chemical unit composed of two amino acid units linked by a peptide bond.

diploid chromosome number The number of chromosomes found characteristically in the cells (except gametes) of sexually reproducing species.

disaccharidase Any disaccharide-hydrolyzing enzyme.

disaccharide A type of carbohydrate known also as a "double sugar"; all disaccharides have the molecular formula $C_{12}H_{22}O_{11}$.

disjunction The separation of homologous chromosome pairs at the end of the first meiotic division.

disposal problems Problems, due to technological oversight, that result when commercial and technological activities produce solid and/or chemical wastes that must be disposed of.

dissecting microscope A tool of biological study that magnifies the image of a biological specimen up to 20 times normal size for purposes of gross dissection.

dominance A pattern of genetic inheritance in which the effects of a dominant allele mask those of a recessive allele.

dominant allele (gene) An allele (gene) whose effect masks that of its recessive allele.

double-stranded chromosome The two-stranded structure that results from chromosomal replication.

Down's syndrome In human beings, a condition, characterized by mental and physical retardation, that may be caused by the nondisjunction of chromosome number 21.

Drosophila The common fruit fly, an organism that has served as an object of genetic research in the development of the gene-chromosome theory.

ductless gland See **endocrine gland.**

ecology The science that studies the interactions of living things with each other and with the nonliving environment.

ecosystem The basic unit of study in ecology, including the plant and animal community in interaction with the nonliving environment.

ectoderm An embryonic tissue that differentiates into skin and nerve tissue in the adult animal.

effector An organ specialized to produce a response to an environmental stimulus: effectors may be muscles or glands.

egestion The process by which undigested food materials are eliminated from the body.

electron microscope A tool of biological study that uses a focused beam of electrons to produce an image of a biological specimen magnified up to 25,000 times normal size.

element The simplest form of matter; an element is a substance (e.g., nitrogen) made up of a single type of atom.

embryo An organism in the early stages of development following fertilization.

embryonic development A series of complex processes by which animal and plant embryos develop into adult organisms.

emphysema A disorder of the human respiratory system in which lung tissue deteriorates, leaving the lung with diminished capacity and efficiency.

emulsification A process by which fat globules are surrounded by bile to form fat droplets.

endocrine ("ductless") gland A gland (e.g., thyroid, pituitary) specialized for the production of hormones and their secretion directly into the bloodstream; such glands lack ducts.

endoderm An embryonic tissue that differentiates into the digestive and respiratory tract lining in the adult animal.

endoplasmic reticulum (ER) A cell organelle known to function in the transport of cell products from place to place within the cell.

environmental laws Federal, state, and local legislation enacted in an attempt to protect environmental resources—a positive aspect of human involvement with the environment.

enzymatic hydrolysis An enzyme-controlled reaction by which complex food molecules are broken down chemically into simpler subunits.

enzyme An organic catalyst that controls the rate of metabolism of a single type of substrate; enzymes are protein in nature.

enzyme-substrate complex A physical association between an enzyme molecule and its substrate within which the substrate is metabolized.

epicotyl A portion of the plant embryo that specializes to become the upper stem, leaves, and flowers of the adult plant.

epidermis The outermost cell layer in a plant or animal.

epiglottis In a human being, a flap of tissue that covers the upper end of the trachea during swallowing and prevents inhalation of food.

esophagus A structure in the upper portion of the digestive tract that conducts the food from the pharynx to the midgut.

essential amino acid An amino acid that cannot be synthesized by the human body, but must be obtained by means of the diet.

estrogen A hormone, secreted by the ovary, that regulates the production of female secondary sex characteristics.

evolution Any process of gradual change through time.

excretion The life function by which living things eliminate metabolic wastes from their cells.

exoskeleton A chitinous material that covers the outside of the bodies of most arthropods and provides protection for internal organs and anchorage for muscles.

exploitation of organisms Systematic removal of animals and plants with commercial value from their environments, for sale—a negative aspect of human involvement with the environment.

extensor A skeletal muscle that extends (opens) a joint.

external development Embryonic development that occurs outside the body of the female parent (e.g., birds).

external fertilization Fertilization that occurs outside the body of the female parent (e.g., fish).

extracellular digestion Digestion that occurs outside the cell.

fallopian tube See **oviduct.**

fatty acid An organic molecule that is a component of certain lipids.

fauna The animal species comprising an ecological community.

feces The semisolid material that results from the solidification of undigested foods in the large intestine.

fertilization The fusion of gametic nuclei in the process of sexual reproduction.

filament The portion of the stamen that supports the anther.

flagella Microscopic, whiplike structures found on certain cells that aid in locomotion and circulation.

flexor A skeletal muscle that flexes (closes) a joint.

flora The plant species comprising an ecological community.

flower The portion of a flowering plant that is specialized for sexual reproduction.

fluid-mosaic model A model of the structure of the cell membrane in which large protein molecules are thought to be embedded in a bilipid layer.

follicle One of many areas within the ovary that serve as sites for the periodic maturation of ova.

follicle stage The stage of the menstrual cycle in which an ovum reaches its final maturity under the influence of the hormone FSH.

follicle-stimulating hormone (FSH) A pituitary hormone that regulates the maturation of, and the secretion of estrogen by, the ovarian follicle.

food chain A series of nutritional relationships in which food energy is passed from producer to herbivore to carnivore to decomposer; a segment of a food web.

food web A construct showing a series of interrelated food chains and illustrating the complex nutritional interrelationships that exist in an ecosystem.

fossil The preserved direct or indirect remains of an organism that lived in the past, as found in the geologic record.

fraternal twins In human beings, twin offspring that result from the simultaneous fertilization of two ova by two sperm; such twins are not genetically identical.

freshwater biome An aquatic biome made up of many separate freshwater systems that vary in size and stability and may be closely associated with terrestrial biomes.

fruit Any plant structure that contains seeds; a mechanism of seed dispersal.

Fungi One of the five biological kingdoms; it includes organisms unable to manufacture their own organic foods (e.g., mushroom).

gallbladder An accessory organ that stores bile.

gallstones A disorder of the human digestive tract in which deposits of hardened cholesterol lodge in the gallbladder.

gamete A specialized reproductive cell produced by organisms of sexually reproducing species. (See **sperm; ovum; pollen; ovule.**)

gametogenesis The process of cell division by which gametes are produced. (See **meiosis; spermatogenesis; oogenesis.**)

ganglion An area of bunched nerve cells that acts as a switching point for nerve impulses traveling from receptors and to effectors.

garden pea The research organism used by Mendel in his early scientific work in genetic inheritance.

gastric cecum A gland in the grasshopper that secretes digestive enzymes.

gastrula A stage of embryonic development in animals in which the embryo assumes a tube-within-a-tube structure and distinct embryonic tissues (ectoderm, mesoderm, endoderm) begin to differentiate.

gastrulation The process by which a blastula becomes progressively more indented, forming a gastrula.

gene A unit of heredity; a discrete portion of a chromosome thought to be responsible for the production of a single type of polypeptide; the "factor" responsible for the inheritance of a genetic trait.

gene frequency The proportion (percentage) of each allele for a particular trait that is present in the gene pool of a population.

gene linkage A pattern of inheritance in which genes located along the same chromosome are prevented from assorting independently, but are linked together in their inheritance.

gene mutation An alteration of the chemical nature of a gene that changes its ability to control the production of a polypeptide chain.

gene pool The sum total of all the inheritable genes for the traits in a given sexually reproducing population.

gene-chromosome theory A theory of genetic inheritance that is based on current understanding of the relationships between the biochemical control of traits and the process of cell division.

genetic counseling Clinical discussions concerning inheritance patterns that are designed to inform prospective parents of the potential for expression of a genetic disorder in their offspring.

genetic engineering The use of various techniques to move genes from one organism to another.

genetic screening A technique for the detection of human genetic disorders in which bodily fluids are analyzed for the presence of certain marker chemicals.

genotype The particular combination of genes in an allele pair.

genus A level of biological classification that represents a subdivision of the phylum level; having fewer organisms with great similarity (e.g., *Drosophila,* paramecium).

geographic isolation The separation of species populations by geographical barriers, facilitating the evolutionary process.

geologic record A supporting item of evidence of organic evolution, supplied within the earth's rock and other geological deposits.

germination The growth of the pollen tube from a pollen grain; the growth of the embryonic root and stem from a seed.

gestation The period of prenatal development of a placental mammal; human gestation requires approximately 9 months.

gizzard A portion of the digestive tract of certain organisms, including the earthworm and the grasshopper, in which food is ground into smaller fragments.

glomerulus A capillary network lying within Bowman's capsule of the nephron.

glucagon A hormone, secreted by the islets of Langerhans, that regulates the release of blood sugar from stored glycogen.

glucose A monosaccharide produced commonly in photosynthesis and used by both plants and animals as a "fuel" in the process of respiration.

glycerol An organic compound that is a component of certain lipids.

glycogen A polysaccharide synthesized in animals as a means of storing glucose; glycogen is stored in the liver and in the muscles.

goiter A disorder of the human regulatory system in which the thyroid gland enlarges because of a deficiency of dietary iodine.

Golgi complex Cell organelles that package cell products and move them to the plasma membrane for secretion.

gonad An endocrine gland that produces the hormones responsible for the production of various secondary sex characteristics. (See **ovary; testis.**)

gout A disorder of the human excretory system in which uric acid accumulates in the joints, causing severe pain.

gradualism A theory of the time frame required for organic evolution which assumes that evolutionary change is slow, gradual, and continuous.

grafting A technique of plant propagation in which the stems of desirable plants are attached (grafted) to rootstocks of related varieties to produce new plants for commercial purposes.

grana The portion of the chloroplast within which chlorophyll molecules are concentrated.

grassland A terrestrial biome characterized by wide variation in temperature and a climax flora that includes grasses.

growth A process by which cells increase in number and size, resulting in an increase in size of the organism.

growth-stimulating hormone (GSH) A pituitary hormone regulating the elongation of the long bones of the body.

guanine A nitrogenous base found in both DNA and RNA molecules.

guard cell One of a pair of cells that surround the leaf stomate and regulate its size.

habitat The environment or set of ecological conditions within which an organism lives.

Hardy-Weinberg principle A hypothesis, advanced by G. H. Hardy and W. Weinberg, which states that the gene pool of a population should remain stable as long as a set of "ideal" conditions is met.

heart In human beings, a four-chambered muscular pump that facilitates the movement of blood throughout the body.

helix Literally a spiral; a term used to describe the "twisted ladder" shape of the DNA molecule.

hemoglobin A type of protein specialized for the transport of respiratory oxygen in certain organisms, including earthworms and human beings.

herbivore A heterotrophic organism that consumes plant matter as its primary source of nutrition. (See **primary consumer.**)

hermaphrodite An animal organism that produces both male and female gametes.

heterotroph An organism that typically carries on heterotrophic nutrition.

heterotroph hypothesis A scientific hypothesis devised to explain the probable origin and early evolution of life on earth.

heterotrophic nutrition A type of nutrition in which organisms must obtain their foods from outside sources of organic nutrients.

heterozygous A term used to refer to an allele pair in which the alleles have different contrasting effects (e.g., *Aa, RW*).

high blood pressure A disorder of the human transport system in which systolic and diastolic pressures register higher than normal because of narrowing of the artery opening.

histamine A chemical product of the body that causes irritation and swelling of the mucous membranes.

homeostasis The condition of balance and dynamic stability that characterizes living systems under normal conditions.

homologous chromosomes A pair of chromosomes that carry corresponding genes for the same traits.

homologous structures Structures present within different species that can be shown to have had a common origin, but that may or may not share a common function.

homozygous A term used to refer to an allele pair in which the alleles are identical in terms of effect (e.g., *AA, aa*).

hormone A chemical product of an endocrine gland which has a regulatory effect on the cell's metabolism.

host The organism that is harmed in a parasitic relationship.

hybrid A term used to describe a heterozygous genotype. (See **heterozygous.**)

hybridization A technique of plant/animal breeding in which two varieties of the same species are crossbred in the hope of producing offspring with the favorable traits of both varieties.

hydrogen bond A weak electrostatic bond that holds together the twisted strands of DNA and RNA molecules.

hydrolysis The chemical process by which a complex food molecule is split into simpler components through the addition of a molecule of water to the bonds holding it together.

hypocotyl A portion of the plant embryo that specializes to become the root and lower stem of the adult plant.

hypothalamus An endocrine gland whose secretions affect the pituitary gland.

identical twins In human beings, twin offspring resulting from the separation of the embryonic cell mass of a single fertilization into two separate masses; such twins are genetically identical.

importation of organisms The introduction of nonactive plants and animals into new areas where they compete strongly with native species—a negative aspect of human involvement with the environment.

in vitro fertilization A laboratory technique in which fertilization is accomplished outside the mother's body using mature ova and sperm extracted from the parents' bodies.

inbreeding A technique of plant/animal breeding in which a "purebred" variety is bred only with its own members, so as to maintain a set of desired characteristics.

independent assortment A pattern of inheritance in which genes on different, nonhomologous chromosomes are free to be inherited randomly and regardless of the inheritance of the others.

ingestion The mechanism by which an organism takes in food from its environment.

inorganic compound A chemical compound that lacks the element carbon or hydrogen (e.g., table salt: NaCl).

insulin A hormone, secreted by the islets of Langerhans, that regulates the storage of blood sugar as glycogen.

intercellular fluid (ICF) The fluid that bathes cells and fills intercellular spaces.

interferon A substance, important in the fight against human cancer, that may now be produced in large quantities through techniques of genetic engineering.

intermediate inheritance Any pattern of inheritance in which the offspring expresses a phenotype different from the phenotypes of its parents and usually representing a form intermediate between them.

internal development Embryonic development that occurs within the body of the female parent.

internal fertilization Fertilization that occurs inside the body of the female parent.

interneuron A type of neuron, located in the central nervous system, that is responsible for the interpretation of impulses received from sensory neurons.

intestine A portion of the digestive tract in which chemical digestion and absorption of digestive end-products occur.

intracellular digestion A type of chemical digestion carried out within the cell.

iodine A chemical stain used in cell study; an indicator used to detect the presence of starch. (See **staining.**)

islets of Langerhans An endocrine gland, located within the pancreas, that produces the hormones insulin and glucagon.

karyotype An enlarged photograph of the paired homologous chromosomes of an individual cell that is used in the detection of certain genetic disorders involving chromosome mutation.

karyotyping A technique for the detection of human genetic disorders in which a karyotype is analyzed for abnormalities in chromosome structure or number.

kidney The excretory organ responsible for maintaining the chemical composition of the blood. (See **nephron.**)

kidney failure A disorder of the human excretory system in which there is a general breakdown of the kidney's ability to filter blood components.

kingdom A level of biological classification that includes a broad grouping of organisms displaying general structural similarity; five kingdoms have been named by scientists.

lacteal A small extension of the lymphatic system, found inside the villus, that absorbs fatty acids and glycerol resulting from lipid hydrolysis.

lactic acid fermentation A type of anaerobic respiration in which glucose is converted to two lactic acid molecules.

large intestine A portion of the digestive tract in which undigested foods are solidified by means of water absorption to form feces.

lateral meristem A plant growth region located under the epidermis or bark of a stem. (See **cambium.**)

Latin The language used in biological classification for naming organisms by means of binomial nomenclature.

lenticel A small pore in the stem surface that permits the absorption and release of respiratory gases within stem tissues.

leukemia A disorder of the human transport system in which the bone marrow produces large numbers of abnormal white blood cells. (See **cancer.**)

lichen A symbiosis of alga and fungus that frequently acts as a pioneer species on bare rock.

limiting factor Any abiotic or biotic condition that places limits on the survival of organisms and on the growth of species populations in the environment.

lipase Any lipid-hydrolyzing enzyme.

lipid An organic compound composed of carbon, hydrogen, and oxygen in which hydrogen and oxygen are *not* in a 2:1 ratio (e.g., a wax, plant oil); many lipids are constructed of a glycerol and three fatty acids.

liver An accessory organ that stores glycogen, produces bile, destroys old red blood cells, deaminates amino acids, and produces urea.

lock-and-key model A theoretical model of enzyme action that attempts to explain the concept of enzyme specificity.

lung The major organ of respiratory gas exchange.

luteinizing hormone (LH) A pituitary hormone that regulates the conversion of the ovarian follicle into the corpus luteum.

lymph Intercellular fluid (ICF) that has passed into the lymph vessels.

lymph node One of a series of structures in the body that act as reservoirs of lymph and also contain white blood cells as part of the body's immune system.

lymph vessel One of a branching series of tubes that collect ICF from the tissues and redistribute it as lymph.

lymphatic circulation The movement of lymph throughout the body.

lymphocyte A type of white blood cell that produces antibodies.

lysosome A cell organelle that houses hydrolytic enzymes used by the cell in the process of chemical digestion.

Malpighian tubules In arthropods (e.g., grasshopper), an organ specialized for the removal of metabolic wastes.

maltase A specific enzyme that catalyzes the hydrolysis (and dehydration synthesis) of maltose.

maltose A type of disaccharide; a maltose molecule is composed of two units of glucose joined together by dehydration synthesis.

marine biome An aquatic biome characterized by relatively stable conditions of moisture, salinity, and temperature.

marsupial mammal See **nonplacental mammal.**

mechanical digestion Any of the processes by which foods are broken apart physically into smaller particles.

medulla The portion of the human brain responsible for regulating the automatic processes of the body.

meiosis The process by which four monoploid nuclei are formed from a single diploid nucleus.

meningitis A disorder of the human regulatory system in which the membranes of the brain or spinal cord become inflamed.

menstrual cycle A hormone-controlled process responsible for the monthly release of mature ova.

menstruation The stage of the menstrual cycle in which the lining of the uterus breaks down and is expelled from the body via the vaginal canal.

meristem A plant tissue specialized for embryonic development. (See **apical meristem; lateral meristem; cambium.**)

mesoderm An embryonic tissue that differentiates into muscle, bone, the excretory system, and most of the reproductive system in the adult animal.

messenger RNA (m-RNA) A type of RNA that carries the genetic code from the nuclear DNA to the ribosome for transcription.

metabolism All of the chemical processes of life considered together; the sum total of all the cell's chemical activity.

methylene blue A chemical stain used in cell study. (See **staining.**)

microdissection instruments Tools of biological study that are used to remove certain cell organelles from within cells for examination.

micrometer (μm) A unit of linear measurement equal in length to 0.001 millimeter (0.000001 meter), used for expressing the dimensions of cells and cell organelles.

mitochondrion A cell organelle that contains the enzymes necessary for aerobic respiration.

mitosis A precise duplication of the contents of a parent cell nucleus, followed by an orderly separation of these contents into two new, identical daughter nuclei.

mitotic cell division A type of cell division that results in the production of two daughter cells identical to each other and to the parent cell.

Monera One of the five biological kingdoms; it includes simple unicellular forms lacking nuclear membranes (e.g., bacteria).

monohybrid cross A genetic cross between two organisms both heterozygous for a trait controlled by a single allele pair. The phenotypic ratio resulting is 3:1; the genotypic ratio is 1:2:1.

monoploid chromosome number The number of chromosomes commonly found in the gametes of sexually reproducing species.

monosaccharide A type of carbohydrate known also as a "simple sugar"; all monosaccharides have the molecular formula $C_6H_{22}O_6$.

motor neuron A type of neuron that carries "command" impulses from the central nervous system to an effector organ.

mucus A protein-rich mixture that bathes and moistens the respiratory surfaces.

multicellular Having a body that consists of large groupings of specialized cells (e.g., human being).

multiple alleles A pattern of inheritance in which the existence of more than two alleles is hypothesized, only two of which are present in the genotype of any one individual.

muscle A type of tissue specialized to produce movement of body parts.

mutagenic agent Any environmental condition that initiates or accelerates genetic mutation.

mutation Any alteration of the genetic material, either a chromosome or a gene, in an organism.
mutualism A type of symbiosis beneficial to both organisms in the relationship.

nasal cavity A series of channels through which outside air is admitted to the body interior and is warmed and moistened before entering the lung.
natural selection A concept, central to Darwin's theory of evolution, to the effect that the individuals best adapted to their environment tend to survive and to pass their favorable traits on to the next generation.
negative feedback A type of endocrine regulation in which the effects of one gland may inhibit its own secretory activity, while stimulating the secretory activity of another gland.
nephridium An organ found in certain organisms, including the earthworm, specialized for the removal of metabolic wastes.
nephron The functional unit of the kidney. (See **glomerulus; Bowman's capsule.**)
nerve A structure formed from the bundling of neurons carrying sensory or motor impulses.
nerve impulse An electrochemical change in the surface of the nerve cell.
nerve net A network of "nerve" cells in coelenterates such as the hydra.
neuron A cell specialized for the transmission of nerve impulses.
neurotransmitter A chemical substance secreted by a neuron that aids in the transmission of the nerve impulse to an adjacent neuron.
niche The role that an organism plays in its environment.
nitrifying bacteria In the nitrogen cycle, bacteria that absorb ammonia and convert it into nitrate salts.
nitrogen cycle The process by which nitrogen is recycled and made available for use by other organisms.
nitrogen-fixing bacteria A type of bacteria responsible for absorbing atmospheric nitrogen and converting it to nitrate salts in the soil.
nitrogenous base A chemical unit composed of carbon, hydrogen, and nitrogen that is a component part of the nucleotide unit.
nitrogenous waste Any of a number of nitrogen-rich compounds that result from the metabolism of proteins and amino acids in the cell. (See **ammonia; urea; uric acid**.)
nondisjunction A type of chromosome mutation in which the members of one or more pairs of homologous chromosomes fail to separate during the disjunction phase of the first meiotic division.
nonplacental mammal A species of mammal in which internal development is accomplished without the aid of a placental connection (marsupial mammals).

nucleic acid An organic compound composed of repeating units of nucleotide.

nucleolus A cell organelle located within the nucleus that is known to function in protein synthesis.

nucleotide The repeating unit making up the nucleic acid polymer (e.g., DNA, RNA).

nucleus A cell organelle that contains the cell's genetic information in the form of chromosomes.

nutrition The life function by which living things obtain food and process it for their use.

omnivore A heterotrophic organism that consumes both plant and animal matter as sources of nutrition.

one gene-one polypeptide A scientific hypothesis concerning the role of the individual gene in protein synthesis.

oogenesis A type of meiotic cell division in which one ovum and three polar bodies are produced from each primary sex cell.

open transport system A type of circulatory system in which the transport fluid is *not* always enclosed within blood vessels (e.g., grasshopper).

oral cavity In human beings, the organ used for the ingestion of foods.

oral groove The ingestive organ of the paramecium.

organ transplant An application of the study of immunity in which an organ or tissue of a donor is transplanted into a compatible recipient.

organelle A small, functional part of a cell specialized to perform a specific life function (e.g., nucleus, mitochondrion).

organic compound A chemical compound that contains the elements carbon and hydrogen (e.g., carbohydrate, protein).

organic evolution The mechanism thought to govern the changes in living species over geologic time.

osmosis A form of passive transport by which water is absorbed or released by cells.

ovary A female gonad that secretes the hormone estrogen, which regulates female secondary sex characteristics; the ovary also produces ova, which are used in reproduction.

overcropping A negative aspect of human involvement with the environment in which soil is overused for the production of crops, leading to exhaustion of soil nutrients.

overgrazing The exposure of soil to erosion due to the loss of stabilizing grasses when it is overused by domestic animals—a negative aspect of human involvement with the environment.

overhunting A negative aspect of human involvement with the environment in which certain species have been greatly reduced or made extinct by uncontrolled hunting practices.

oviduct A tube that serves as a channel for conducting mature ova from the ovary to the uterus; the site of fertilization and the earliest stages of embryonic development.

ovulation The stage of the menstrual cycle in which the mature ovum is released from the follicle into the oviduct.

ovule A structure located within the flower ovary that contains a monoploid egg nucleus and serves as the site of fertilization.

ovum A type of gamete produced as a result of oogenesis in female animals; the egg, the female sex cell.

oxygen-18 A radioactive isotope of oxygen that is used to trace the movement of this element in biochemical reaction sequences.

oxygenated blood Blood that contains a high percentage of oxyhemoglobin.

oxyhemoglobin Hemoglobin that is loosely bound to oxygen for purposes of oxygen transport.

palisade layer A cell layer found in most leaves that contains high concentrations of chloroplasts.

pancreas An accessory organ which produces enzymes that complete the hydrolysis of foods to soluble end-products; also the site of insulin and glucagon production.

parasitism A type of symbiosis from which one organism in the relationship benefits, while the other (the "host") is harmed, but not ordinarily killed.

parathormone A hormone of the parathyroid gland that regulates the metabolism of calcium in the body.

parathyroid gland An endocrine gland whose secretion, parathormone, regulates the metabolism of calcium in the body.

passive immunity A temporary immunity produced as a result of the injection of preformed antibodies.

passive transport Any process by which materials are absorbed into the cell interior from an area of high concentration to an area of low concentration, without the expenditure of cell energy (e.g., osmosis, diffusion).

penis A structure that permits internal fertilization through direct implantation of sperm into the female reproductive tract.

peptide bond A type of chemical bond that links the nitrogen atom of one amino acid with the terminal carbon atom of a second amino acid in the formation of a dipeptide.

peripheral nerves Nerves in the earthworm and grasshopper that branch from the ventral nerve cord to other parts of the body.

peripheral nervous system A major subdivision of the nervous system that consists of all the nerves of all types branching through the body. (See **autonomic nervous system; somatic nervous system.**)

peristalsis A wave of contraction of the smooth muscle lining; the digestive tract that causes ingested food to pass along the food tube.

petal An accessory part of the flower that is thought to attract pollinating insects.

pH A chemical unit used to express the concentration of hydrogen ion (H^+), or the acidity, of a solution.

phagocyte A type of white blood cell that engulfs and destroys bacteria.

phagocytosis The process by which the ameba surrounds and ingests large food particles for intracellular digestion.

pharynx The upper part of the digestive tube that temporarily stores food before digestion.

phenotype The observable trait that results from the action of an allele pair.

phenylketonuria (PKU) A genetically related human disorder in which the homozygous combination of a particular mutant gene prevents the normal metabolism of the amino acid phenylalanine.

phloem A type of vascular tissue through which water and dissolved sugars are transported in plants from the leaf downward to the roots for storage.

phosphate group A chemical group made up of phosphorus and oxygen that is a component part of the nucleotide unit.

phosphoglyceraldehyde (PGAL) An intermediate product formed during photosynthesis that acts as the precursor of glucose formation.

photochemical reactions A set of biochemical reactions in photosynthesis in which light is absorbed and water molecules are split. (See **photolysis**.)

photolysis The portion of the photochemical reactions in which water molecules are split into hydrogen atoms and made available to the carbon fixation reactions.

photosynthesis A type of autotrophic nutrition in which green plants use the energy of sunlight to convert carbon dioxide and water into glucose.

phylum A level of biological classification that is a major subdivision of the kingdom level, containing fewer organisms with greater similarity (e.g., Chordata).

pinocytosis A special type of absorption by which liquids and particles too large to diffuse through the cell membrane may be taken in by vacuoles formed at the cell surface.

pioneer autotrophs The organisms supposed by the heterotroph hypothesis to have been the first to evolve the ability to carry on autotrophic nutrition.

pioneer species In an ecological succession, the first organisms to inhabit a barren environment.

pistil The female sex organ of the flower. (See **stigma; style; ovary**.)

pituitary gland An endocrine gland that produces hormones regulating the secretions of other endocrine glands; the "master gland."

placenta In placental mammals, a structure composed of both embryonic and maternal tissues that permits the diffusion of soluble substances to and from the fetus for nourishment and the elimination of fetal waste.

placental mammal A mammal species in which embryonic development occurs internally with the aid of a placental connection to the female parent's body.

Plant One of the five biological kingdoms; it includes multicellular organisms whose cells are bounded by cell walls and which are capable of photosynthesis (e.g., maple tree).

plasma The liquid fraction of blood, containing water and dissolved proteins.

plasma membrane A cell organelle that encloses the cytoplasm and other cell organelles and regulates the passage of materials into and out of the cell.

platelet A cell-like component of the blood that is important in clot formation.

polar body One of three nonfunctional cells produced during oogenesis that contain monoploid nuclei and disintegrate soon after completion of the process.

polio A disorder of the human regulatory system in which viral infection of the central nervous system may result in severe paralysis.

pollen The male gamete of the flowering plant.

pollen tube A structure produced by the germinating pollen grain that grows through the style to the ovary and carries the sperm nucleus to the ovule for fertilization.

pollination The transfer of pollen grains from anther to stigma.

pollution control The development of new procedures to reduce the incidence of air, water, and soil pollution—a positive aspect of human involvement with the environment.

polyploidy A type of chromosome mutation in which an entire set of homologous chromosomes fail to separate during the disjunction phase of the first meiotic division.

polysaccharide A type of carbohydrate composed of repeating units of monosaccharide that form a polymeric chain.

polyunsaturated fat A type of fat in which many bonding sites are unavailable for the addition of hydrogen atoms.

population All the members of a particular species in a given geographical location at a given time.

population control The use of various practices to slow the rapid growth in the human population—a positive aspect of human interaction with the environment.

population genetics A science that studies the genetic characteristics of a sexually reproducing species and the factors that affect its gene frequencies.

postnatal development The growth and maturation of an individual from birth, through aging, to death.

prenatal development The embryonic development that occurs within the uterus before birth. (See **gestation.**)

primary consumer Any herbivorous organism that receives food energy from the producer level (e.g., mouse); the second stage of a food chain.

primary sex cell The diploid cell that undergoes meiotic cell division to produce monoploid gametes.

producer Any autotrophic organism capable of trapping light energy and converting it to the chemical bond energy of food (e.g., green plants); the organisms forming the basis of the food chain.

progesterone A hormone produced by the corpus luteum and/or placenta that has the effect of maintaining the uterine lining and suppressing ovulation during gestation.

protease Any protein-hydrolyzing enzyme.

protein A complex organic compound composed of repeating units of amino acid.

Protista One of the five biological kingdoms; it includes simple unicellular forms whose nuclei are surrounded by nuclear membranes (e.g., ameba, paramecium).

pseudopod A temporary, flowing extension of the cytoplasm of an ameba that is used in nutrition and locomotion.

pulmonary artery One of two arteries that carry blood from the heart to the lungs for reoxygenation.

pulmonary circulation Circulation of blood from the heart through the lungs and back to the heart.

pulmonary vein One of four veins that carry oxygenated blood from the lungs to the heart.

pulse Rhythmic contractions of the artery walls that help to push the blood fluid through the capillary networks of the body.

punctuated equilibrium A theory of the time frame required for evolution which assumes that evolutionary change occurs in "bursts" with long periods of relative stability intervening.

pyramid of biomass A construct used to illustrate the fact that the total biomass available in each stage of a food chain diminishes from producer level to consumer level.

pyramid of energy A construct used to illustrate the fact that energy is lost at each trophic level in a food chain, being most abundant at the producer level.

pyruvic acid An intermediate product in the aerobic or anaerobic respiration of glucose.

receptor An organ specialized to receive a particular type of environmental stimulus.

recessive allele (gene) An allele (gene) whose effect is masked by that of its dominant allele.

recombinant DNA DNA molecules that have been moved from one cell to another in order to give the recipient cell a genetic characteristic of the donor cell.

recombination The process by which the members of segregated allele pairs are randomly recombined in the zygote as a result of fertilization.

rectum The portion of the digestive tract in which digestive wastes are stored until they can be released to the environment.

red blood cell Small, nonnucleated cells in the blood that contain hemoglobin and carry oxygen to bodily tissues.

reduction division See **meiosis.**

reflex A simple, inborn, involuntary response to an environmental stimulus.

reflex arc The complete path, involving a series of three neurons (sensory, interneuron, and motor), working together, in a reflex action.

regeneration A type of asexual reproduction in which new organisms are produced from the severed parts of a single parent organism; the replacement of lost or damaged tissues.

regulation The life process by which living things respond to changes within and around them, and by which all life processes are coordinated.

replication An exact self-duplication of the chromosome during the early stages of cell division; the exact self-duplication of a molecule of DNA.

reproduction The life process by which new cells arise from preexisting cells by cell division.

reproductive isolation The inability of species varieties to interbreed and produce fertile offspring, because of variations in behavior or chromosome structure.

respiration The life function by which living things convert the energy of organic foods into a form more easily used by the cell.

response The reaction of an organism to an environmental stimulus.

rhizoid A rootlike fiber produced by fungi that secrete hydrolytic enzymes and absorb digested nutrients.

ribonucleic acid (RNA) A type of nucleic acid that operates in various ways to facilitate protein synthesis.

ribose A five-carbon sugar found as a component part of the nucleotides of RNA molecules only.

ribosomal RNA (r-RNA) The type of RNA that makes up the ribosome.

ribosome A cell organelle that serves as the site of protein synthesis in the cell.

root A plant organ specialized to absorb water and dissolved substances from the soil, as well as to anchor the plant to the soil.

root hair A small projection of the growing root that serves to increase the surface area of the root for absorption.

roughage A variety of undigestible carbohydrates that add bulk to the diet and facilitate the movement of foods through the intestine.

runner A type of vegetative propagation in which an above-ground stem (runner) produces roots and leaves and establishes new organisms with identical characteristics.

saliva A fluid secreted by salivary glands that contains hydrolytic enzymes specific to the digestion of starches.

salivary gland The gland that secretes saliva, important in the chemical digestion of certain foods.

salt A chemical composed of a metal and a nonmetal joined by means of an ionic bond (e.g., sodium chloride).

saprophyte A heterotrophic organism that obtains its nutrition from the decomposing remains of dead plant and animal tissues (e.g., fungus, bacteria).

saturated fat A type of fat molecule in which all available bonding sites on the hydrocarbon chains are taken up with hydrogen atoms.

scrotum A pouch extending from the wall of the lower abdomen that houses the testes at a temperature optimum for sperm production.

secondary consumer Any carnivorous animal that derives its food energy from the primary consumer level (e.g., a snake); the third level of a food chain.

secondary sex characteristics The physical features, different in males and females, that appear with the onset of sexual maturity.

seed A structure that develops from the fertilized ovule of the flower and germinates to produce a new plant.

seed dispersal Any mechanism by which seeds are distributed in the environment so as to widen the range of a plant species. (See **fruit.**)

segregation The random separation of the members of allele pairs that occurs during meiotic cell division.

self-pollination A type of pollination in which the pollen of a flower pollinates another flower located on the same plant organism.

sensory neuron A type of neuron specialized for receiving environmental stimuli, which are detected by receptor organs.

sepal An accessory part of the flower that functions to protect the bud during development.

sessile A term that relates to the "unmoving" state of certain organisms, including the hydra.

seta One of several small, chitinous structures (setae) that aid the earthworm in its locomotor function.

sex chromosomes A pair of homologous chromosomes carrying genes that determine the sex of an individual; these chromosomes are designated as X and Y.

sex determination A pattern of inheritance in which the conditions of maleness and femaleness are determined by the inheritance of a pair of sex chromosomes (XX = female; XY = male).

sex linkage A pattern of inheritance in which certain nonsex genes are located on the X sex chromosome, but have no corresponding alleles on the Y sex chromosome.

sex-linked trait A genetic trait whose inheritance is controlled by the genetic pattern of sex linkage (e.g., color blindness).

sexual reproduction A type of reproduction in which new organisms are formed as a result of the fusion of gametes from two parent organisms.

shell An adaptation for embryonic development in many terrestrial, externally developing species that protects the developing embryo from drying and physical damage (e.g., birds).

sickle-cell anemia A genetically related human disorder in which the homozygous combination of a mutant gene leads to the production of abnormal hemoglobin and crescent-shaped red blood cells.

skeletal muscle A type of muscle tissue associated with the voluntary movements of skeletal levers in locomotion.

small intestine In human beings, the longest portion of the food tube, in which final digestion and absorption of soluble end-products occur.

smooth muscle See **visceral muscle.**

somatic nervous system A subdivision of the peripheral nervous system that is made up of nerves associated with voluntary actions.

speciation The process by which new species are thought to arise from previously existing species.

species A biological grouping of organisms so closely related that they are capable of interbreeding and producing fertile offspring (e.g., human being).

species presentation The establishment of game lands and wildlife refuges that have permitted the recovery of certain endangered species—a positive aspect of human involvement with the environment.

sperm A type of gamete produced as a result of spermatogenesis in male animals; the male reproductive cell.

spermatogenesis A type of meiotic cell division in which four sperm cells are produced for each primary sex cell.

spinal cord The part of the central nervous system responsible for reflex action, as well as impulse conduction between the peripheral nervous system and the brain.

spindle apparatus A network of fibers that form during cell division and to which centromeres attach during the separation of chromosomes.

spiracle One of several small pores in arthropods, including the grasshopper, that serve as points of entry of respiratory gases from the atmosphere to the tracheal tubes.

spongy layer A cell layer found in most leaves that is loosely packed and contains many air spaces to aid in gas exchange.

spore A specialized asexual reproductive cell produced by certain plants.

sporulation A type of asexual reproduction in which spores released from special spore cases on the parent plant germinate and grow into new adult organisms of the species.

staining A technique of cell study in which chemical stains are used to make cell parts more visible for microscopic study.

stamen The male reproductive structure in a flower. (See **anther; filament.**)

starch A type of polysaccharide produced and stored by plants.

stem A plant organ specialized to support the leaves and flowers of a plant, as well as to conduct materials between the roots and the leaves.

stigma The sticky upper portion of the pistil, which serves to receive pollen.

stimulus Any change in the environment to which an organism responds.

stomach A muscular organ that acts to liquefy food and that produces gastric protease for the hydrolysis of protein.

stomate A small opening that penetrates the lower epidermis of a leaf and through which respiratory and photosynthetic gases diffuse.

strata The layers of sedimentary rock that contain fossils, whose ages may be determined by studying the patterns of sedimentation.

stroke A disorder of the human regulatory system in which brain function is impaired because of oxygen starvation of brain centers.

stroma An area of the chloroplast within which the carbon-fixation reactions occur; stroma lie between pairs of grana.

style The portion of the pistil that connects the stigma to the ovary.

substrate A chemical that is metabolized by the action of a specific enzyme.

succession A situation in which an established ecological community is gradually replaced by another until a climax community is established.

survival of the fittest The concept, frequently associated with Darwin's theory of evolution, that in the intraspecies competition among naturally occurring species the organisms best adapted to the particular environment will survive.

sweat glands In human beings, the glands responsible for the production of perspiration.

symbiosis A term which refers to a variety of biotic relationships in which organisms of different species live together in close physical association.

synapse The gap that separates the terminal branches of one neuron from the dendrites of an adjacent neuron.

synapsis The intimate, highly specific pairing of homologous chromosomes that occurs in the first meiotic division, forming tetrads.

synthesis The life function by which living things manufacture the complex compounds required to sustain life.

systemic circulation The circulation of blood from the heart through the body tissues (except the lungs) and back to the heart.

systole The higher pressure registered during blood pressure testing. (See **diastole**.)

taiga A terrestrial biome characterized by long, severe winters and climax flora that includes coniferous trees.

Tay-Sachs A genetically related human disorder in which fatty deposits in the cells, particularly of the brain, inhibit proper functioning of the nervous system.

technological oversight A term relating to human activities that adversely affect environmental quality due to failure to adequately assess the environmental impact of a technological development.

teeth Structures located in the mouth that are specialized to aid in the mechanical digestion of foods.

temperate deciduous forest A terrestrial biome characterized by moderate climatic conditions and climax flora that includes deciduous trees.

template A pattern or design provided by the DNA molecule for the synthesis of protein molecules.

tendon A type of connective tissue that attaches a skeletal muscle to a bone.

tendonitis A disorder of the human locomotor system in which the junction between a tendon and a bone becomes irritated and inflamed.

tentacle A grasping structure in certain organisms, including the hydra, that contains stinging cells and is used for capturing prey.

terminal branch A cytoplasmic extension of the neuron that transmits a nerve impulse to adjacent neurons via the secretion of neurotransmitters.

terrestrial biome A biome that comprises primarily land ecosystems, the characteristics of which are determined by the major climate zone of the earth.

test cross A genetic cross accomplished for the purpose of determining the genotype of an organism expressing a dominant phenotype; the unknown is crossed with a homozygous recessive.

testis A gonad in human males that secretes the hormone testosterone, which regulates male secondary sex characteristics; the testis also produces sperm cells for reproduction.

testosterone A hormone secreted by the testis that regulates the production of male secondary sex characteristics.

tetrad A grouping of four chromatids that results from synapsis.

thymine A nitrogenous base found only in DNA.

thyroid gland An endocrine gland that regulates the body's general rate of metabolism through secretion of the hormone thyroxin.

thyroid-stimulating hormone (TSH) A pituitary hormone that regulates the secretions of the thyroid gland.

thyroxin A thyroid hormone that regulates the body's general metabolic rate.

tongue A structure that aids in the mechanical digestion of foods.

trachea A cartilage-ringed tube that conducts air from the mouth to the bronchi.

tracheal tube An adaptation in arthropods (e.g., grasshopper) which functions to conduct respiratory gases from the environment to the moist internal tissues.

Tracheophyta A phylum of the Plant Kingdom whose members (tracheophytes) contain vascular tissues and true roots, stems, and leaves (e.g., geranium, fern, bean, maple tree, corn).

transfer RNA (t-RNA) A type of RNA that functions to transport specific amino acids from the cytoplasm to the ribosome for protein synthesis.

translocation A type of chromosome mutation in which a section of a chromosome is transferred to a nonhomologous chromosome.

transpiration The evaporation of water from leaf stomates.

transpiration pull A force that aids the upward conduction of materials in the xylem by means of the evaporation of water (transpiration) from leaf surfaces.

transport The life function by which substances are absorbed, circulated, and released by living things.

triplet codon A group of three nitrogenous bases that provide information for the placement of amino acids in the synthesis of proteins.

tropical forest A terrestrial biome characterized by a warm, moist climate and a climax flora that includes many species of broadleaf trees.

tropism A plant growth response to an environmental stimulus.

tuber A type of vegetative propagation in which an underground stem (tuber) produces new tubers, each of which is capable of producing new organisms with identical characteristics.

tundra A terrestrial biome characterized by permanently frozen soil and climax flora that includes lichens and mosses.

tympanum A receptor organ in arthropods (e.g., grasshopper) which is specialized to detect vibrational stimuli.

ulcer A disorder of the human digestive tract in which a portion of its lining erodes and becomes irritated.

ultracentrifuge A tool of biological study that uses very high speeds of centrifugation to separate cell parts for examination.

umbilical cord In placental mammals, a structure containing blood vessels that connects the placenta to the embryo.

unicellular Having a body that consists of a single cell (e.g., paramecium).

uracil A nitrogenous base that is a component part of the nucleotides of RNA molecules only.

urea A type of nitrogenous waste with moderate solubility and moderate toxicity.

ureter In human beings, a tube that conducts urine from the kidney to the urinary bladder.

urethra In human beings, a tube that conducts urine from the urinary bladder to the exterior of the body.

uric acid A type of nitrogenous waste with low solubility and low toxicity.

urinary bladder An organ responsible for the temporary storage of urine.

urine A mixture of water, salts, and urea excreted from the kidney.

use and disuse A term associated with the evolutionary theory of Lamarck, since proved incorrect.

uterus In female placental mammals, the organ within which embryonic development occurs.

vaccination An inoculation of dead or weakened disease organisms that stimulates the body's immune system to produce active immunity.

vacuole A cell organelle that contains storage materials (e.g., starch, water) housed inside the cell.

vagina In female placental mammals, the portion of the reproductive tract into which sperm are implanted during sexual intercourse and through which the baby passes during birth.

variation A concept, central to Darwin's theory of evolution, that refers to the range of adaptation which can be observed in all species.

vascular tissues Tubelike plant tissues specialized for the conduction of water and dissolved materials within the plant. (See **xylem; phloem.**)

vegetative propagation A type of asexual reproduction in which new plant organisms are produced from the vegetative (nonfloral) parts of the parent plant.

vein (human) A relatively thin-walled blood vessel that carries blood from capillary networks back toward the heart.

vein (plant) An area of vascular tissues located in the leaf that aid the upward transport of water and minerals through the leaf and the transport of dissolved sugars to the stem and roots.

vena cava One of two major arteries that return blood to the heart from the body tissues.

ventral nerve cord The main pathway for nerve impulses between the brain and peripheral nerves of the grasshopper and earthworm.

ventricle One of two thick-walled, muscular chambers of the heart that pump blood out to the lungs and body.

villi Microscopic projections of the lining of the small intestine that absorb the soluble end-products of digestion. (See **lacteal.**)

visceral muscle A type of muscle tissue associated with the involuntary movements of internal organs (e.g., peristalsis in the small intestine).

vitamin a type of nutrient that acts as a coenzyme in various enzyme-controlled reactions.

water cycle The mechanism by which water is made available to living things in the environment through the processes of precipitation, evaporation, runoff, and percolation.

water pollution A type of technological oversight that involves the addition of some unwanted factor (e.g., sewage, heavy metals, heat, toxic chemicals) to our water resources.

Watson-Crick model A model of DNA structure devised by J. Watson and F. Crick that hypothesizes a "twisted ladder" arrangement for the DNA molecule.

white blood cell A type of blood cell that functions in disease control. (See **phagocyte; lymphocyte.**)

xylem A type of vascular tissue through which water and dissolved minerals are transported upward through a plant from the root to the stems and leaves.

yolk A food substance, rich in protein and lipid, found in the eggs of many animal species.

yolk sac The membrane that surrounds the yolk food supply of the embryos of many animal species.

zygote The single diploid cell that results from the fusion of gametes in sexual reproduction; a fertilized egg.

Regents Examinations, Answers, and Student Self-Appraisal Charts

Examination
August 2007
Living Environment

PART A

Answer all questions in this part. [30]

Directions (1–30): For *each* statement or question, select the word or expression that, of those given, best completes the statement or answers the question. Record your answers in the spaces provided.

1 Which condition would most likely upset the stability of an ecosystem?

  1 a cycling of elements between organisms and the environment

  2 energy constantly entering the environment

  3 green plants incorporating sunlight into organic compounds

  4 a greater mass of animals than plants   1_____

2 In 1910, Thomas Morgan discovered a certain pattern of inheritance in fruit flies known as sex linkage. This discovery extended the ideas of inheritance that Gregor Mendel had discovered while working with garden peas in 1865. Which principle of scientific inquiry does this illustrate?

  1 A control group must be part of a valid experiment.

  2 Scientific explanations can be modified as new evidence is found.

  3 The same experiment must be repeated many times to validate the results.

  4 Values can be used to make ethical decisions about scientific discovery.

        2_____

3 As a human red blood cell matures, it loses its nucleus. As a result of this loss, a mature red blood cell lacks the ability to

  1 take in material from the blood

  2 release hormones to the blood

  3 pass through artery walls

  4 carry out cell division

        3_____

4 Enzyme molecules normally interact with substrate molecules. Some medicines work by blocking enzyme activity in pathogens. These medicines are effective because they

  1 are the same size as the enzyme

  2 are the same size as the substrate molecules

  3 have a shape that fits into the enzyme

  4 have a shape that fits into all cell receptors

        4_____

5 The diagram below represents three human body systems.

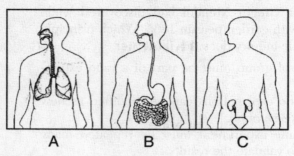

A   B   C   5_____

Which row in the chart below correctly shows what systems A, B, and C provide for the human body?

Row	System A	System B	System C
(1)	blood cells	glucose	hormones
(2)	oxygen	absorption	gametes
(3)	gas exchange	nutrients	waste removal
(4)	immunity	coordination	carbon dioxide

6 Which statement describes one function of the placenta in mammals?

1 It allows blood of the mother to mix with the blood of the fetus.

2 It contains fluid that protects the embryo from harm.

3 It removes waste products that are produced in the cells of the fetus.

4 It synthesizes food for the embryo.   6_____

7 The arrows in the diagram below indicate the movement of materials into and out of a single-celled
organism.

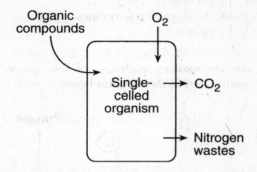

The movements indicated by all the arrows are
directly involved in

1 the maintenance of homeostasis
2 respiration, only
3 excretion, only
4 the digestion of proteins   7_____

8 The brown summer feathers of ptarmigans, small
Arctic birds, are replaced by white feathers after
winter arrives. Which statement best explains this
observation?

1 The expression of genes can be modified by the
environment.
2 Holes in the ozone layer vary in size depending
on the season.
3 Acids in rain bleach the brown feathers of the
birds.
4 Mutations occur only during certain seasons.   8_____

9 A child has brown hair and brown eyes. His father has brown hair and blue eyes. His mother has red hair and brown eyes. The best explanation for the child having brown hair and brown eyes is that

1 a gene mutation occurred that resulted in brown hair and brown eyes
2 gene expression must change in each generation so evolution can occur
3 the child received genetic information from each parent
4 cells from his mother's eyes were present in the fertilized egg

9_____

10 The diagram below represents a portion of a type of organic molecule present in the cells of organisms.

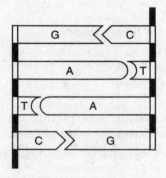

What will most likely happen if there is a change in the base sequence of this molecule?

1 The molecule will be converted into an inorganic compound.
2 The amino acid sequence may be altered during protein synthesis.
3 The chromosome number will decrease in future generations.
4 The chromosome number may increase within the organisms.

10_____

11 All cells in an embryo have the same DNA. However, the embryonic cells form organs, such as the brain and the kidneys, which have very different structures and functions. These differences are the result of

  1 having two types of cells, one type from each parent

  2 rapid mitosis causing mutations in embryo cells

  3 new combinations of cells resulting from meiosis

  4 certain genes being expressed in some cells and not in others   11_____

12 Viruses frequently infect bacteria and insert new genes into the genetic material of the bacteria. When these infected bacteria reproduce asexually, which genes would most likely be passed on?

  1 only the new genes

  2 only the original genes

  3 both the original and the new genes

  4 neither the original nor the new genes   12_____

13 A mutation changes a gene in a cell in the stomach of an organism. This mutation could cause a change in

  1 both the organism and its offspring

  2 the organism, but not its offspring

  3 its offspring, but not the organism itself

  4 neither the organism nor its offspring   13_____

14 A certain protein is found in mitochondria, chloroplasts, and bacteria. This provides evidence that
plants and bacteria

1 have some similar DNA base sequences
2 can use carbon dioxide to make proteins
3 digest proteins into simple sugars
4 contain certain pathogenic microbes   14_____

15 Extinction of a species could result from

1 evolution of a type of behavior that produces
greater reproductive success
2 synthesis of a hormone that controls cellular
communication
3 limited genetic variability in the species
4 fewer unfavorable mutations in the species   15_____

16 The rate at which all organisms obtain, transform,
and transport materials depends on an immediate
supply of

1 ATP and enzymes
2 solar energy and carbon dioxide
3 carbon dioxide and enzymes
4 ATP and solar energy   16_____

17 As women age, their reproductive cycles stop due to
decreased

1 digestive enzyme production
2 production of ATP
3 levels of specific hormones
4 heart rate   17_____

18 The evolutionary pathways of five species are represented in the diagram below.

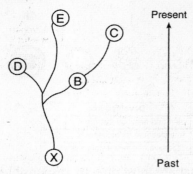

Which statement is supported by the diagram?

1 Species *C* is the ancestor of species *B*.
2 Species *D* and *E* evolved from species *B*.
3 Species *X* evolved later than species *D* but before species *B*.
4 Both species *C* and species *D* are related to species *X*.   18_____

19 Which cell is normally produced as a direct result of meiosis?

1 a uterine cell having half the normal species number of chromosomes
2 an egg having the full species number of chromosomes
3 a zygote having the full species number of chromosomes
4 a sperm having half the normal species number of chromosomes   19_____

20 The diagram and chart below represent some of the changes a zygote undergoes during its development.

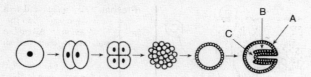

Layer	Develops Into
A	skin and nervous system
B	muscles and blood vessels
C	digestive and respiratory systems

The processes that are most directly responsible for these changes are

1 sorting and recombination of genetic information
2 mitosis and differentiation
3 meiosis and adaptation
4 fertilization and cycling of materials   20____

21 The diagram below shows the growth pattern of some skin cells in the human body after they have been exposed to ultraviolet radiation.

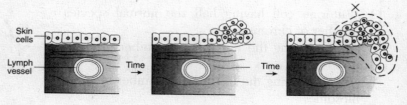

The cells in area X are most likely

1 red blood cells
2 cancer cells
3 white blood cells
4 sex cells   21____

22 The diagram below represents a cross section of part of a leaf.

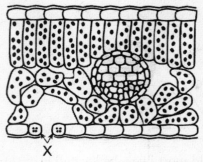

X

Which life functions are directly regulated through feedback mechanisms associated with the actions of the structures labeled *X*?

1 excretion and immunity
2 digestion and coordination
3 circulation and reproduction
4 respiration and photosynthesis   22_____

23 One irreversible effect of both deforestation and water pollution on the environment is the

1 extinction of species
2 thinning of the ozone shield
3 depletion of atmospheric carbon dioxide levels
4 increase in renewable resources   23_____

24 An energy pyramid containing autotrophs and other organisms from a food chain is represented below.

Carnivores would most likely be located in

1 level I, only
2 level I and level II
3 level III, only
4 level II and level III

24_____

25 Which statement describes a situation that leads to stability within an ecosystem?

1 Carbon dioxide and water are released only by abiotic sources in the ecosystem.
2 Interactions between biotic and abiotic components regulate carbon dioxide and water levels.
3 Animals provide the oxygen used by plants, and plants provide the nitrogen needed by animals.
4 Organisms provide all the necessary energy for the maintenance of this ecosystem.

25_____

26 Worms that had been invaded by bacteria were eaten by a species of bird. Many of these birds died as a result. The most likely explanation for this is that the

1 bacteria interfered with normal life functions of the birds
2 disease that killed the birds was inherited
3 gene alterations in the bacterial cells killed the birds
4 birds produced antigens in response to the bacteria   26_____

27 Which action illustrates an increased understanding and concern by humans for ecological interrelationships?

1 importing organisms in order to stabilize existing ecosystems
2 eliminating pollution standards for industries that promote technology
3 removing natural resources at a rate equal to the needs of the population
4 implementing laws to regulate the number of animals hunted and killed each year   27_____

28 The graph below represents the growth of a population of flies in a jar.

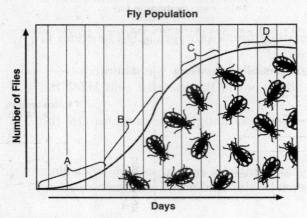

Fly Population

Which letter indicates the part of the graph that represents the carrying capacity of the environment in the jar?

(1) *A*   (3) *C*
(2) *B*   (4) *D*   28_____

29 One likely reason some experimental automobiles have been developed to use electricity rather than gasoline is that

1 gasoline is made from petroleum, a nonrenewable resource
2 Earth has an unlimited supply of fossil fuels
3 the use of electricity will eliminate the need for all antipollution laws
4 the use of electricity will increase the manufacture of antipollution devices for cars   29_____

30 Ecosystems will have a greater chance of maintaining equilibrium over a long period of time if they have

  1 organisms imported by humans from other environments

  2 a sudden change in climate

  3 a diversity of organisms

  4 predators eliminated from the food chains   30_____

PART B–1

Answer all questions in this part.   [10]

Directions (31–40): For *each* statement or question, select the word or expression that, of those given, best completes the statement or answers the question. Record your answers in the spaces provided.

31 The diagram below represents a food pyramid.

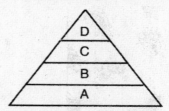

The concentration of the pesticide DDT in individual organisms at level *D* is higher than the concentration in individuals at level *A* because DDT is

1 synthesized by organisms at level *D*
2 excreted by organisms at level *A* as a toxic waste
3 produced by organisms at level *C* which are eaten by organisms at level *D*
4 passed through levels *A*, *B*, and *C* to organisms at level *D*

31 _____

32 Which concept is represented in the graph below?

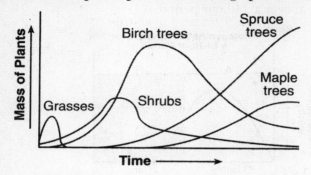

1 ecological succession in a community
2 cycling of carbon and nitrogen in a forest
3 energy flow in a food chain over time
4 negative human impact on the environment   32_____

33 The graph below shows photosynthetic activity in an ecosystem over a 24-hour period.

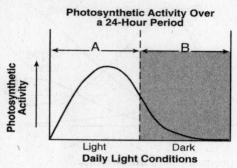

Photosynthetic Activity Over a 24-Hour Period

Data for a study on respiration in this ecosystem should be collected during

1 interval *A*, from only the producers in the ecosystem

2 intervals *A* and *B*, from only the consumers in the ecosystem

3 intervals *A* and *B*, from both the producers and consumers in the ecosystem

4 interval *A* only, from abiotic but not biotic components of the ecosystem

33 _____

Base your answers to questions 34 and 35 on the information and data table below and on your knowledge of biology.

A student studied the location of single-celled photosynthetic organisms in a lake for a period of several weeks. The depth at which these organisms were found at different times of the day varied greatly. Some of the data collected are shown in the table below.

Data Table

Light Conditions at Different Times of the Day	Average Depth of Photosynthetic Organisms (cm)
full light	150
moderate light	15
no light	10

34 A valid inference based on these data is that

1 most photosynthetic organisms live below a depth of 150 centimeters
2 oxygen production increases as photosynthetic organisms move deeper in the lake
3 photosynthetic organisms respond to changing light levels
4 photosynthetic organisms move up and down to increase their rate of carbon dioxide production   34_____

35 Which materials would the student most likely have used in this investigation?

   1 microscope, pipette, and slides with coverslips
   2 graduated cylinder, triple-beam balance, and chromatography paper
   3 thermometer, electric balance, and biological stains
   4 computer, pH paper, and gel electrophoresis apparatus

35_____

36 A student prepared a slide of pollen grains from a flower. First the pollen was viewed through the low-power objective lens and then, without moving the slide, viewed through the high-power objective lens of a compound light microscope.

Which statement best describes the relative number and appearance of the pollen grains observed using these two objectives?

   1 low power: 25 small pollen grains
      high power: 100 large pollen grains
   2 low power: 100 small pollen grains
      high power: 25 large pollen grains
   3 low power: 25 large pollen grains
      high power: 100 small pollen grains
   4 low power: 100 large pollen grains
      high power: 25 small pollen grains

36_____

Base your answers to questions 37 and 38 on the diagram below, which represents a sequence of events in a biological process that occurs within human cells and on your knowledge of biology.

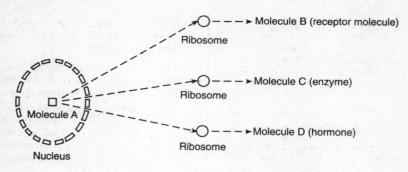

37 Molecule *A* contains the

1 starch necessary for ribosome synthesis in the cytoplasm
2 organic substance that is broken down into molecules *B*, *C*, and *D*
3 proteins that form the ribosome in the cytoplasm
4 directions for the synthesis of molecules *B*, *C*, and *D*   37_____

38 Molecules *B*, *C*, and *D* are similar in that they are usually

1 composed of genetic information
2 involved in the synthesis of antibiotics
3 composed of amino acids
4 involved in the diffusion of oxygen into the cell   38_____

39 A technique used to produce new plants is represented in the diagram below.

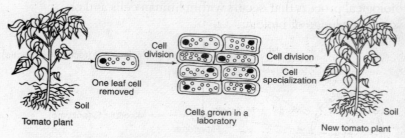

Which statement is best supported by the information in the diagram?

1 The one leaf cell removed formed a zygote that developed into a new plant by mitotic cell division.

2 This procedure is used to produce new tomato plants that are clones of the original tomato plant.

3 The cell taken from the leaf produced eight cells, each having one-half of the genetic information of the original leaf cell.

4 The new tomato plant will not be able to reproduce sexually because it was produced by mitotic cell division.

39____

Base your answer to question 40 on the information below and on your knowledge of biology.

Students cut 20 rod-shaped pieces of potato of the same diameter and length. Five pieces of potato were placed into each of four beakers containing different concentrations of sugar solutions. Each potato piece was measured again after 24 hours. The table below shows the results of their experiment.

Change in Length

Concentration of Sugar Solution (grams per liter)	Original Length of Potato Pieces (mm)	Average Length After 24 Hours (mm)
0	50.0	52.0
5	50.0	44.0
8	50.0	43.5
10	50.0	42.5

40 Which graph best represents the information in the
data table?

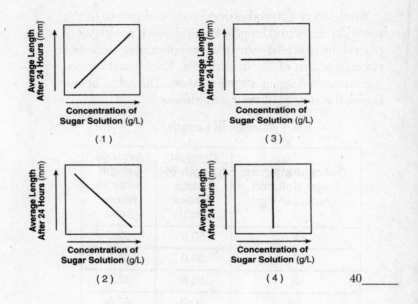

Concentration of
Sugar Solution (g/L)

(1)

Concentration of
Sugar Solution (g/L)

(3)

Concentration of
Sugar Solution (g/L)

(2)

Concentration of
Sugar Solution (g/L)

(4)

40_____

PART B–2

Answer all questions in this part.   [15]

Directions (41–54): **For those questions that are followed by four choices, record your answers in the spaces provided. For all other questions in this part, record your answers in accordance with the directions.**

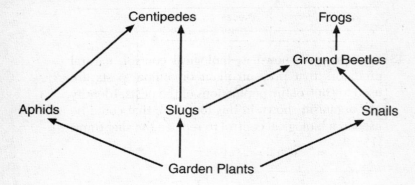

Gardeners sometimes use slug traps to capture and kill slugs. These traps were tested in a garden with a large slug population. Organisms found in the trap after one week are shown in the table below.

Organisms in Trap

Organism	Number in Trap
slugs	8
snails	1
aphids	13
centipedes	1
ground beetles	98

41 How many organisms in the trap were herbivores?

    (1) 5   (3) 22

    (2) 9   (4) 99   41_____

42 State *one* reason the slug traps are *not* the best method to control slugs.   [1]

_____

_____

43 In a process known as biological control, natural predators that prey on plant or animal pests are used to control the populations of the pests. Identify *one* organism shown in this food web that could be used as a biological control to replace the slug traps. [1]

_____

_____

44 Write the structures listed below in order from least complex to most complex.   [1]

    organ
    cell
    organism
    organelle
    tissue

    Least complex: _____

                     _____

                     _____

                     _____

    Most complex: _____

45 To prevent harm to the fetus, women should avoid tobacco, alcohol, and certain medications during pregnancy. State *one* specific way that *one* of these substances could harm the fetus. [1]

_____

_____

Base your answers to questions 46 through 48 on the information below and on your knowledge of biology.

Arsenic and Old Glucocorticoids

Constant exposure to small amounts of arsenic in drinking water has been found to increase the risk of cancer and other diseases. In January of 2001, the EPA (Environmental Protection Agency) lowered the acceptable levels of arsenic in drinking water from 50 ppb (parts per billion) to 10 ppb.

Researchers are now trying to determine how arsenic affects the body. Recent experiments suggest that arsenic may block the activity of hormones. One group of hormones affected by arsenic is glucocorticoids, which are responsible for activating many genes that appear to suppress cancer.

Rat tumor cells were used to determine the effect of arsenic on glucocorticoids. One group of cells was treated with a solution of synthetic glucocorticoid and arsenic, another with a solution of synthetic glucocorticoid and water, and a third group with a solution containing only water. Researchers then measured the activity of one of the genes that is usually activated by glucocorticoids. The genes in the cells treated with the hormone and arsenic mixture and those treated with just water did not become activated. The genes in the cells treated with the hormone and water mixture were activated. Researchers concluded that arsenic blocked the normal activity of the hormone. They are now extending their studies to determine if arsenic acts in a similar manner in other types of cells and in entire organisms.

46 Research suggests that a buildup of arsenic in the cells of humans may be harmful because

　1 synthetic arsenic can be formed by the breakdown of glucocorticoids in the body
　2 arsenic prevents the action of genes that are important in reactions that suppress cancer
　3 arsenic prevents the reaction in which water and hormones bond and attach to cancer cells
　4 glucocorticoids can build up in tissues and cause an increase in the absorption of arsenic   46____

47 State *one* reason this study should be extended to other cells or to other complex organisms.   [1]

_____

_____

48 Identify *one* specific hormone in the body, other than glucocorticoid. Explain how disruption of the activity of the hormone you identified might upset a feedback mechanism in the body.   [2]

_____

_____

_____

Base your answers to questions 49 and 50 on the diagram below, which illustrates some steps in genetic engineering and on your knowledge of biology.

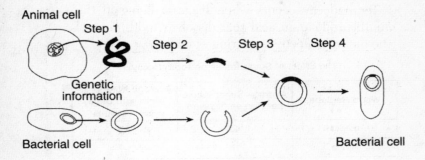

49 What is the result of step 3?

1 a new type of molecular base is formed
2 different types of minerals are joined together
3 DNA from the bacterial cell is cloned
4 DNA from different organisms is joined together   49____

50 State *one* way that enzymes are used in step 2. [1]

_____

_____

Base your answers to questions 51 through 54 on the information and data table below and on your knowledge of biology.

Tooth decay occurs when bacteria living in the mouth produce an acid that dissolves tooth enamel (the outer, protective covering of a tooth).

The Effect of Sugar Intake on Tooth Decay

World Regions	Average Sugar Intake per Person (kg/year)	Average Number of Teeth with Decay per Person
Americas	40	3.0
Africa	18	1.7
Southeast Asia	14	1.6
Europe	36	2.6

Directions (51–53): Using the information in the data table, construct a bar graph on the grid provided on the next page, following the directions below.

51 Mark an appropriate scale on the axis labeled "Average Sugar Intake per Person." [1]

52 Construct vertical bars in the bracketed area for each world region to represent the "Average Sugar Intake per Person." Place the bars on the left side of each bracketed region and shade the bars as shown below. (The bar for Americas has been done for you.) [1]

53 Construct vertical bars in the bracketed area for each world region to represent the "Average Number of Teeth with Decay per Person." Place the bars on the right side of each bracketed region and shade in each bar as shown below. [1]

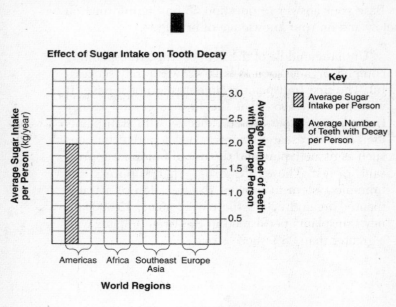

54 Which statement is a valid conclusion regarding tooth decay?

1 As sugar intake increases, the acidity in the mouth decreases, reducing tooth decay.

2 As sugar intake increases, tooth decay increases in Europe and the Americas, but not in Africa and Southeast Asia.

3 The greater the sugar intake, the greater the average number of decayed teeth.

4 The greater the sugar intake, the faster a tooth decays.

54____

PART C

Answer all questions in this part. [17]

Directions (55–62): **Record your answers in the spaces provided.**

Base your answer to question 55 on the information below and on your knowledge of biology.

Until the middle of the 20th century, transplanting complex organs, such as kidneys, was rarely successful. The first transplant recipients did not survive. It was not until 1954 that the first successful kidney transplant was performed. Success with transplants increased as research scientists developed techniques such as tissue typing and the use of immunosuppressant drugs. These are drugs that suppress the immune system to prevent the rejection of a transplanted organ. In 2002, there were nearly 15,000 kidney transplants performed in the United States with a greater than 95% success rate.

55 Describe the relationship of the immune system to organ transplants and the use of immunosuppressant drugs to prevent the rejection of a transplanted organ. In your answer be sure to:

- state *one* way the immune system is involved in the rejection of transplanted organs   [1]

- explain why the best source for a donated kidney would be the identical twin of the recipient   [1]

- explain why immunosuppressant drugs might be needed to prevent rejection of a kidney received from a donor other than an identical twin   [1]

- state *one* reason a person may get sick more easily when taking an immunosuppressant drug   [1]

_____

_____

_____

_____

_____

_____

_____

_____

_____

Base your answers to questions 56 through 58 on the information below and on your knowledge of biology.

A population of gray squirrels lived in the trees surrounding four houses in a city. The houses and trees were removed, and a tall office building was constructed in their place. Some of the squirrels were able to survive by relocating to the trees in a park nearby.

56 State *one* specific way the relocated squirrels would most likely interact with a gray squirrel population that has lived in the park for many years.   [1]

_____

_____

57 State *one* specific way the relocated squirrels will change an abiotic factor in the park ecosystem. [1]

_____

_____

58 State *one* specific natural factor in the park ecosystem that will limit the growth of the squirrel population and support your answer. [1]

_____

_____

59 An individual has placed an editorial in the community newspaper stating that the local recycling
program should be discontinued. Respond to this
editorial by explaining the importance of the local
recycling program for the environment. In your
explanation be sure to:

- state *one* effect the increasing human population
  will have on the availability of natural resources
  [1]

- state *one* reason why recycling is important   [1]

- identify *two* natural resources or products made
  from natural resources that can be recycled   [2]

_____

_____

_____

_____

_____

_____

_____

_____

_____

_____

60 An insect pest known as the medfly significantly reduced the orange crop in California. Pesticides were used to control the medfly. Using the concept of natural selection, explain how the continued use of a certain pesticide may become ineffective in controlling this fly. Your answer must include the concepts of:

- variation   [1]
- adaptive value of a variation (adaptation)   [1]
- survival   [1]
- reproduction   [1]

_____

_____

_____

_____

_____

_____

_____

_____

_____

Base your answers to questions 61 and 62 on the passage below and on your knowledge of biology.

Human activities have had a major impact on biodiversity. Scientists cannot solve this problem alone. Concerned individuals need to be involved in restoring and maintaining biodiversity.

61 Explain how a loss of biodiversity today can affect the survival of humans in the future. [1]

_____

_____

_____

62 State *one* specific action that you as a student can take in your community to help maintain or increase biodiversity. [1]

_____

_____

PART D

Directions (63–75): **Answer all questions in this part.** [13]

For those questions that are followed by four choices, record your answers in the spaces provided. For all other questions in this part, record your answers in accordance with the directions given in the question.

63 The data in the table below were collected during a reaction-time experiment conducted in five biology classes. Average reaction times for each class were determined first at room temperature and then after cooling each student's hand in cold water for two minutes.

Average Reaction Times to Grab a Falling Ruler

Class	At Room Temperature (seconds)	After Cooling (seconds)
1	.42	.48
2	.36	.41
3	.35	.47
4	.43	.58
5	.44	.47
Averages	.40	.48

Which statement is best supported by the data?

1 Cooling the hand increases the reaction time.
2 Cooling the hand does not affect the reaction time.
3 Cooling the hand affects only some subjects.
4 Two minutes of cooling is not enough to affect reaction time.

63_____

64 A student hypothesized that the pulse rate in humans would increase 1 hour after eating a meal. Pulse rates were obtained from nine classmates 1 hour after eating lunch. The data in beats per minute were recorded as: 60, 64, 56, 68, 72, 76, 72, 80, and 68. State *one* error in this experiment. [1]

_____

_____

Base your answers to questions 65 and 66 on the information below and on your knowledge of biology.

A student read a magazine article that claimed people who exercise for 30 minutes are able to solve more math problems than if they had not exercised. The student convinced four of his friends to test this claim. First, he gave them 15 minutes to do 50 math problems. The number each person solved is shown in the trial 1 graph. Next, all four of the students exercised for 30 minutes. At the end of the 30 minutes, they were given another 50 math problems of equal difficulty for the same amount of time. The number of math problems each student solved is shown in the trial 2 graph.

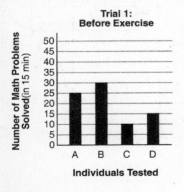

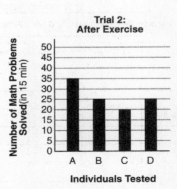

65 Explain why exercise could influence the ability of a student to solve math problems. [1]

_____

_____

66 State whether or not exercising for 30 minutes improved the ability of students to solve math problems. Support your answer using data from the graphs. [1]

_____

_____

_____

Base your answer to question 67 on the information and data table below and on your knowledge of biology.

Body Structures and Reproductive Characteristics of Four Organisms

Organism	Body Structures	Reproductive Characteristics
pigeon	feathers, scales 2 wings, 2 legs	lays eggs
A	scales 4 legs	lays eggs
B	fur 2 leathery wings, 2 legs	gives birth to live young provides milk for offspring
C	fur 4 legs	lays eggs provides milk for offspring

67 Explain why it would be difficult to determine which one of the other three organisms from the table should be placed in box 1. [1]

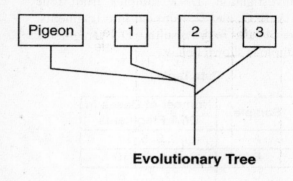

Evolutionary Tree

_____

_____

_____

68 If frog eggs taken from a freshwater pond are placed in a saltwater aquarium, what will most likely happen?

1 Water will leave the eggs.
2 Salt will leave the eggs.
3 Water will neither enter nor leave the eggs.
4 The eggs will burst. 68____

Base your answers to questions 69 through 71 on the information below and on your knowledge of biology.

In an investigation, DNA samples from four organisms, *A*, *B*, *C*, and *D*, were cut into fragments. The number of bases in the resulting DNA fragments for each sample is shown below.

Data Table

Sample	Number of Bases in DNA Fragments
A	3, 9, 5, 14
B	8, 4, 12, 10
C	11, 7, 6, 8
D	4, 12, 8, 11

69 The diagram below represents the gel-like material through which the DNA fragments moved during gel electrophoresis. Draw lines to represent the position of the fragments from each DNA sample when electrophoresis is completed.   [1]

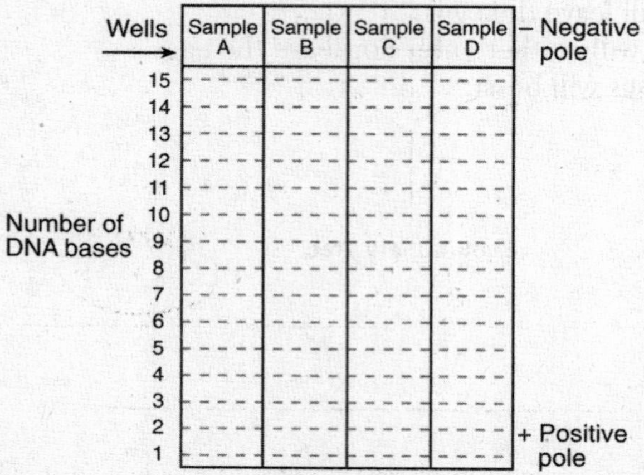

Wells →	Sample A	Sample B	Sample C	Sample D	− Negative pole

Number of DNA bases: 15, 14, 13, 12, 11, 10, 9, 8, 7, 6, 5, 4, 3, 2, 1

+ Positive pole

70 Which *two* DNA samples are the most similar? Support your answer using data from this investigation. [1]

Samples _____ and _____

_____

_____

71 State *one* specific use for the information obtained from the results of gel electrophoresis. [1]

_____

_____

Base your answers to questions 72 through 74 on the information below and on your knowledge of biology.

In the *Beaks of Finches* laboratory activity, students were each assigned a tool to use to pick up seeds. In round one, students acting as birds used their assigned tools to pick up small seeds from their own large dishes (the environment) and place them in smaller dishes (their stomachs). The seeds collected by each student were counted. Some students were able to collect many seeds, while others collected just a few.

In round two, students again used their assigned tools to collect seeds. This time several students were picking up seeds from the same dish of seeds.

72 Explain how this laboratory activity illustrates the process of natural selection. [1]

_____

_____

73 One factor that influences the evolution of a species that was *not* part of this laboratory activity is

1 struggle for survival
2 variation
3 competition
4 overproduction

73____

74 Identify *one* trait, other than beak characteristics, that could contribute to the ability of a finch to feed successfully. [1]

_____

_____

75 A student fills a dialysis membrane bag with a mixture of red dye, yellow dye, and water. He soaks the bag in pure water for 24 hours and then observes that the water outside the bag turns yellow. Which statement best explains the results of this experiment?

1 Water diffused into the membrane bag.
2 The dialysis membrane actively transported yellow dye molecules.
3 Only red dye diffused through the membrane.
4 The yellow dye molecules are smaller than the red dye molecules.

75____

Answers
August 2007
Living Environment

Answer Key

PART A

1. 4	**7.** 1	**13.** 2	**19.** 4	**25.** 2
2. 2	**8.** 1	**14.** 1	**20.** 2	**26.** 1
3. 4	**9.** 3	**15.** 3	**21.** 2	**27.** 4
4. 3	**10.** 2	**16.** 1	**22.** 4	**28.** 4
5. 3	**11.** 4	**17.** 3	**23.** 1	**29.** 1
6. 3	**12.** 3	**18.** 4	**24.** 3	**30.** 3

PART B–1

31. 4	**34.** 3	**37.** 4	**40.** 2
32. 1	**35.** 1	**38.** 3	
33. 3	**36.** 2	**39.** 2	

PART B–2

41. 3
42. *See* Answers Explained.
43. *See* Answers Explained.
44. *See* Answers Explained.
45. *See* Answers Explained.
46. 2
47. *See* Answers Explained.
48. *See* Answers Explained.
49. 4
50. *See* Answers Explained.
51. *See* Answers Explained.
52. *See* Answers Explained.
53. *See* Answers Explained.
54. 3

PART C

55. *See* Answers Explained.
56. *See* Answers Explained.
57. *See* Answers Explained.
58. *See* Answers Explained.
59. *See* Answers Explained.
60. *See* Answers Explained.
61. *See* Answers Explained.
62. *See* Answers Explained.

PART D

63. 1
64. *See* Answers Explained.
65. *See* Answers Explained.
66. *See* Answers Explained.
67. *See* Answers Explained.
68. 1
69. *See* Answers Explained.
70. *See* Answers Explained.
71. *See* Answers Explained.
72. *See* Answers Explained.
73. 4
74. *See* Answers Explained.
75. 4

Answers Explained

PART A

1. **4** *A greater mass of animals than plants* is the condition that would most likely upset the stability of an ecosystem. Solar energy captured by producer organisms (plants) is inefficiently passed on to the consumer level in a food pyramid, with as much as 90% lost to the environment as heat at each exchange. For this reason, any food pyramid must have a greater biomass of producer organisms than of consumers (animals) that directly or indirectly depend on the producers for food. This ecosystem will most likely experience a rapid reduction in consumer biomass until the proper balance is restored.

WRONG CHOICES EXPLAINED

(1) *A cycling of elements between organisms and the environment* is *not* the condition that would most likely upset the stability of an ecosystem. Elements essential to life (such as carbon, hydrogen, oxygen, and nitrogen) cycle naturally in healthy ecosystems with the help of biological processes. Without such cycling, these elements would no longer be available to support life, so life would soon cease. An ecosystem with cycling of elements would tend to be stable.

(2) *Energy constantly entering the environment* is *not* the condition that would most likely upset the stability of an ecosystem. Energy constantly enters the ecosystem in the form of sunlight, which is captured by green plants and converted to chemical bond energy. Solar energy captured by producer organisms (plants) is inefficiently passed on to the consumer level in a food pyramid, so must be constantly replenished. An ecosystem with a constant source of energy would tend to be stable.

(3) *Green plants incorporating sunlight into organic compounds* is *not* the condition that would most likely upset the stability of an ecosystem. Solar energy captured by producer organisms (plants) is used in the process of photosynthesis to create chemical bonds found in molecules of glucose. Glucose is then converted to more complex organic molecules that are used as food by consumer organisms, forming the basis of the food pyramid. An ecosystem in which plants incorporate sunlight into organic compounds would tend to be stable.

2. **2** *Scientific explanations can be modified as new evidence is found* is the principle of scientific inquiry illustrated by this discovery. As each new scientific discovery is made, it adds to the body of knowledge about biological phenomena and provides researchers with additional information on which to base new experiments. In this illustration, Mendel's work with garden peas identi-

fied, but could not explain, a discrete inheritance pattern; Morgan's work with fruit flies was able to explain the genetic basis of Mendel's observations.

WRONG CHOICES EXPLAINED

(1), (3), (4) *A control group must be part of a valid experiment, The same experiment must be repeated many times to validate the results*, and *Values can be used to make ethical decisions about scientific discovery* are *not* the principles of scientific inquiry illustrated by this discovery. Although each of these principles is a true statement about valid scientific experiments, none of them is directly illustrated by the discovery described.

3. **4** As a result of the loss of its nucleus, a mature red blood cell lacks the ability to *carry out cell division*. The cell nucleus contains genetic material that carries the blueprint of cellular biochemistry. During cell division in nucleated cells, this genetic material is passed on to the daughter cells. Red blood cells that lack a nucleus cannot reproduce by cell division and eventually die.

WRONG CHOICES EXPLAINED

(1) As a result of the loss of its nucleus, a mature red blood cell does *not* lack the ability to *take in material from the blood*. As is true of all living cells, a mature red blood cell takes in water, nutrients, and inorganic materials from the blood plasma by simple diffusion. In fact, the principal function of mature red blood cells is to take in and transport oxygen gas that enters blood plasma via the alveoli of the lung.

(2) As a result of the loss of its nucleus, a mature red blood cell does *not* lack the ability to *release hormones to the blood*. Red blood cells cannot perform this function under any circumstance. Hormones are released into the blood from specialized endocrine tissues, not from red blood cells.

(3) As a result of the loss of its nucleus, a mature red blood cell does *not* lack the ability to *pass through artery walls*. Red blood cells cannot perform this function under any circumstance. The muscular walls of normal arteries are too thick and impervious to allow red blood cells to pass through them.

4. **3** These medicines are effective because they *have a shape that fits into the enzyme*. The "lock-and-key" model of enzyme specificity explains that the chemical shape of an enzyme is complementary to the chemical shape of its substrate, allowing the two molecules to fit together at the active site in a manner similar to a key that fits only one specific lock. This chemical fit allows the enzyme to catalyze a chemical reaction involving the substrate. The enzyme-blocking medicine molecule has the same chemical shape as the substrate molecule and, once attached to the enzyme at the active site, blocks attachment of substrate molecules and prevents the chemical reaction from occurring.

WRONG CHOICES EXPLAINED

(1), (2) It is *not* true that these medicines are effective because they *are the same size as the enzyme* or because they *are the same size as the substrate molecules*. The critical factor in the medicines' effectiveness is not their size relative to these molecules, but their chemical shape in relation to the enzyme's active site.

(4) It is *not* true that these medicines are effective because they *have a shape that fits into all cell receptors*. Cell receptors are specialized proteins embedded in the cell membrane that function to communicate with other cells and/or to actively transport specific molecules into and out of the cell. It would be dangerous and unethical to produce a medicine that blocks all cell receptors because such a medicine could shut down all such cell communication and transport mechanisms, potentially imperiling life. The medicines described are designed to block only one kind of enzyme.

5. **3** Row 3 in the chart correctly shows what systems *A*, *B*, and *C* provide for the human body. System *A* is the respiratory system, which provides moist interior membranes that allow the absorption of oxygen gas into, and the excretion of carbon dioxide gas out of, the body. System *B* is the digestive system, which produces hydrolytic enzymes for the breakdown of complex foods and moist interior membranes specialized for the absorption of simple nutrient molecules such as glucose, amino acids, and fatty acids. System *C* is the excretory system, which is specialized for the removal of toxic waste molecules such as urea and salts from the blood and for their elimination to the outside environment.

WRONG CHOICES EXPLAINED

(1), (2), (4) Rows *1*, *2*, and *4* in the chart do *not* correctly show what systems *A*, *B*, and *C* provide for the human body. Each of these rows contains one or more incorrect items of information. [NOTE: Care must be taken to correctly identify system *C* as the excretory system and *not* as the female reproductive system, which it resembles in this diagram.]

6. **3** *It removes waste products that are produced in the cells of the fetus* is the statement that describes one function of the placenta in mammals. As the maternal blood circulates through the placenta, fetal wastes such as carbon dioxide and urea diffuse across a tissue boundary into the maternal bloodstream. The mother's blood then circulates through her body, where the carbon dioxide and urea are removed by diffusion in the lung and kidney, respectively.

WRONG CHOICES EXPLAINED

(1) *It allows blood of the mother to mix with the blood of the fetus* is *not* the statement that describes one function of the placenta in mammals. In a normal placenta, the maternal and fetal bloodstreams are kept separate by a

membrane made up of maternal and fetal tissues. This placental membrane permits diffusion of essential molecules between the maternal and fetal bloodstreams but prevents blood tissue from passing between mother and fetus.

(2) *It contains fluid that protects the embryo from harm* is *not* the statement that describes one function of the placenta in mammals. The embryo (fetus) is surrounded by a fluid-filled sac known as the amnion, which provides protection from mechanical shock. The placenta does not provide this function in mammals.

(4) *It synthesizes food for the embryo* is *not* the statement that describes one function of the placenta in mammals. Humans are heterotrophic organisms, meaning that they must obtain their food by consuming animal and plant matter. The human placenta transmits food molecules from the maternal blood to the fetal blood via diffusion through membranes.

7. **1** The movements indicated by all the arrows are directly involved in *the maintenance of homeostasis*. Homeostasis is a state of dynamic equilibrium, or "steady state," that is achieved in healthy living things and that provides the optimum environment for the performance of life activities. Each arrow in the diagram indicates the direction in which essential materials are moving into or out of a healthy single-celled organism in order to promote homeostasis in that organism.

WRONG CHOICES EXPLAINED

(2) The movements indicated by all the arrows are *not* directly involved in *respiration, only*. The arrows indicating the movement of carbon dioxide and nitrogen wastes are involved in excretion, and the arrow indicating the movement of organic compounds is involved in nutrition.

(3) The movements indicated by all the arrows are *not* directly involved in *excretion, only*. The arrow indicating the movement of oxygen is involved in respiration, and the arrow indicating the movement of organic compounds is involved in nutrition.

(4) The movements indicated by all the arrows are *not* directly involved in *the digestion of proteins*. The arrows indicating the movement of carbon dioxide and nitrogen wastes are involved in excretion, and the arrow indicating the movement of oxygen is involved in respiration.

8. **1** *The expression of genes can be modified by the environment* is the statement that best explains the observation described. The ptarmigan's feather color change under wintry conditions is an indication that the genes for feather pigmentation are expressed differently under different environmental conditions.

WRONG CHOICES EXPLAINED

(2) *Holes in the ozone layer vary in size depending on the season* is *not* the statement that best explains the observation described. There is no informa-

tion given in the observation relating to the presence or effect of the ozone layer.

(3) *Acids in rain bleach the brown feathers of the birds* is *not* the statement that best explains the observation described. There is no information given in the observation relating to the presence or effect of acid rain. In addition, the observation includes information that the color change is due to feather replacement, not bleaching.

(4) *Mutations occur only during certain seasons* is *not* the statement that best explains the observation described. Mutations occur not just when seasons change, but randomly in cells at all times because of external and internal mutagenic factors.

9. **3** *The child received genetic information from each parent* is the best explanation for the child having brown hair and brown eyes. During the processes of gamete formation and fertilization, chromosomes of both parents randomly assort and recombine to produce unique combinations of traits expressed by both parents. The fact that this child has received the brown eye color of the mother and the brown hair color of the father is evidence of this random assortment and recombination.

WRONG CHOICES EXPLAINED

(1) *A gene mutation occurred that resulted in brown hair and brown eyes* is *not* the best explanation for the child having brown hair and brown eyes. Only mutations received in gamete-producing cells may be passed on to successive generations and therefore contribute to the alteration of phenotypes. It is unlikely that such a mutation could be responsible for the observation described.

(2) *Gene expression must change in each generation so evolution can occur* is *not* the best explanation for the child having brown hair and brown eyes. Organic evolution is a process involving mutation, variation, and natural selection of favorable traits. Genes are expressed as a function of their control over biochemical processes in the cells of an individual.

(4) *Cells from the mother's eyes were present in the fertilized egg* is *not* the best explanation for the child having brown hair and brown eyes. This is a nonsense distracter. A fertilized egg is a single cell that forms from the fusion of two monoploid gametes and so cannot contain mature differentiated cells of either parent.

10. **2** *The amino acid sequence may be altered during protein synthesis* is what will most likely happen if there is a change in the base sequence of this molecule. The molecule illustrated is a portion of a DNA strand showing four base pairs. The base pairs provide a code for the synthesis of proteins in the cell. If the sequence of base pairs in this molecule changes, then the sequence of amino acids in the manufactured protein may change as well.

WRONG CHOICES EXPLAINED

(1) *The molecule will be converted into an inorganic compound* is *not* what will most likely happen if there is a change in the base sequence of this molecule. Organic compounds are defined as those containing the elements carbon and hydrogen. Changing the sequence of base pairs will not alter the fact that DNA is an organic molecule.

(3), (4) *The chromosome number will decrease in future generations* and *The chromosome number may increase within the organisms* are *not* what will most likely happen if there is a change in the base sequence of this molecule. Chromosomes are complex cell organelles made up of thousands of complete strands of DNA. Altering the base pair sequence in one of the DNA molecules will not affect the total number of chromosomes present in the cells.

11. **4** These differences are the result of *certain genes being expressed in some cells and not in others*. In the process of differentiation, highly specialized cell tissues develop that display specific characteristics. As cells differentiate, genes needed for the proper operation of these cells are switched "on," whereas genes not needed in the specialized cells are switched "off." This allows specialized cells to operate efficiently, producing only those substances essential for cell operation.

WRONG CHOICES EXPLAINED

(1) These differences are *not* the result of *having two types of cells, one type from each parent*. The two types of cells referenced are probably sperm cells and egg cells. These cells are necessary to complete the fertilization process, but their presence in the fertilization process does not directly relate to the differentiation of cells that occurs after fertilization.

(2) These differences are *not* the result of *rapid mitosis causing mutations in embryo cells*. Mitosis is a type of cell division in which a diploid ($2n$) parent cell gives rise to genetically identical diploid ($2n$) daughter cells. Mitosis does not cause mutations in embryo cells.

(3) These differences are *not* the result of *new combinations of cells resulting from meiosis*. Meiosis is a process by which monoploid (n) male and female gametes are formed from diploid ($2n$) primary sex cells. This process precedes the process of fertilization and does not directly relate to the differentiation of cells that occurs after fertilization.

12. **3** *Both the original and the new genes* are those that would be most likely to be passed on when these infected bacteria reproduce asexually. Because the process of mitotic cell division results in exact duplication of the genetic material in the parent cell, all genes will be replicated and passed on when the bacteria reproduce by asexual means. The process of mitosis cannot differentiate between bacterial and viral genes when replicating new strands of DNA.

WRONG CHOICES EXPLAINED

(1), (2), (4) It is *not* true that *only the new genes*, *only the original genes*, or *neither the original nor the new genes* are those that would be most likely to be passed on when these infected bacteria reproduce asexually. All genes, both original and new, will be replicated and passed on during asexual reproduction. See correct answer above.

13. **2** This mutation could cause a change in *the organism, but not its offspring*. Only mutations received in gamete-producing cells may be passed on to successive generations and therefore contribute to the alteration of phenotypes. The mutation described may cause an alteration in the stomach tissues of this individual organism, but this alteration will not be passed on to subsequent generations because it occurred in somatic, not gametic, tissue.

WRONG CHOICES EXPLAINED

(1), (3), (4) This mutation could *not* cause a change in *both the organism and its offspring*; *its offspring, but not the organism itself*; or *neither the organism nor its offspring*. A mutation in a somatic (body) cell cannot be passed on to offspring. See correct answer above.

14. **1** The fact that a certain protein is found in mitochondria, chloroplasts, and bacteria provides evidence that plants and bacteria *have some similar DNA base sequences*. Specific DNA base sequences provide codes for the production of specific proteins in living cells. When the same specific protein is found in different species, this fact provides evidence that DNA sequences are similar in those species and implies a genetic relationship between them.

WRONG CHOICES EXPLAINED

(2) The fact that a certain protein is found in mitochondria, chloroplasts, and bacteria does *not* provide evidence that plants and bacteria *can use carbon dioxide to make proteins*. Carbon dioxide is an inorganic by-product of the process of cellular respiration. Carbon dioxide is taken up by plants in the process of photosynthesis and is used in the production of glucose, not proteins. Most bacteria do not use carbon dioxide for any purpose.

(3) The fact that a certain protein is found in mitochondria, chloroplasts, and bacteria does *not* provide evidence that plants and bacteria *digest proteins into simple sugars*. When proteins are digested within living cells, the simple subunits that immediately result are amino acids, not simple sugars.

(4) The fact that a certain protein is found in mitochondria, chloroplasts, and bacteria does *not* provide evidence that plants and bacteria *contain certain pathogenic microbes*. Pathogenic (disease-causing) microbes such as viruses may invade the living cells of any organism. The presence of a specific protein in the cells of different species may be an indication of the presence of a specific microbe, though it is unlikely that a plant and a bacterium would be susceptible to the same types of microbes.

15. **3** Extinction of a species could result from *limited genetic variability in the species*. A species lacking variations that provide adaptive advantages under various conditions may not be able to adapt readily to a changing environment. Environmental pressures may eliminate susceptible members of the species, reducing its numbers drastically over time. Eventually, such a non-adaptive species reaches a point at which it can no longer sustain a breeding population and may become extinct.

WRONG CHOICES EXPLAINED

(1) Extinction of a species could *not* result from *evolution of a behavior that produces greater reproductive success*. Evolution of such a behavior represents an adaptive advantage that would make the species less likely, not more likely, to become extinct.

(2) Extinction of a species could *not* result from *synthesis of a hormone that controls cellular communication*. Such a hormone may be routinely produced in the species and probably represents a favorable adaptation that would make the species less likely, not more likely, to become extinct.

(4) Extinction of a species could *not* result from *fewer unfavorable mutations in the species*. Mutations are random events whose effects may be favorable or unfavorable. In the event that chance events made these mutations more favorable, this would make the species less likely, not more likely, to become extinct.

16. **1** The rate at which all organisms obtain, transform, and transport materials depends on an immediate supply of *ATP and enzymes*. ATP (adenosine triphosphate) is a molecule that stores the chemical bond energy released from the oxidation of glucose during respiration. ATP is used in cells to provide the energy needed for certain cell processes that require activation energy. Enzymes are molecules that catalyze specific chemical reactions in cells, thereby speeding the rate of cell processes. Both ATP and enzymes affect the rate at which these cell processes work.

WRONG CHOICES EXPLAINED

(2), (3) The rate at which all organisms obtain, transform, and transport materials does *not* depend on an immediate supply of *solar energy and carbon dioxide* or *carbon dioxide and enzymes*. Carbon dioxide is a low-energy by-product of the process of respiration that is not directly related to the rates of most cell processes (except photosynthesis). Solar energy is the ultimate source of energy available to the majority of living things. While a constant source of solar energy is essential to life, its effect on the rate of cell processes is made through the process of photosynthesis and so is not immediate.

(4) The rate at which all organisms obtain, transform, and transport materials does *not* depend on an immediate supply of *ATP and solar energy*. Solar energy is the ultimate source of energy available to the majority of living things. While a constant source of solar energy is essential to life, its effect on

the rate of cell processes is made through the process of photosynthesis and so is not immediate.

17. **3** As women age, their reproductive cycles stop because of decreased *levels of specific hormones*. The female reproductive (menstrual) cycle is controlled by a series of hormones produced in the pituitary gland and in the reproductive organs. The levels of these hormones are dependent on the amount of estrogen produced by the ovary. As the production of this hormone drops off, the reproductive cycle becomes erratic and eventually ceases in menopause.

WRONG CHOICES EXPLAINED

(1) As women age, their reproductive cycles do *not* stop because of decreased *digestive enzyme production*. Digestive enzymes are produced in the digestive organs, including the mouth, stomach, small intestine, and pancreas. Digestive enzymes have no direct control over the female reproductive cycle.

(2) As women age, their reproductive cycles do *not* stop because of decreased *production of ATP*. ATP (adenosine triphosphate) is a molecule that stores the chemical bond energy released from the oxidation of glucose during respiration. ATP has no direct control over the female reproductive cycle.

(4) As women age, their reproductive cycles do *not* stop because of decreased *heart rate*. The heart rate is controlled by nerves in the brain and spinal cord and may be influenced by the production of certain hormones (e.g., adrenaline). The heart rate has no direct control over the female reproductive cycle.

18. **4** *Both species C and species D are related to species X* is the statement that is supported by the diagram. In this representation, species X is a common ancestor of species B, C, D, and E. Although species X may be extinct, an analysis of its DNA would reveal a close similarity to species C and D, indicating a genetic relationship among these three species.

WRONG CHOICES EXPLAINED

(1) *Species C is the ancestor of species B* is *not* the statement that is supported by the diagram. As represented in the diagram, species B predates species C in a direct evolutionary line. Therefore, species C is the descendant, not the ancestor, of species B.

(2) *Species D and E evolved from species B* is *not* the statement that is supported by the diagram. The evolutionary line that gave rise to species D and E branched off from the line that gave rise to species B before any of these species appeared in the geological record. Although species B, D, and E share a common ancestor (species X), species D and E could not have evolved from species B.

(3) *Species X evolved later than species D but before species B* is *not* the statement that is supported by the diagram. As represented in the diagram, species X predates both species D and species B. Therefore, both species D and species B evolved later than species X.

19. **4** *A sperm having half the normal species number of chromosomes* is the cell that is normally produced as a direct result of meiosis. Meiosis is a process by which monoploid (*n*) male and female gametes are formed from primary sex cells containing the diploid (2*n*), or normal species number, of chromosomes. In human males, meiosis results in the formation of monoploid sperm cells.

WRONG CHOICES EXPLAINED
(1) *A uterine cell having half the normal species number of chromosomes* is *not* the cell that is normally produced as a direct result of meiosis. A uterine cell is a somatic (body) cell that normally contains the diploid (2*n*), or normal species number, of chromosomes.
(2) *An egg having the full species number of chromosomes* is *not* the cell that is normally produced as a direct result of meiosis. An egg cell resulting from meiosis contains the monoploid (*n*) number of chromosomes, or half the normal species number.
(3) *A zygote having the full species number of chromosomes* is *not* the cell that is normally produced as a direct result of meiosis. A zygote is a diploid (2*n*) cell that results from the fusion of a monoploid (*n*) egg and a monoploid (*n*) sperm in the process of fertilization.

20. **2** *Mitosis and differentiation* are the processes that are most directly responsible for the changes illustrated in the diagram. The diagram illustrates the stages in the development of a zygote as it divides by mitosis to produce two, then four, and eventually a mass of undifferentiated embryonic cells. The last stage of the diagram illustrates the embryonic mass as its cells begin to differentiate into the specialized layers that will give rise to the tissues and organs of the mature organism.

WRONG CHOICES EXPLAINED
(1) *Sorting and recombination of genetic information* are *not* the processes that are most directly responsible for the changes illustrated in the diagram. Sorting and recombination are processes that occur during the process of meiosis (gamete formation), which precedes the events illustrated in the diagram.
(3) *Meiosis and adaptation* are *not* the processes that are most directly responsible for the changes illustrated in the diagram. Meiosis is a process by which monoploid (*n*) male and female gametes are formed from primary sex cells containing the diploid (2*n*) number of chromosomes. This process precedes the events illustrated in the diagram. Adaptation is a term relating to the

success of a species variety as it responds to a changing set of environmental conditions. This term has no direct relationship to the process illustrated in the diagram.

(4) *Fertilization and cycling of materials* are *not* the processes that are most directly responsible for the changes illustrated in the diagram. Fertilization is a process in which a monoploid (n) sperm cell fuses with a monoploid (n) egg cell to produce a diploid $(2n)$ zygote. This process immediately precedes the first stage shown in the illustration. Cycling of materials relates to a process that occurs in healthy environments in which elements essential to life (e.g., carbon, hydrogen, oxygen, and nitrogen) are cycled between living things and the nonliving environment. This term has no direct relationship to the process illustrated in the diagram.

21. **2** The cells in area X are most likely *cancer cells*. Cancer is the uncontrolled growth of abnormal cells that crowd out the normal, healthy tissues of a living organism. Cancer is set off in individual cells that have been exposed to mutagenic agents such as ultraviolet radiation.

WRONG CHOICES EXPLAINED

(1), (3), (4) The cells in area X are *not* most likely *red blood cells*, *white blood cells*, or *sex cells*. These are specialized types of cells found in the human body, none of which are located in skin tissue.

22. **4** *Respiration and photosynthesis* are the life functions that are directly regulated through feedback mechanisms associated with the actions of the structures labeled X. The structures labeled X are guard cells that surround a stomate on the bottom surface of a leaf. The guard cells control the opening and closing of the stomate and thereby regulate the rate of gas exchange between the atmosphere and the moist interior of the leaf. Atmospheric gases admitted through the stomate include oxygen and carbon dioxide gas, which are important in the life functions of respiration and photosynthesis, respectively.

WRONG CHOICES EXPLAINED

(1) *Excretion and immunity* are *not* the life functions that are directly regulated through feedback mechanisms associated with the actions of the structures labeled X. Excretion of excess oxygen and carbon dioxide gases through the stomate is regulated by the guard cells. Immunity is not a life function as such, but a process in which an organism gains a level of protection against a specific disease. Immunity is not regulated by the guard cells.

(2) *Digestion and coordination* are *not* the life functions that are directly regulated through feedback mechanisms associated with the actions of the structures labeled X. Digestion is not a life function as such, but an aspect of the life function of nutrition in which complex foods are broken down into simple organic molecules. Coordination is not a life function as such, but an

aspect of the life function of regulation in which life processes are controlled for the mutual benefit of other processes. Neither digestion nor coordination is regulated by the guard cells.

(3) *Circulation and reproduction* are *not* the life functions that are directly regulated through feedback mechanisms associated with the actions of the structures labeled *X*. Circulation is not a life function as such, but an aspect of the life function of transport in which materials are moved throughout the body of an organism. Reproduction is a life function in which new members of a species are produced by the fusion of gametes and the growth of an embryo. Neither circulation nor reproduction is regulated by the guard cells.

23. **1** *Extinction of species* is one irreversible effect of both deforestation and water pollution on the environment. When natural habitats are altered significantly, as by deforestation or water pollution, environmental conditions favorable to some species in that habitat may change so markedly that the species may no longer be able to survive. If the habitat is destroyed over a wide area, the species may be unable to recover and may die out completely (become extinct). Extinction is an irreversible condition.

WRONG CHOICES EXPLAINED

(2) *Thinning of the ozone shield* is probably *not* one irreversible effect of both deforestation and water pollution on the environment. The ozone layer is located high in the atmosphere, where it provides a protective shield by filtering out harmful and mutagenic ultraviolet radiation. Theoretically, the ozone layer is being thinned by the use of certain aerosol chemicals that pollute the atmosphere. While this is not directly linked to deforestation and water pollution, scientists acknowledge that every event affecting the environment is linked to every other event, so it is impossible to say that there is no cause-and-effect relationship in this case.

(3) *Depletion of atmospheric carbon dioxide levels* is *not* one irreversible effect of both deforestation and water pollution on the environment. Deforestation and water pollution are both responsible for the destruction of photosynthetic plants and algae. The destruction of these organisms leads to an increase, not a depletion, of carbon dioxide in the atmosphere.

(4) *Increase in renewable resources* is *not* one irreversible effect of both deforestation and water pollution on the environment. Wood and other plant matter are renewable natural resources that are decreased, not increased, as a result of deforestation and water pollution.

24. **3** Carnivores would most likely be located in *level III, only*. In this energy pyramid, level I contains the autotrophs (green plants and other producer organisms), level II contains the herbivores (plant eaters or primary consumers), and level III contains the carnivores (meat eaters or secondary consumers). The tapering of the pyramid from level I through level III represents the fact that solar energy captured by producer organisms at level I is

inefficiently passed on to the consumer levels (II and III) in an energy pyramid, with as much as 90% of the energy lost to the environment as heat at each exchange.

WRONG CHOICES EXPLAINED

(1), (2), (4) Carnivores would *not* most likely be located in *level I, only, level I and level II,* or *level II and level III.* Carnivores are located exclusively in level III. See correct answer above.

25. **2** *Interactions between biotic and abiotic components regulate carbon dioxide and water levels* is the statement that describes a situation that leads to stability within an ecosystem. In any ecosystem, it is the delicate balance of biotic (living) and abiotic (nonliving) factors that promote the stability of the environment and work together to support the many species of living things contained within it.

WRONG CHOICES EXPLAINED

(1) *Carbon dioxide and water are released only by abiotic sources in the ecosystem* is *not* the statement that describes a situation that leads to stability within an ecosystem. Although abiotic factors may be responsible for the release of carbon dioxide and water in the environment, these materials are also released as a result of the life process of respiration, which is carried on by all living species.

(3) *Animals provide the oxygen used by plants, and plants provide the nitrogen needed by animals* is *not* the statement that describes a situation that leads to stability within an ecosystem. In a natural ecosystem, oxygen gas is released by plants as a result of the process of photosynthesis. Nitrogen is cycled in the environment by means of a complex interaction among plants, animals, bacteria, and the abiotic environment.

(4) *Organisms provide all the necessary energy for the maintenance of this ecosystem* is *not* the statement that describes a situation that leads to stability within an ecosystem. Any stable ecosystem must have a reliable source of energy in the form of sunlight or heat in order to sustain itself. An ecosystem dependent solely on the energy contained in living organisms will quickly exhaust the available energy and become incapable of supporting life.

26. **1** The most likely explanation for this observation is that the *bacteria interfered with normal life functions of the birds.* Although additional experimentation would be necessary to confirm this hypothesis, it seems likely that the bacteria may have had some negative effect on the birds and upset the homeostatic balance necessary to sustain life. It is not clear from the observation exactly what caused this negative effect.

WRONG CHOICES EXPLAINED

(2) The most likely explanation for this observation is *not* that the *disease that killed the birds was inherited.* Given the information shared in the obser-

vation, it appears likely that the disease was acquired through infection, not inherited from a prior generation.

(3) The most likely explanation for this observation is *not* that the *gene alterations in the bacterial cells killed the birds*. Although unlikely, it is possible that the strain of bacteria ingested by the birds had received one or more mutations that made its presence in the birds lethal.

(4) The most likely explanation for this observation is *not* that the *birds produced antigens in response to the bacteria*. Antigens are substances (usually proteins) that are produced by and are unique to an organism. When the bacteria invaded the birds, the birds most likely produced antibodies, not antigens, in an attempt to neutralize the antigens produced by the bacteria.

27. **4** *Implementing laws to regulate the number of animals hunted and killed each year* is the action that illustrates an increased understanding and concern by humans for ecological interrelationships. Improving the interaction between human activities and wild species is key to improving the overall health of the ecosystem. Hunting to control growing populations of game species such as deer is a human activity that takes the place of natural predation once carried on by wolves and cougars (killed off by humans). Overhunting any species can have long-term negative effects on that species' population (and on future hunting opportunities). Laws that regulate the number of animals killed via hunting help to stabilize the ecosystem for all species, including humans.

WRONG CHOICES EXPLAINED

(1) *Importing organisms in order to stabilize existing ecosystems* is *not* the action that illustrates an increased understanding and concern by humans for ecological interrelationships. When humans import nonnative species and release them into the native ecosystem, these species may become invasive, crowding out native species and altering the natural community substantially. This action illustrates decreased, not increased, understanding and concern by humans for ecological interrelationships.

(2) *Eliminating pollution standards for industries that promote technology* is *not* the action that illustrates an increased understanding and concern by humans for ecological interrelationships. Industries that pollute degrade the natural ecosystem for all species, including humans. Pollution disrupts natural cycles and contaminates the air, water, and soil on which we depend for human survival. This action illustrates decreased, not increased, understanding and concern by humans for ecological interrelationships.

(3) *Removing natural resources at a rate equal to the needs of the population* is *not* the action that illustrates an increased understanding and concern by humans for ecological interrelationships. Renewable resources (such as wood) have a finite capacity to create replacement resources for themselves. Nonrenewable resources (such as fossil fuels and minerals) have no capacity to re-create themselves. Once these nonrenewable resources are used up,

there will be no more to extract from the environment. This action illustrates decreased, not increased, understanding and concern by humans for ecological interrelationships.

28. **4** Letter *D* indicates the part of the graph that represents the carrying capacity of the environment in the jar. The carrying capacity of an environment is the number of organisms of any given species that can be supported for an extended period given limited quantities of food, oxygen, light, minerals, water, and similar resources. When the carrying capacity of a population is reached, the population levels off; if it is exceeded, the population declines. The level part of the graph (letter *D*) indicates that the carrying capacity was reached after about 11 days.

WRONG CHOICES EXPLAINED

(1), (2), (3) Letters *A*, *B*, and *C* do *not* indicate the part of the graph that represents the carrying capacity of the environment in the jar. These parts of the graph indicate the growth phases of the population when the numbers of individuals remained below the carrying capacity because resources were not yet limited.

29. **1** One reason some experimental automobiles have been developed to use electricity rather than gasoline is that *gasoline is made from petroleum, a nonrenewable resource*. Fossil fuels such as coal, natural gas, and oil are nonrenewable energy sources formed deep in the Earth over millions of years. Once these resources are used up, they cannot be replaced. It is imperative that humans work purposefully to develop electricity production capabilities through the use of renewable sources such as water, wind, solar, hydrogen, and fusion technologies.

WRONG CHOICES EXPLAINED

(2) It is *not* true that one reason some experimental automobiles have been developed to use electricity rather than gasoline is that *Earth has an unlimited supply of fossil fuels*. Fossil fuels such as coal, natural gas, and oil are nonrenewable energy sources formed deep in the Earth over millions of years. Once these resources are used up, they cannot be replaced.

(3) It is *not* true that one reason some experimental automobiles have been developed to use electricity rather than gasoline is that *the use of electricity will eliminate the need for all antipollution laws*. Antipollution laws have been developed to curb the human tendency to release contaminants into the natural environment, thereby endangering the lives of the plants and animals that share our fragile ecosystems with us. Electricity is produced by a variety of means, including the burning of fossil fuels that would release particulates, hydrocarbons, and greenhouse gases into the atmosphere if the power industry were not regulated by antipollution laws.

(4) It is *not* true that one reason some experimental automobiles have been developed to use electricity rather than gasoline is that *the use of electricity will increase the manufacture of antipollution devices for cars*. Antipollution devices on cars have been developed to reduce the amount of damaging air pollutants released from the burning of gasoline. The use of electricity as a power source for cars would decrease, not increase, the need for such devices.

30. **3** Ecosystems will have a greater chance of maintaining equilibrium over a long period of time if they have *a diversity of organisms*. Each species population within an ecosystem has a specific role (niche) in the environment. Balanced, stable environments contain a large number of different kinds of species whose niches complement each other such that essential roles are played out for the mutual benefit of all members of the ecosystem.

WRONG CHOICES EXPLAINED

(1) Ecosystems will *not* have a greater chance of maintaining equilibrium over a long period of time if they have *organisms imported by humans from other environments*. When humans import nonnative species and release them into the native ecosystem, these species may become invasive, crowding out native species and altering the natural community substantially. Such an event would tend to upset, not maintain, the equilibrium of the ecosystem.

(2) Ecosystems will *not* have a greater chance of maintaining equilibrium over a long period of time if they have *a sudden change in climate*. Each ecosystem is built around the abiotic (nonliving) aspects of the environment, including its climate. A sudden climatic change would make survival of established plants and animals difficult. Such an event would tend to upset, not maintain, the equilibrium of the ecosystem.

(4) Ecosystems will *not* have a greater chance of maintaining equilibrium over a long period of time if they have *predators eliminated from the food chains*. Predators serve an important role in the natural environment by controlling populations of herbivores. Eliminating predators from food chains would allow herbivores to increase their numbers beyond the carrying capacity of the environment. Such an event would tend to upset, not maintain, the equilibrium of the ecosystem.

PART B-1

31. **4** The concentration of the pesticide DDT in individual organisms at level D is higher than the concentration in individuals at level A because DDT is *passed through levels A, B, and C to organisms at level D*. At each level of the food pyramid, the available biomass and energy decrease because of inefficiencies in energy and material transfer. Insoluble contaminants such as DDT found at low concentrations in plants (level A) become progressively

more concentrated as the plants are consumed by herbivores (level B) and the herbivores are consumed by carnivores (levels C and D).

WRONG CHOICES EXPLAINED

(1), (2), (3) It is *not* true that the concentration of the pesticide DDT in individual organisms at level D is higher than the concentration in individuals at level A because DDT is *synthesized by organisms at level D, excreted by organisms at level A as a toxic waste,* or *produced by organisms at level C which are eaten by organisms at level D.* DDT is a synthetic chemical manufactured by humans and was used as an insecticide in the early mid-20th century. DDT is not naturally synthesized or excreted as a metabolic waste by living organisms at any level of the food pyramid.

32. **1** *Ecological succession in a community* is the concept represented in the graph. Ecological succession is a term that describes the replacement of one plant community by other, progressively more complex plant communities until a stable climax community is established. In the graph, a typical upland forest succession is represented in which grasses invade a barren area as the pioneer species, followed by shrubs, birches, spruces, and maples. At the end of the time shown, a mixed climax community of birch, spruce, and maple trees has become established.

WRONG CHOICES EXPLAINED

(2) *Cycling of carbon and nitrogen in a forest* is *not* the concept represented in the graph. Such a graph would depict the interaction between biotic (living) and abiotic (nonliving) factors in the environment as they take in, convert, and pass on various forms of carbon and nitrogen compounds.

(3) *Energy flow in a food chain over time* is *not* the concept represented in the graph. Such a graph would depict organisms at different trophic (feeding) levels as they interact in various nutritional relationships to pass energy from lower to higher levels of the food chain.

(4) *Negative human impact on the environment* is *not* the concept represented in the graph. Such a graph would depict a variety of human activities and demonstrate how these activities negatively impact living things and upset the natural equilibrium to the detriment of the entire ecosystem.

33. **3** Data for a study on respiration in this ecosystem should be collected during *intervals A and B, from both the producers and consumers in the ecosystem.* Respiration is a life function that is carried on by all living things at all times of the day and night. A study of this life function should be designed to collect data from the full range of organisms involved and the full range of time during which the process occurs.

WRONG CHOICES EXPLAINED

(1), (2), (4) Data for a study on respiration in this ecosystem should *not* be collected during *interval A, from only the producers in the ecosystem; inter-*

vals A and B, from only the consumers in the ecosystem; or *interval A only, from abiotic but not biotic components of the ecosystem.* For the reasons stated above, a study of respiration should be conducted across both intervals of time from all the living organisms in the ecosystem. Respiration is a life function carried on by the biotic (living) components of the ecosystem. Abiotic (nonliving) components do not carry on respiration.

34. **3** *Photosynthetic organisms respond to changing light levels* is a valid inference based on these data. The information provided in the chart indicates that the depth at which photosynthetic organisms suspend themselves is directly proportional to the intensity of the light conditions. The stronger the light intensity, the deeper these organisms are found (up to a depth of 150 cm).

WRONG CHOICES EXPLAINED
(1) *Most photosynthetic organisms live below a depth of 150 centimeters* is *not* a valid inference based on these data. In fact, the information provided indicates that these photosynthetic organisms dwell at a depth above, not below, 150 cm.

(2) *Oxygen production increases as photosynthetic organisms move deeper in the lake* is *not* a valid inference based on these data. Because the production of oxygen gas is associated with the process of photosynthesis, it is reasonable to infer that oxygen production is greatest near the lake surface. However, there were no data presented explicitly on the production of oxygen in the lake.

(4) *Photosynthetic organisms move up and down to increase their rate of carbon dioxide production* is *not* a valid inference based on these data. Photosynthetic organisms produce carbon dioxide during the process of respiration. However, there were no data presented explicitly on the production of carbon dioxide in the data chart.

35. **1** *Microscope, pipette, and slides with coverslips* are the materials that the student most likely would have used in this investigation. The student most likely would have used the pipette to draw water samples from varying depths and then used the slides and coverslips to prepare the water samples for examination under the microscope.

WRONG CHOICES EXPLAINED
(2), (3), (4) *Graduated cylinder, triple-beam balance, and chromatography paper; thermometer, electric balance, and biological stains;* and *computer, pH paper, and gel electrophoresis apparatus* are *not* the materials that the student most likely would have used in this investigation. While the computer might have been useful in this experiment, none of the other materials listed directly relate to the study as described.

36. **2** *Low power: 100 small pollen grains; high power: 25 large pollen grains* is the statement that best describes the relative number and appearance of the pollen grains observed using these two objectives. Using low power (likely 100×), the student observes a field of view that encompasses 100 pollen grains whose individual sizes appear small because of the relatively low magnification used. Switching to high power (likely 200×), the student observes a field of view only 25% (by area) that of the low-power field, encompassing only 25 pollen grains whose individual sizes appear two times larger than under low-power magnification.

WRONG CHOICES EXPLAINED

(1), (3), (4) *Low power: 25 small pollen grains, high power: 100 large pollen grains*; *low power: 25 large pollen grains, high power: 100 small pollen grains*; and *low power: 100 large pollen grains, high power: 25 small pollen grains* are *not* the statements that best describe the relative number and appearance of the pollen grains observed using these two objectives. Using the mathematical logic cited above, any data that indicate fewer, larger pollen grains observed under low power than under high power are illogical. The field of view of low-power magnification encompasses a larger area of the slide and magnifies the objects less than does the high-power magnification. Each of the incorrect responses violates one or both of these rules.

37. **4** Molecule A contains the *directions for the synthesis of molecules B, C, and D*. Molecule A is DNA, which contains a complex sequence of nitrogenous bases that code for the production of specific proteins. The DNA code is "read" by messenger-RNA molecules, which then carry it from the nucleus to ribosomes located in the cell cytoplasm. At the ribosomes, amino acid molecules are linked in a specific sequence guided by the code on the messenger-RNA. Each specific amino acid sequence represents a functional cell protein (molecules B, C, and D).

WRONG CHOICES EXPLAINED

(1) Molecule A does *not* contain the *starch necessary for ribosome synthesis in the cytoplasm*. Molecule A is DNA. Starch is a complex carbohydrate polymer composed of repeating glucose subunits. By contrast, DNA is a complex nucleic acid polymer composed of nucleotide subunits. Starch is not used in the synthesis of ribosomes in the cytoplasm.

(2) Molecule A does *not* contain the *organic substance that is broken down into molecules B, C, and D*. Molecule A is DNA. Molecules B, C, and D do not represent breakdown products of another organic substance. Rather, they are complex proteins that result from a synthesis process linking amino acid units together in a highly specific sequence.

(3) Molecule A does *not* contain the *proteins that form the ribosome in the cytoplasm*. Molecule A is DNA. DNA is not a protein molecule but a complex

nucleic acid polymer. DNA does not contain the substances that make up the ribosome.

38. **3** Molecules *B*, *C*, and *D* are similar in that they are usually *composed of amino acids*. Molecules *B*, *C*, and *D* are complex proteins that result from a synthesis process linking amino acid units together in a highly specific sequence. At the ribosomes, amino acid molecules are linked in this specific sequence guided by the genetic code.

WRONG CHOICES EXPLAINED

(1) It is *not* true that molecules *B*, *C*, and *D* are similar in that they are usually *composed of genetic information*. Genetic information is housed in molecules of DNA in the nucleus of the cell. This genetic information is used to determine the structure of specific proteins (molecules *B*, *C*, and *D*). These molecules contain no genetic information.

(2) It is *not* true that molecules *B*, *C*, and *D* are similar in that they are usually *involved in the synthesis of antibiotics*. Antibiotics are protein substances synthesized naturally by molds and fungi. As proteins, these molecules are produced in a manner similar to the way that molecules *B*, *C*, and *D* are produced. These molecules do not synthesize antibiotics.

(4) It is *not* true that molecules *B*, *C*, and *D* are similar in that they are usually *involved in the diffusion of oxygen into the cell*. Diffusion is a process by which molecules of a substance move from a region of high relative concentration to a region of low relative concentration of that substance. The diffusion of oxygen into a cell is accomplished without direct assistance from cell proteins and without the expenditure of cell energy. These molecules are not involved in oxygen diffusion.

39. **2** *This procedure is used to produce new tomato plants that are clones of the original tomato plant* is the statement that is best supported by the information in the diagram. The diagram illustrates a process by which a single living cell is removed from a parent tomato plant and placed in a laboratory culture to promote its growth by mitotic cell division into a mass of tomato cells. These cells are maintained in the laboratory culture until specialization (differentiation) occurs and they can be transplanted to soil for development into a new mature tomato plant. This new plant will be a genetic duplicate (clone) of the original parent tomato plant. This method of producing new plants is known as cloning.

WRONG CHOICES EXPLAINED

(1) *The one leaf cell removed formed a zygote that developed into a new plant by mitotic cell division* is *not* the statement that is best supported by the information in the diagram. A zygote is a cell that results from the fusion of an egg cell nucleus and a sperm cell nucleus. The cell removed from the parent tomato plant is not capable of forming a zygote.

(3) *The cell taken from the leaf produced eight cells, each having one-half of the genetic information of the original leaf cell* is *not* the statement that is best supported by the information in the diagram. Cloning is an asexual reproductive process in which all cells involved retain the genetic information of the original parent plant. No cells that contain only one-half this information are produced in cloning.

(4) *The new tomato plant will not be able to reproduce sexually because it was produced by mitotic cell division* is *not* the statement that is best supported by the information in the diagram. The new tomato plant will be a genetic duplicate (clone) of the original parent tomato plant and share all its characteristics and capabilities, including the ability to reproduce by sexual means.

40. **2** Graph 2 best represents the information in the data table. A review of the data in the table reveals that as the sugar concentration (independent variable) increases, the length of the potato rods (dependent variable) decreases. This phenomenon is due to the diffusion of water out of the potato cells in response to the higher concentrations of sugar in the containers. This inverse relationship between the variables is best illustrated by graph 2.

WRONG CHOICES EXPLAINED
(1) Graph 1 does *not* best represent the information in the data table. Graph 1 illustrates a direct relationship between the variables. This graph would be appropriate if the length of the potato rods increased and the sugar concentration increased.

(3) Graph 3 does *not* best represent the information in the data table. Graph 3 illustrates a neutral relationship between the variables. This graph would be appropriate if the length of the potato rods was unaffected by changes in the sugar concentration.

(4) Graph 4 does *not* best represent the information in the data table. Graph 4 is a nonsense distracter that reverses the dependent and independent variable, thereby implying that the length of the potato rods has a neutral effect on the sugar concentration. See correct answer above.

PART B-2

41. **3** A total of 22 organisms in the trap were herbivores. Herbivores are defined as animals that feed on plant matter. A review of the food web indicates that aphids, slugs, and snails are the organisms identified as feeding on garden plants. Counting the numbers of these organisms in the trap (slugs, 8; snails, 1; aphids, 13) yields a total of 22.

WRONG CHOICES EXPLAINED
(1) It is *not* true that a total of 5 organisms in the trap were herbivores. No combination of animals in the trap yields this number.

(2) It is *not* true that a total of 8 organisms in the trap were herbivores. This number can be derived by adding the number of slugs (8) and the number of snails (1), but this method excludes aphids.

(4) It is *not* true that a total of 99 organisms in the trap were herbivores. This number can be derived by adding the number of centipedes (1) and the number of ground beetles (98), but these organisms are carnivores, which consume other animals for food.

42. One credit is allowed for correctly stating *one* reason the slug traps are *not* the best method to control slugs. Acceptable responses include, but are not limited to: [1]

- *Many other organisms are caught in the slug traps.*
- *Traps kill off natural predators of slugs.*
- *Traps may disrupt the food web.*
- *Traps catch more ground beetles than slugs.*
- *Traps cover only a limited amount of the slugs' range.*
- *Traps may not be set where slugs travel.*

43. One credit is allowed for correctly identifying *one* organism shown in this food web that could be used as a biological control to replace the slug traps. Acceptable responses include: [1]

- *Ground beetles*
- *Centipedes*

44. One credit is allowed for correctly placing these terms in order of increasing complexity. Acceptable responses include: [1]

Least complex	organelle
	cell
	tissue
	organ
Most complex	organism

45. One credit is allowed for correctly stating *one* specific way that *one* of these substances could harm the fetus. Acceptable responses include, but are not limited to: [1]

- *Tobacco (alcohol, drugs) could interfere with development.*
- *Alcohol (tobacco, drugs) could cause low birth weight.*
- *Drugs (alcohol) could cause death of the fetus.*
- *Alcohol could cause fetal alcohol syndrome.*
- *Drugs (alcohol) could cause learning defects.*
- *The baby could be born drug- (alcohol-) dependent.*

ANSWERS August 2007

46. **2** Research indicates that a buildup of arsenic in the cells of humans may be harmful because *arsenic prevents the action of genes that are important in reactions that suppress cancer*. This information is found in the second paragraph of the passage and suggests that arsenic interferes with the action of a group of hormones known as glucocorticoids, which are known to activate genes that suppress cancer.

WRONG CHOICES EXPLAINED

(1), (3), (4) Research does *not* indicate that a buildup of arsenic in the cells of humans may be harmful because *synthetic arsenic can be formed by the breakdown of glucocorticoids in the body*, because *arsenic prevents the reaction in which water and hormones bond and attach to the cancer cells*, or because *glucocorticoids can build up in tissues and cause an increase in the absorption of arsenic*. There is no information contained in the passage that supports any of these statements. Students are advised to read the passage carefully for valid information before selecting the correct response.

47. One credit is allowed for correctly stating *one* reason this study should be extended to other cells or to other complex organisms. Acceptable responses include, but are not limited to: [1]

- *The reactions in rat cells could be different from those in other organisms.*
- *To increase the validity of the experimental results.*
- *The results of the experiment indicate what happens only in cells outside the organism.*
- *Rats and humans are both mammals, so similar results could be found in humans.*
- *Animal testing can yield information that provides a direct health benefit to humans.*

48. Two credits are allowed, 1 credit for correctly identifying one specific hormone in the body other than glucocorticoid, and 1 credit for explaining how disruption of the activity of the hormone you identified might upset a feedback mechanism in the body. Acceptable responses include, but are not limited to: [2]

- *Insulin—prevent regulation of glucose levels in the blood*
- *Estrogen (testosterone)—interfere with the message for development of sex characteristics*
- *Adrenaline—block the increase in heart rate needed to respond to emergencies*
- *Pituitary growth hormone—slow the elongation of bones and stunt body growth*

- *Progesterone—prevent maintenance of the uterine lining during pregnancy*
- *Thyroxin—interfere with the metabolism of iodine, leading to goiter formation*

49. **4** *DNA from different organisms is joined together* is the result of step 3. The diagram illustrates the basic steps in a form of genetic engineering in which a gene of one organism is inserted into the genome of a different organism. As a result, the recipient organism is able to produce one or more proteins that it would not normally be able to produce.

WRONG CHOICES EXPLAINED
(1) *A new type of base is formed* is *not* the result of step 3. DNA utilizes four types of nitrogenous bases (A, T, G, and C); RNA utilizes U in place of T. If a new base were formed, it would take many evolutionary steps for it to be incorporated into the genetic code.

(2) *Different types of minerals are joined together* is *not* the result of step 3. Minerals are inorganic compounds composed of a metallic component and a nonmetallic component. Minerals are not normally joined together by biological processes such as the one illustrated.

(3) *DNA from the bacterial cell is cloned* is *not* the result of step 3. Cloning is an asexual reproductive process in which a somatic (body) cell is cultured to create a new organism that retains the genetic information of the original parent organism. The diagram indicates that genetic information is being altered in this process.

50. One credit is allowed for correctly stating *one* way that enzymes are used in step 2. Acceptable responses include, but are not limited to: [1]

- *Enzymes are used to cut the DNA.*
- *Researchers use enzymes to cut the genetic material.*
- *Scientists use enzymes to insert the desired animal gene into the bacterial genome.*
- *Restriction enzymes are used to "snip" the DNA at specific locations.*

51. One credit is allowed for correctly marking an appropriate scale on the axis labeled "Average Sugar Intake per Person." [1]

52. One credit is allowed for correctly constructing vertical bars to represent the "Average Sugar Intake per Person" and shading the bars according to the key. [1]

53. One credit is allowed for correctly constructing vertical bars to represent the "Average Number of Teeth with Decay per Person" and shading the bars according to the key. [1]

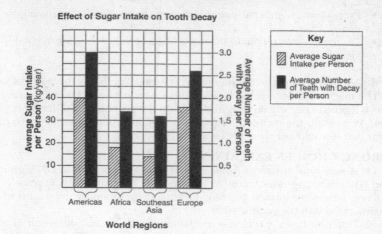

54. **3** *The greater the sugar intake, the greater the average number of decayed teeth* is the statement that is a valid conclusion regarding tooth decay. The graphed data illustrate a clear direct relationship between sugar intake (independent variable) and tooth decay (dependent variable) in this experiment. Areas with the highest sugar intake per person (Americas and Europe) have nearly twice as much tooth decay as areas with the lowest sugar intake per person (Africa and Southeast Asia).

WRONG CHOICES EXPLAINED

(1) *As sugar intake increases, the acidity in the mouth decreases, reducing tooth decay* is *not* the statement that is a valid conclusion regarding tooth decay. The table title relates the fact that bacteria living on the surface of teeth release an acid that dissolves tooth enamel. This statement implies that these bacteria thrive in a sugar-rich environment in the mouth, causing mouth acidity to increase, not decrease, and causing tooth decay to increase as well.

(2) *As sugar intake increases, tooth decay increases in Europe and the Americas, but not in Africa or Southeast Asia* is *not* the statement that is a valid conclusion regarding tooth decay. The graphed data illustrate clearly that sugar intake is responsible for increased tooth decay in all areas, although the total amount of tooth decay is lower in Asia and Africa than in America and Europe.

(4) *The greater the sugar intake, the faster a tooth decays* is *not* the statement that is a valid conclusion regarding tooth decay. There are no data available in the table or in the graph concerning the relationship between sugar intake and the rate of tooth decay.

PART C

55. Four credits are allowed for analyzing the experiment that produced the data shown in the table. In your answer be sure to:

- State *one* way the immune system is involved in the rejection of transplanted organs. [1]
- Explain why the best source for a donated kidney would be the identical twin of the recipient. [1]
- Explain why immunosuppressant drugs might be needed to prevent rejection of a kidney received from a donor other than an identical twin. [1]
- State *one* reason a person may get sick more easily when taking an immunosuppressant drug. [1]

Examples of acceptable responses: [4]

The immune system is designed to recognize and fight off foreign invaders in the body by producing antibodies to attack foreign antigens. [1] Identical twins have 100% of their genes in common, including those for the production of specific antigens, so the donor's tissues would not be rejected by the recipient's antibodies. [1] When a transplanted organ comes from a donor other than an identical twin, immunosuppressants can be used that will stop the immune system from attacking the donated organ. [1] Unfortunately, the immunosuppressants may also depress the immune system, making the recipient more susceptible to infectious diseases. [1]

A healthy immune system recognizes chemical markers in a transplanted organ and attempts to reject it by producing antibodies. [1] If the transplanted organ is donated by an identical twin, the immune system doesn't detect foreign chemical markers and so doesn't try to reject it. [1] Immunosuppressant drugs are designed to block the production of antibodies by the recipient. [1] Immunosuppressant drugs may weaken the immune system, leaving the body open to infection by pathogenic organisms. [1]

56. One credit is allowed for correctly stating *one* specific way the relocated squirrels would most likely interact with a gray squirrel population that has lived in the park for many years. Acceptable responses include, but are not limited to: [1]

- *Relocated squirrels compete with park squirrels for food.*
- *Relocated squirrels can mate with park squirrels.*
- *The relocated squirrels will attempt to fill the environmental niche of the park squirrels.*
- *Relocated squirrels compete with park squirrels for nesting sites.*

57. One credit is allowed for correctly stating *one* specific way the relocated squirrels will change an abiotic factor in the park ecosystem. Acceptable responses include, but are not limited to: [1]

- *The soil will be enriched by added wastes.*
- *Relocated squirrels will use water from the park pond.*
- *Relocated squirrels will take up space.*
- *Relocated squirrels will use oxygen from, and add carbon dioxide to, the park's air.*
- *Relocated squirrels will aerate the soil when burying nuts.*

58. One credit is allowed for correctly stating *one* specific natural factor in the park ecosystem that will limit the growth of the squirrel population and support your answer. Acceptable responses include, but are not limited to: [1]

- *Predators will consume the squirrels for food.*
- *Competition with other gray squirrels will keep the population from increasing.*
- *Spread of disease due to the denser population will eliminate the weaker squirrels.*
- *The squirrels' food supply may become limited, forcing some squirrels to migrate or starve.*
- *The trees may not contain adequate space in the branches for nesting sites for all the squirrels.*

59. Four credits are allowed for responding to this editorial by explaining the importance of the local recycling program for the environment. In your answer, be sure to:

- State *one* effect the increasing human population will have on the availability of natural resources. [1]
- State *one* reason why recycling is important. [1]
- Identify *two* natural resources or products made from natural resources that can be recycled. [2]

Examples of acceptable responses: [4]

As the human population continues to grow, it will place increasing demands on limited natural resources to produce the goods needed for maintaining a standard of living. [1] Recycling is important because it slows the demand for newly extracted natural resources by reusing old ones. [1] Examples of products made from natural resources include plastic (made from petroleum) and aluminum (made from bauxite). [2]

The human population continues to grow at a rate faster than the environment is able to respond. As a result, humans use tremendous quantities of mineral resources that are being depleted and will soon run out completely. [1] Recycling the products made from these mineral resources is essential if we are to have these material resources for our future manufacturing operations. [1] It is important that we recycle paper (made from wood) and steel (made from iron ore) if these materials are to be available in our future. [2]

60. Four credits are allowed for explaining how the continued use of a certain pesticide may become ineffective in controlling the medfly. In your answer, be sure to:

- Include the concept of variation. [1]
- Include the concept of adaptive value of a variation (adaptation). [1]
- Include the concept of survival. [1]
- Include the concept of reproduction. [1]

Examples of acceptable responses: [4]

Some medflies have a variation that provides resistance to pesticides. [1] When a pesticide is present, flies with the favorable variation will survive [1] to reproduce and pass the variation to their offspring. [1] A variety of medfly resistant to the pesticide will result. [1]

Like other species, medflies contain many genetic variations within their populations, some favorable, some neutral, and others unfavorable. [1] The use of chemical pesticides to control medfly outbreaks may provide short-term gains but can have long-term negative consequences for farmers. Whenever a pesticide is used, most medflies in the population are killed, but a few resistant ones survive. [1] Those that survive may be genetically resistant to the pesticide and may pass on their resistance to the next generation by means of reproduction. [1] The result over a few generations may be a superresistant medfly population that cannot be killed off by the use of standard pesticides. [1]

61. One credit is allowed for correctly explaining how a loss of biodiversity today can affect the survival of humans in the future. Acceptable responses include, but are not limited to: [1]

- *A loss of biodiversity can result in a shortage of food.*
- *A lack of materials for building may result.*
- *A loss of medical research on rare species of plants and animals that could lead to new medicines may result.*
- *Single-variety crops are more susceptible to disease and drought.*

62. One credit is allowed for correctly stating one specific action that you as a student can take in your community to help maintain or increase biodiversity. Acceptable responses include, but are not limited to: [1]

- *Plant native trees and wildflowers*
- *Preserve habitats*
- *Recycle resources*
- *Consume fewer products*
- *Work to reduce pollution*
- *Don't import invasive or foreign species*
- *Don't buy products from companies that exploit fragile ecosystems and endangered populations*

PART D

63. **1** *Cooling the hand increases the reaction time* is the statement that is best supported by the data. A review of the data in the table shows a longer average reaction time in all five classes. This delayed reaction time is most likely due to deadened motor nerve endings in the hand not being able to respond quickly to commands delivered from the brain to the effector muscles.

WRONG CHOICES EXPLAINED

(2) *Cooling the hand does not affect the reaction time* is *not* the statement that is best supported by the data. This statement is not supported by the data, which demonstrate a measurable decrease in average reaction time.

(3) *Cooling the hand affects only some subjects* is *not* the statement that is best supported by the data. This statement is not supported by the data, which demonstrate a consistent slowing of reaction time in all five classes.

(4) *Two minutes of cooling is not enough to affect reaction time* is *not* the statement that is best supported by the data. This statement is not supported by the data, which demonstrate a measurable decrease in average reaction time.

64. One credit is allowed for correctly stating *one* error in this experiment. Acceptable responses include, but are not limited to: [1]

- *The student did not obtain pulse rates before lunch.*
- *The sample size chosen was too small.*
- *The student did not design and measure a control group.*
- *There was no indication of what the nine students were doing just before the data were collected.*

65. One credit is allowed for correctly explaining why exercise could influence the ability of a student to solve math problems. Acceptable responses include, but are not limited to: [1]

- *The blood will bring more oxygen to the brain.*
- *Increased blood flow will remove wastes from the brain.*
- *Increased blood flow will bring more glucose to the brain.*

66. One credit is allowed for correctly stating whether or not exercising for 30 minutes improved the ability of students to solve math problems and support your answer by referencing the data. Acceptable responses include, but are not limited to: [1]

- *Yes, because three of the four students solved more problems after exercising.*
- *No, because one student did fewer problems.*
- *Cannot tell because there are results from only four students.*
- *Inconclusive because there are no data for a separate control group.*

67. One credit is allowed for correctly explaining why it would be difficult to determine which one of the other three organisms from the table should be placed in box 1. Acceptable responses include, but are not limited to: [1]

- *The pigeon shares characteristics with all of the other organisms.*
- *Organisms A and C also lay eggs.*
- *None of the other three organisms share all characteristics in common with the pigeon.*
- *The evolutionary origins of the various structures and characteristics are unclear.*

68. **1** If frog eggs taken from a freshwater pond are placed in a saltwater aquarium, it is most likely that *water will leave the eggs*. Frog eggs from a freshwater pond have a water concentration equal to that of pond water. If these eggs are moved into a saltwater environment, a concentration gradient will be established that will cause water to move by osmosis from a region of relatively high water concentration (the egg interior) to a region of relatively low water concentration (the saltwater environment).

WRONG CHOICES EXPLAINED

(2) If frog eggs taken from a freshwater pond are placed in a saltwater aquarium, it is *not* most likely that *salt will leave the eggs*. In this situation, salt is in relatively high concentration outside the eggs in the saltwater environment and so will be more likely to diffuse into the eggs than out of them.

(3) If frog eggs taken from a freshwater pond are placed in a saltwater aquarium, it is *not* most likely that *water will neither enter nor leave the eggs*. If water were to neither enter nor leave the eggs, they would be in equilibri-

um with their environment, as they are in a freshwater pond. Once they are moved to the saltwater environment, the equilibrium would be disturbed and water would move toward lower concentrations outside the eggs.

(4) If frog eggs taken from a freshwater pond are placed in a saltwater aquarium, it is *not* most likely that *the eggs will burst*. This situation would occur if the eggs were moved from the freshwater pond into distilled (100%) water. In this case water would move from the distilled water container through the cell membranes and into the egg cytoplasm by osmosis. This water movement would continue until the cytoplasm burst the membranes of the egg.

69. One credit is allowed for correctly drawing lines to represent the positions of the fragments from each DNA sample when electrophoresis is completed: [1]

70. One credit is allowed for correctly stating which two samples are the most similar and support your answer using data from this investigation. Acceptable responses include, but are not limited to: [1]

- *Band D—because they have the most fragments in common.*
- *Band D—these samples are the most similar because they both show DNA fragments of 4, 8, and 12 bases.*

71. One credit is allowed for correctly stating *one* specific use for the information obtained from the results of gel electrophoresis. Acceptable responses include, but are not limited to: [1]

- *Determining the identities of criminals*
- *Determining the parents of a child*
- *Determining the identity of a crime victim*
- *Determining the identities of victims of natural disasters*
- *Determining evolutionary relationships*

72. One credit is allowed for correctly explaining how this laboratory activity illustrates the process of natural selection. Acceptable responses include, but are not limited to: [1]

- *The tools represent types of beaks, some of which are more successful for gathering seeds and so are more favorable for survival.*
- *Students with favorable "beaks" survived.*
- *Picking the seeds out of the same dish with different types of beaks mimics the process of competition.*
- *The "most fit" students were the ones with "beaks" that could pick up the most seeds the fastest.*

73. **4** *Overproduction* is a factor that influences the evolution of a species that was *not* part of this laboratory activity. A key point of Darwin's theory of natural selection is the concept that naturally occurring species have a tendency to produce more offspring than can possibly survive. This concept, known as overproduction, is not directly reflected in the *Beaks of Finches* laboratory activities.

WRONG CHOICES EXPLAINED

(1), (3) *Struggle for survival* and *competition* are factors that influence the evolution of a species that were part of this laboratory activity. These concepts are represented by the students when they tried to gather more seeds in their dishes than their partners. Collecting more seeds was considered to be more successful and worthy of survival than collecting fewer seeds.

(2) *Variation* is a factor that influences the evolution of a species that was part of this laboratory activity. The different tools used represented the variations in beak type that might occur naturally in a bird species.

74. One credit is allowed for correctly identifying one trait, other than beak characteristics, that could contribute to the ability of a finch to feed successfully. Acceptable responses include, but are not limited to: [1]

- *Strength*
- *Vision*
- *Coordination*
- *Ability to produce digestive enzymes*
- *Ability to fly*
- *Territorial behavior*

75. **4** *The yellow dye molecules are smaller than the red dye molecules* is the statement that best explains the results of this experiment. In this experiment both red dye molecules and yellow dye molecules were placed inside the dialysis membrane. The water outside the membrane turned yellow, indicating that the yellow dye molecules were small enough to pass through the microscopic pores in the membrane. The fact that the red dye remained inside the membrane is an indication that the red dye molecules were too large to pass through the membrane pores.

WRONG CHOICES EXPLAINED

(1) *Water diffused into the membrane bag* is *not* the statement that best explains the results of this experiment. It is probable that water molecules are small enough to pass readily back and forth through the microscopic pores in the dialysis membrane. This fact does not affect the experimental results.

(2) *The dialysis membrane actively transported yellow dye molecules* is *not* the statement that best explains the results of this experiment. Active transport is a type of cross-membrane transport in which the energy of ATP molecules is used by a living cell to move molecules of a substance against the concentration gradient from a region of relatively low concentration to a region of relatively high concentration of that substance. Dialysis tubing is not a living membrane and therefore cannot perform this type of active transport.

(3) *Only red dye diffused through the membrane* is *not* the statement that best explains the results of this experiment. This statement is directly contradicted by the information given. If only the red dye molecules had diffused through the membrane, then the beaker water would have turned red instead of yellow.

STANDARDS/KEY IDEAS	AUGUST 2007 QUESTION NUMBERS	NUMBER OF CORRECT RESPONSES
STANDARD 1		
Key Idea 1: The central purpose of scientific inquiry is to develop explanations of natural phenomena in a continuing and creative process.	2, 54	
Key Idea 2: Beyond the use of reasoning and consensus, scientific inquiry involves the testing of proposed explanations involving the use of conventional techniques and procedures and usually requiring considerable ingenuity.	35	
Key Idea 3: The observations made while testing proposed explanations, when analyzed using conventional and invented methods, provide new insights into natural phenomena.	40, 51, 52, 53	
Laboratory Checklist	36	
STANDARD 4		
Key Idea 1: Living things are both similar to and different from each other and from nonliving things.	1, 3, 5, 7, 43, 44, 46, 47, 56, 57, 58,	
Key Idea 2: Organisms inherit genetic information in a variety of ways that result in continuity of structure and function between parents and offspring.	8, 9, 10, 11, 12, 37, 38, 49, 50	
Key Idea 3: Individual organisms and species change over time.	13, 14, 15, 18, 60	
Key Idea 4: The continuity of life is sustained through reproduction and development.	6, 17, 19, 20, 39, 45	
Key Idea 5: Organisms maintain a dynamic equilibrium that sustains life.	4, 16, 21, 22, 26, 33, 34, 48, 55	
Key Idea 6: Plants and animals depend on each other and their physical environment.	24, 25, 28, 30, 31, 32, 41, 61	

STANDARDS/KEY IDEAS	AUGUST 2007 QUESTION NUMBERS	NUMBER OF CORRECT RESPONSES
STANDARD 4		
Key Idea 7: Human decisions and activities have a profound impact on the physical and living environment.	23, 27, 29, 42, 59, 62	
REQUIRED LABORATORIES		
Lab 1: "Relationships and Biodiversity"	67, 69, 70, 71	
Lab 2: "Making Connections"	63, 64, 65, 66	
Lab 3: "The Beaks of Finches"	72, 73, 74	
Lab 5: "Diffusion Through a Membrane"	68, 75	

Examination
June 2008
Living Environment

PART A

Answer all questions in this part. [30]

Directions (1–30): For *each* statement or question, select the word or expression that best completes the statement or answers the question. Record your answers in the spaces provided.

1 The chart below contains both autotrophic and heterotrophic organisms.

A	owl	cat	shark
B	mouse	corn	dog
C	squirrel	bluebird	alga

Organisms that carry out only heterotrophic nutrition are found in

1 row *A*, only   3 rows *A* and *B*
2 row *B*, only   4 rows *A* and *C*   1_____

2 A stable pond ecosystem would *not* contain

1 materials being cycled
2 oxygen
3 decomposers
4 more consumers than producers   2_____

3 Although all of the cells of a human develop from one fertilized egg, the human is born with many different types of cells. Which statement best explains this observation?

1 Developing cells may express different parts of their identical genetic instructions.
2 Mutations occur during development as a result of environmental conditions.
3 All cells have different genetic material.
4 Some cells develop before other cells.   3____

4 Humans require organ systems to carry out live processes. Single-celled organisms do not have organ systems and yet they are able to carry out life processes. This is because

1 human organ systems lack the organelles found in single-celled organisms
2 a human cell is more efficient than the cell of a single-celled organism
3 it is not necessary for single-celled organisms to maintain homeostasis
4 organelles present in single-celled organisms act in a manner similar to organ systems   4____

5 Certain poisons are toxic to organisms because they interfere with the function of enzymes in mitochondria. This results directly in the inability of the cell to

1 store information
2 build proteins
3 release energy from nutrients
4 dispose of metabolic wastes   5____

6 At warm temperatures, a certain bread mold can often be seen growing on bread as a dark-colored mass. The same bread mold growing on bread in a cooler environment is red in color. Which statement most accurately describes why this change in the color of the bread mold occurs?

1 Gene expression can be modified by interactions with the environment.

2 Every organism has a different set of coded instructions.

3 The DNA was altered in response to an environmental condition.

4 There is no replication of genetic material in the cooler environment.   6_____

7 Asexually reproducing organisms pass on hereditary information as

1 sequences of A, T, C, and G

2 chains of complex amino acids

3 folded protein molecules

4 simple inorganic sugars   7_____

8 Species of bacteria can evolve more quickly than species of mammals because bacteria have

1 less competition

2 more chromosomes

3 lower mutation rates

4 higher rates of reproduction   8_____

9 The diagram below represents the synthesis of a portion of a complex molecule in an organism.

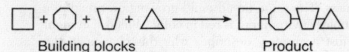

Building blocks   Product

Which row in the chart could be used to identify the building blocks and product in the diagram?

Row	Building Blocks	Product
(1)	starch molecules	glucose
(2)	amino acid molecules	part of protein
(3)	sugar molecules	ATP
(4)	DNA molecules	part of starch

9_____

10 Which diagram best represents the relative locations of the structures in the list below?

A—chromosome
B—nucleus
C—cell
D—gene

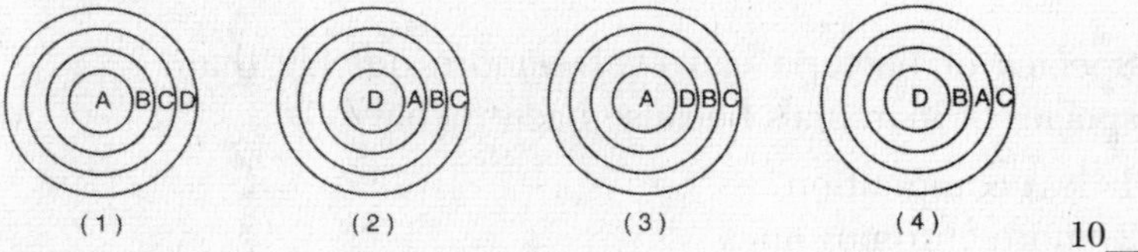

(1)   (2)   (3)   (4)

10_____

11 Which nuclear process is represented below?

A DNA molecule untwists. → The two strands of DNA separate. → Molecular bases pair up. → Two identical DNA molecules are produced.

1 recombination   3 replication
2 fertilization   4 mutation

11_____

12 For centuries, certain animals have been crossed to produce offspring that have desirable qualities. Dogs have been mated to produce Labradors, beagles, and poodles. All of these dogs look and behave very differently from one another. This technique of producing organisms with specific qualities is known as

1 gene replication   3 random mutation
2 natural selection   4 selective breeding   12_____

13 Certain insects resemble the bark of the trees on which they live. Which statement provides a possible biological explanation for this resemblance?

1 The insects needed camouflage so they developed protective coloration.
2 Natural selection played a role in the development of this protective coloration.
3 The lack of mutations resulted in the protective coloration.
4 The trees caused mutations in the insects that resulted in protective coloration.   13_____

14 When is extinction of a species most likely to occur?

1 when environmental conditions remain the same and the proportion of individuals within the species that lack adaptive traits increases
2 when environmental conditions remain the same and the proportion of individuals within the species that possess adaptive traits increases
3 when environmental conditions change and the adaptive traits of the species favor the survival and reproduction of some of its members
4 when environmental conditions change and the members of the species lack adaptive traits to survive and reproduce   14_____

15 In what way are photosynthesis and cellular respiration similar?

1 They both occur in chloroplasts.
2 They both require sunlight.
3 They both involve organic and inorganic molecules.
4 They both require oxygen and produce carbon dioxide.   15____

16 Which process will increase variations that could be inherited?

1 mitotic cell division
2 active transport
3 recombination of genes
4 synthesis of proteins   16____

17 Some cells involved in the process of reproduction are represented in the diagram below.

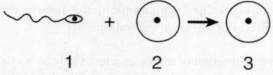

1   **2**   **3**

The process of meiosis formed

1 cell 1, only   3 cell 3, only
2 cells 1 and 2   4 cells 2 and 3   17____

18 Kangaroos are mammals that lack a placenta. Therefore, they must have an alternate way of supplying the developing embryo with

1 nutrients
2 carbon dioxide
3 enzymes
4 genetic information   18____

19 Which substance is the most direct source of the energy that an animal cell uses for the synthesis of materials?

1 ATP   3 DNA
2 glucose   4 starch   19____

20 To increase chances for a successful organ transplant, the person receiving the organ should be given special medications. The purpose of these medications is to

1 increase the immune response in the person receiving the transplant
2 decrease the immune response in the person receiving the transplant
3 decrease mutations in the person receiving the transplant
4 increase mutations in the person receiving the transplant   20____

21 The diagram below represents the cloning of a carrot plant.

Compared to each cell of the original carrot plant, each cell of the new plant will have

1 the same number of chromosomes and the same types of genes
2 the same number of chromosomes, but different types of genes
3 half the number of chromosomes and the same types of genes
4 half the number of chromosomes, but different types of genes   21____

22 The development of an embryo is represented in the diagram below.

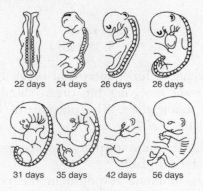

(Not drawn to scale)

These changes in the form of the embryo are a direct result of

1 uncontrolled cell division and mutations
2 differentiation and growth
3 antibodies and antigens inherited from the father
4 meiosis and fertilization

22____

23 The diagram below represents an event that occurs in the blood.

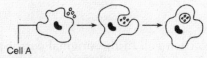

Cell A

Which statement best describes this event?

1 Cell *A* is a white blood cell releasing antigens to destroy bacteria.
2 Cell *A* is a cancer cell produced by the immune system and it is helping to prevent disease.
3 Cell *A* is a white blood cell engulfing disease-causing organisms.
4 Cell *A* is protecting bacteria so they can reproduce without being destroyed by predators.

23____

24 In an ecosystem, the growth and survival of organisms are dependent on the availability of the energy from the Sun. This energy is available to organisms in the ecosystem because

1 producers have the ability to store energy from light in organic molecules
2 consumers have the ability to transfer chemical energy stored in bonds to plants
3 all organisms in a food web have the ability to use light energy
4 all organisms in a food web feed on autotrophs   24____

25 Which factor has the greatest influence on the type of ecosystem that will form in a particular geographic area?

1 genetic variation in the animals
2 climate conditions
3 number of carnivores
4 percentage of nitrogen gas in the atmosphere   25____

26 Farming reduces the natural biodiversity of an area, yet farms are necessary to feed the world's human population. This situation is an example of

1 poor land use   3 conservation
2 a trade-off   4 a technological fix   26____

27 A food chain is represented below.

Grass → Cricket → Frog → Owl

This food chain contains

(1) 4 consumers and no producers
(2) 1 predator, 1 parasite, and 2 producers
(3) 2 carnivores and 2 herbivores
(4) 2 predators, 1 herbivore, and 1 producer   27____

28 A volcanic eruption destroyed a forest, covering the soil with volcanic ash. For many years, only small plants could grow. Slowly, soil formed in which shrubs and trees could grow. These changes are an example of

1 manipulation of genes
2 evolution of a species
3 ecological succession
4 equilibrium

28____

29 A major reason that humans can have such a significant impact on an ecological community is that humans

1 can modify their environment through technology
2 reproduce faster than most other species
3 are able to increase the amount of finite resources available
4 remove large amounts of carbon dioxide from the air

29____

30 Rabbits are herbivores that are not native to Australia. Their numbers have increased steadily since being introduced into Australia by European settlers. One likely reason the rabbit population was able to grow so large is that the rabbits

1 were able to prey on native herbivores
2 reproduced more slowly than the native animals
3 successfully competed with native herbivores for food
4 could interbreed with the native animals

30____

PART B–1

Answer all questions in this part.　[12]

Directions (31–42): For *each* statement or question, select the word or expression that best completes the statement or answers the question. Record your answers in the spaces provided.

31 Which laboratory procedure is represented in the diagram below?

Paper towel

 1 placing a coverslip over a specimen
 2 removing a coverslip from a slide
 3 adding stain to a slide without removing the coverslip
 4 reducing the size of air bubbles under a coverslip　　31____

32 In the United States, there has been relatively little experimentation involving the insertion of genes from other species into human DNA. One reason for the lack of these experiments is that

 1 the subunits of human DNA are different from the DNA subunits of other species
 2 there are many ethical questions to be answered before inserting foreign genes into human DNA
 3 inserting foreign DNA into human DNA would require using techniques completely different from those used to insert foreign DNA into the DNA of other mammals
 4 human DNA always promotes human survival, so there is no need to alter it　　32____

33 The development of an experimental research plan should *not* include a

1 list of safety precautions for the experiment
2 list of equipment needed for conducting the experiment
3 procedure for the use of technologies needed for the experiment
4 conclusion based on data expected to be collected in the experiment

33_____

34 A student performed an experiment to demonstrate that a plant needs chlorophyll for photosynthesis. He used plants that had green leaves with white areas. After exposing the plants to sunlight, he removed a leaf from each plant and processed the leaves to remove the chlorophyll. He then tested each leaf for the presence of starch. Starch was found in the area of the leaf that was green, and no starch was found in the area of the leaf that was white. He concluded that chlorophyll is necessary for photosynthesis.

Which statement represents an assumption the student had to make in order to draw this conclusion?

1 Starch is synthesized from the glucose produced in the green areas of the leaf.
2 Starch is converted to chlorophyll in the green areas of the leaf.
3 The white areas of the leaf do not have cells.
4 The green areas of the leaf are heterotrophic.

34_____

35 The diagram below represents an interaction between parts of an organism.

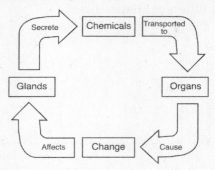

The term *chemicals* in this diagram represents

1 starch molecules   3 hormone molecules
2 DNA molecules   4 receptor molecules   35____

36 The diagram below represents two cells, X and Y.

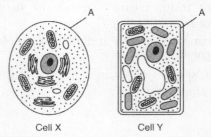

Which statement is correct concerning the structure labeled A?

1 It aids in the removal of metabolic wastes in both cell X and cell Y.
2 It is involved in cell communication in cell X, but not in cell Y.
3 It prevents the absorption of CO_2 in cell X and O_2 in cell Y.
4 It represents the cell wall in cell X and the cell membrane in cell Y.   36____

37 The graph below provides information about the reproductive rates of four species of bacteria, A, B, C, and D, at different temperatures.

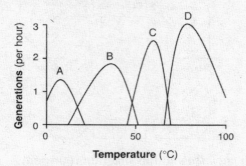

Temperature (°C)

Which statement is a valid conclusion based on the information in the graph?

1 Changes in temperature cause bacteria to adapt to form new species.

2 Increasing temperatures speed up bacterial reproduction.

3 Bacteria can survive only temperatures between 0°C and 100°C.

4 Individual species reproduce within a specific range of temperatures.

37_____

38 The diagram below shows some of the steps in protein synthesis.

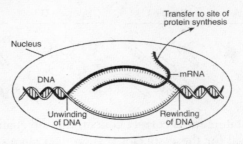

The section of DNA being used to make the strand of mRNA is known as a

(1) carbohydrate   (3) ribosome

(2) gene   (4) chromosome   38_____

39 An energy pyramid is shown below.

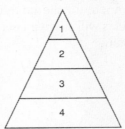

Which graph best represents the relative energy content of the levels of this pyramid?

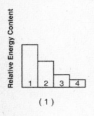
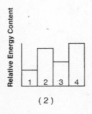
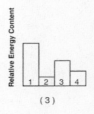
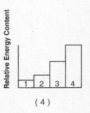

39_____

40 The diagram below represents four different species of bacteria.

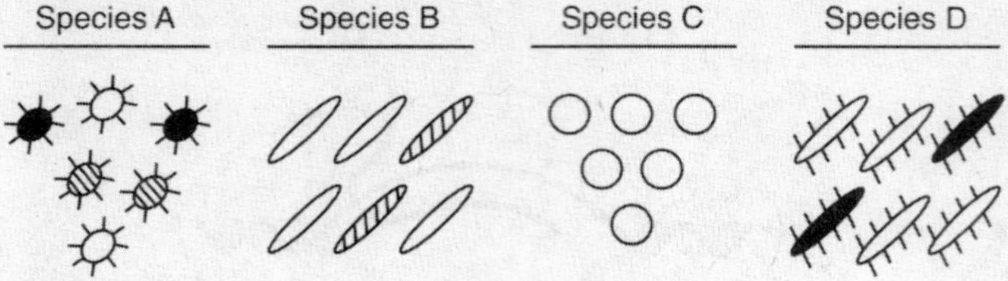

Species A	Species B	Species C	Species D

Which statement is correct concerning the chances of survival for these species if there is a change in the environment?

1 Species A has the best chance of survival because it has the most genetic diversity.
2 Species C has the best chance of survival because it has no gene mutations.
3 Neither species B nor species D will survive because they compete for the same resources.
4 None of the species will survive because bacteria reproduce asexually.

40_____

41 The diagram below represents possible evolutionary relationships between groups of organisms.

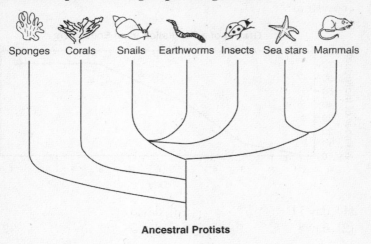

Sponges   Corals   Snails   Earthworms   Insects   Sea stars   Mammals

Ancestral Protists

Which statement is a valid conclusion that can be drawn from the diagram?

1 Snails appeared on Earth before corals.
2 Sponges were the last new species to appear on Earth.
3 Earthworms and sea stars have a common ancestor.
4 Insects are more complex than mammals.   41____

42 On which day did the population represented in the graph below reach the carrying capacity of the ecosystem?

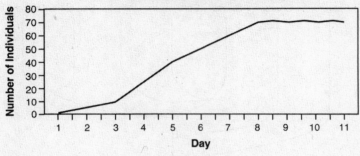

Growth of a Population in an Ecosystem

(1) day 11   (3) day 3
(2) day 8   (4) day 5

42____

PART B–2

Answer all questions in this part.   [13]

Directions (43–55): **For those questions that are followed by four choices, record your answers in the spaces provided. For all other questions in this part, record your answers in accordance with the directions given in the question.**

Base your answers to questions 43 through 47 on the information below and on your knowledge of biology.

Each year, a New York State power agency provides its customers with information about some of the fuel sources used in generating electricity. The table below applies to the period of 2002–2003.

Fuel Sources Used

Fuel Source	Percentage of Electricity Generated
hydro (water)	86
coal	5
nuclear	4
oil	1
solar	0

Directions (43 and 44): Using the information given, construct a bar graph *on the grid on the next page*, following the directions below.

43 Mark an appropriate scale on the axis labeled "Percentage of Electricity Generated."   [1]

44 Construct vertical bars to represent the data. Shade in *each* bar.   [1]

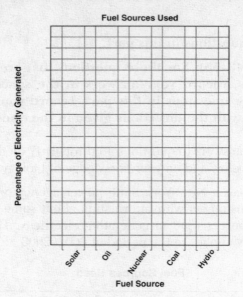

Fuel Sources Used

Percentage of Electricity Generated (y-axis)

Solar   Oil   Nuclear   Coal   Hydro

Fuel Source

45 Identify *one* fuel source in the table that is considered a fossil fuel.   [1]

_____

46 Identify *one* fuel source in the table that is classified as a renewable resource.   [1]

_____

47 State *one* specific environmental problem that can result from burning coal to generate electricity. [1]

_____

_____

_____

Base your answers to questions 48 and 49 on the diagram below that shows some interactions between several organisms located in a meadow environment and on your knowledge of biology.

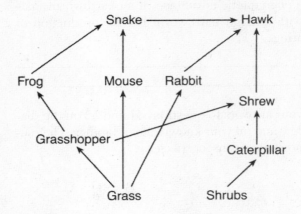

48 A rapid *decrease* in the frog population results in a change in the hawk population. State how the hawk population may change. Support your answer.   [1]

_____

_____

49 Identify *one* cell structure found in a producer in this meadow ecosystem that is *not* found in the carnivores.   [1]

_____

_____

50 Individuals of some species, such as earthworms, have both male and female sex organs. In many cases, however, these individuals do not fertilize their own eggs.

State *one* genetic advantage of an earthworm mating with another earthworm for the production of offspring. [1]

_____

_____

Base your answers to questions 51 and 52 on the diagram below and on your knowledge of biology. The diagram represents six insect species.

Species E   Species F

51 A dichotomous key to these six species is shown below. Complete the missing information for sections 5.*a.* and 5.*b.* so that the key is complete for all *six* species. [1]

Dichotomous Key

1. a. has small wings . . . . go to 2
   b. has large wings . . . . go to 3

2. a. has a single pair of wings . . . . Species A
   b. has a double pair of wings . . . . . . Species B

3. a. has a double pair of wings . . . . go to 4
   b. has a single pair of wings . . . . Species C

4. a. has spots . . . . go to 5
   b. does not have spots . . . . Species D

5. a. _____ . . . . Species E
   b. _____ . . . . Species F

52 Use the key to identify the drawings of species *A*, *B*, *C*, and *D*. Place the letter of *each* species on the line located below the drawing of the species.   [1]

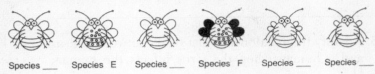

Species ___   Species E   Species ___   Species F   Species ___   Species ___

Base your answers to questions 53 through 55 on the information below and on your knowledge of biology.

Proteins on the surface of a human cell and on a bird influenza virus are represented in the diagram below.

Human Cell   Bird Influenza Virus

53 In the space below, draw a change in the bird influenza virus that would allow it to infect this human cell.   [1]

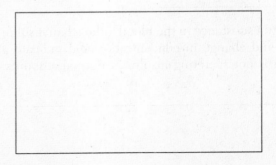

54 Explain how this change in the virus could come about. [1]

_____

_____

55 Identify the relationship that exists between a virus and a human when the virus infects the human. [1]

_____

_____

PART C

Answer all questions in this part. [17]

Directions (56–67): **Record your answers in the spaces provided.**

Base your answers to questions 56 and 57 on the information below and on your knowledge of biology.

Insulin is a hormone that has an important role in the maintenance of homeostasis in humans.

56 Identify the structure in the human body that is the usual source of insulin. [1]

_____

57 Identify a substance in the blood, other than insulin, that could change in concentration and indicate a person is not secreting insulin in normal amounts. [1]

_____

Base your answers to questions 58 and 59 on the information below and on your knowledge of biology.

The hedgehog, a small mammal native to Africa and Europe, has been introduced to the United States as an exotic pet species. Scientists have found that hedgehogs can transfer pathogens to humans and domestic animals. Foot-and-mouth viruses, *Salmonella*, and certain fungi are known pathogens carried by hedgehogs. As more and more of these exotic animals are brought into this country, the risk of infection increases in the human population.

58 State *one negative* effect of importing exotic species to the United States. [1]

_____

_____

59 State *one* way the human immune system might respond to an invading pathogen associated with handling a hedgehog. [1]

_____

_____

Base your answers to questions 60 through 62 on the information below and on your knowledge of biology.

The last known wolf native to the Adirondack Mountains of New York State was killed over a century ago. Several environmental groups have recently proposed reintroducing the wolf to the Adirondacks. These groups claim there is sufficient prey to support a wolf population in this area. These prey include beaver, deer, and moose. Opponents of this proposal state that the Adirondacks already have a dominant predator, the Eastern coyote.

60 State *one* effect the reintroduction of the wolf may have on the coyote population within the Adirondacks. Explain why it would have this effect. [1]

_____

_____

61 Explain why the coyote is considered a limiting factor in the Adirondack Mountains. [1]

_____

_____

62 State *one* ecological reason why some individuals might support the reintroduction of wolves to the Adirondacks. [1]

_____

_____

_____

63 You have been assigned to design an experiment to determine the effects of light on the growth of tomato plants. In your experimental design be sure to:
- state *one* hypothesis to be tested [1]
- identify the independent variable in the experiment [1]
- describe the type of data to be collected [1]

_____

_____

_____

_____

_____

_____

_____

64 In some land plants, guard cells are found only on the lower surfaces of the leaves. In some water plants, guard cells are found only on the upper surfaces of the leaves. Explain how guard cells in both land and water plants help maintain homeostasis. In your answer be sure to:

 • identify *one* function regulated by the guard cells in leaves   [1]

 • explain how guard cells carry out this function [1]

 • give *one* possible evolutionary advantage of the position of the guard cells on the leaves of land plants   [1]

_____

_____

_____

_____

_____

_____

_____

Base your answers to questions 65 and 66 on the information below and on your knowledge of biology.

Scientists are increasingly concerned about the possible effects of damage to the ozone layer.

65 Damage to the ozone layer has resulted in mutations in skin cells that lead to cancer. Will the mutations that caused the skin cancers be passed on to offspring? Support your answer.   [1]

_____

_____

66 State *two* specific ways in which an ocean ecosystem will change (other than fewer photosynthetic organisms) if populations of photosynthetic organisms die off as a result of damage to the ozone layer. [2]

_____

_____

_____

_____

_____

67 Lawn wastes, such as grass clippings and leaves, were once collected with household trash and dumped into landfills. Identify *one* way that this practice was harmful to the environment. [1]

_____

_____

PART D

Answer all questions in this part. [13]

Directions **(68–80): For those questions that are followed by four choices, record your answers in the spaces provided. For all other questions in this part, record your answers in accordance with the directions given in the question.**

68 In preparation for an electrophoresis procedure, enzymes are added to DNA in order to

  1 convert the DNA into gel
  2 cut the DNA into fragments
  3 change the color of the DNA
  4 produce longer sections of DNA   68_____

69 Paper chromatography is a laboratory technique that is used to

  1 separate different molecules from one another
  2 stain cell organelles
  3 indicate the pH of a substance
  4 compare relative cell sizes   69_____

70 A marathon runner frequently experiences muscle cramps while running. If he stops running and rests, the cramps eventually go away. The cramping in the muscles most likely results from

  1 lack of adequate oxygen supply to the muscle
  2 the runner running too slowly
  3 the runner warming up before running
  4 increased glucose production in the muscle   70_____

Base your answers to questions 71 through 73 on the information below and on your knowledge of biology.

A series of investigations was performed on four different plant species. The results of these investigations are recorded in the data table below.

Characteristics of Four Plant Species

Plant Species	Seeds	Leaves	Pattern of Vascular Bundles (structures in stem)	Type of Chlorophyll Present
A	round/small	needle-like	scattered bundles	chlorophyll a and b
B	long/pointed	needle-like	circular bundles	chlorophyll a and c
C	round/small	needle-like	scattered bundles	chlorophyll a and b
D	round/small	needle-like	scattered bundles	chlorophyll b

71 Based on these data, which *two* plant species appear to be most closely related? Support your answer. [1]

Plant species _____ and _____

_____

_____

72 What additional information could be gathered to support your answer to question 71? [1]

_____

_____

73 State *one* reason why scientists might want to know if two plant species are closely related. [1]

_____

_____

Base your answers to questions 74 and 75 on the data table below and on your knowledge of biology.

Dietary Preferences of Finches

Species of Finch	Preferred Foods
A	nuts and seeds
B	worms and insects
C	fruits and seeds
D	insects and seeds
E	nuts and seeds

74 Based on its preferred food, species *B* would be classified as a

1 decomposer
2 producer
3 carnivore
4 parasite   74_____

75 Which two species would most likely be able to live in the same habitat without competing with each other for food?

1 *A* and *C*
2 *B* and *C*
3 *B* and *D*
4 *C* and *E*   75_____

Base your answers to questions 76 and 77 on the experimental setup shown below.

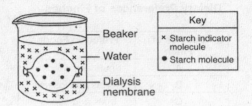

76 On the diagram below, draw in the expected locations of the molecules after a period of one hour. [1]

77 When starch indicator is used, what observation would indicate the presence of starch?   [1]

_____

_____

_____

78 State *one* reason why some molecules can pass through a certain membrane, but other molecules can *not*.   [1]

_____

_____

79 A plant cell in a microscopic field of view is represented below.

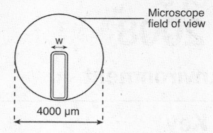

The width (w) of this plant cell is closest to

1 200 μm
2 800 μm
3 1200 μm
4 1600 μm   79_____

80 The diagram below represents a plant cell in tap water as seen with a compound light microscope.

Which diagram best represents the appearance of the cell after it has been placed in a 15% salt solution for two minutes?

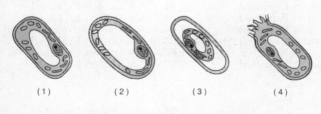

(1)   (2)   (3)   (4)   80_____

Answers
June 2008

Living Environment

Answer Key

PART A

1. 1	6. 1	11. 3	16. 3	21. 1	26. 2
2. 4	7. 1	12. 4	17. 2	22. 2	27. 4
3. 1	8. 4	13. 2	18. 1	23. 3	28. 3
4. 4	9. 2	14. 4	19. 1	24. 1	29. 1
5. 3	10. 2	15. 3	20. 2	25. 2	30. 3

PART B–1

31. 3	33. 4	35. 3	37. 4	39. 4	41. 3
32. 2	34. 1	36. 1	38. 2	40. 1	42. 2

PART B–2

43. *See* Answers Explained.	50. *See* Answers Explained.
44. *See* Answers Explained.	51. *See* Answers Explained.
45. *See* Answers Explained.	52. *See* Answers Explained.
46. *See* Answers Explained.	53. *See* Answers Explained.
47. *See* Answers Explained.	54. *See* Answers Explained.
48. *See* Answers Explained.	55. *See* Answers Explained.
49. *See* Answers Explained.	

PART C

56. *See* Answers Explained.	62. *See* Answers Explained.
57. *See* Answers Explained.	63. *See* Answers Explained.
58. *See* Answers Explained.	64. *See* Answers Explained.
59. *See* Answers Explained.	65. *See* Answers Explained.
60. *See* Answers Explained.	66. *See* Answers Explained.
61. *See* Answers Explained.	67. *See* Answers Explained.

PART D

68. 2	75. 2
69. 1	76. *See* Answers Explained.
70. 1	77. *See* Answers Explained.
71. *See* Answers Explained.	78. *See* Answers Explained.
72. *See* Answers Explained.	79. 2
73. *See* Answers Explained.	80. 3
74. 3	

Answers Explained

PART A

1. **1** Organisms that carry out only heterotrophic nutrition are found in *row A, only*. Heterotrophic organisms are those that derive their nutritional requirements by consuming the bodies of other organisms. Animals are heterotrophic (other-feeding) organisms, whereas most plants and algae are autotrophic (self-feeding) organisms that derive their nutritional requirements from the self-contained process of photosynthesis. Row A lists the owl, cat, and shark, all of which are animals that consume other animals for food and therefore are heterotrophic organisms.

WRONG CHOICES EXPLAINED

(2), (3), (4) Organisms that carry out only heterotrophic nutrition are *not* found in *row B, only*, *rows A and B*, or *rows A and C*. Row B includes corn and row C includes alga, both of which are autotrophic organisms. Therefore, neither row B nor row C contains *only* heterotrophic organisms.

2. **4** A stable pond ecosystem would *not* contain *more consumers than producers*. Any stable ecosystem must have many more producers than consumers because solar energy captured by producer organisms is inefficiently passed on to consumers, with as much as 90% of the energy lost to the environment as heat at each exchange at each trophic (feeding) level. An ecosystem with more consumers than producers would quickly destabilize and collapse until an appropriate balance could be reestablished.

WRONG CHOICES EXPLAINED

(1) A stable pond ecosystem *would* contain *materials being recycled*. Materials needed to build the structural and functional chemical components of living things include carbon, hydrogen, oxygen, and nitrogen, among others. There is a finite quantity of these materials present on Earth, so they must be recycled to allow the continuation of life.

(2) A stable pond ecosystem *would* contain *oxygen*. Molecular oxygen is a gas found commonly in Earth's air, water, and soil environments and is essential for sustaining life in the vast majority of living things that make up a stable ecosystem. Oxygen is needed by these organisms for the process of aerobic respiration.

(3) A stable pond ecosystem *would* contain *decomposers*. Decomposers, such as many bacteria and fungi, obtain their nutritional requirements by consuming the bodies of other organisms. In doing so, they assist in the process of material recycling discussed above.

3. **1** *Developing cells may express different parts of their identical genetic instructions* is the statement that best explains this observation. In the process of differentiation, highly specialized cell tissues develop that display specific characteristics. As cells differentiate, genes needed for the proper operation of these cells are switched "on," whereas genes not needed in the specialized cells are switched "off." This allows specialized cells to operate efficiently, producing only those substances essential for cell operation.

WRONG CHOICES EXPLAINED
(2) *Mutations occur during development as a result of environmental conditions* is *not* the statement that best explains this observation. Mutations of genetic material are random events caused by mutagenic agents (such as chemicals or radiation) present in the environment. The random nature of mutations makes it impossible that they are responsible for cell differentiation.

(3) *All cells have different genetic material* is *not* the statement that best explains this observation. All the cells of a human are derived from a single fertilized egg cell that undergoes mitotic cell division. The result of mitotic cell division is the creation of genetically identical, not different, daughter cells.

(4) *Some cells develop before others* is *not* the statement that best explains this observation. Although new cells develop and specialize continuously during the process of human growth, each new cell takes on the specialized characteristics of the tissue from which it forms.

4. **4** Single-celled organisms do not have organ systems and yet are able to carry out life processes because *organelles present in single-celled organisms act in a manner similar to organ systems*. These specialized subcellular structures contain enzymes needed to carry out essential life functions such as respiration (mitochondria), nutrition (food vacuoles), regulation (nucleus), synthesis (ribosomes), and excretion (contractile vacuoles), among others.

WRONG CHOICES EXPLAINED
(1) It is *not* true that single-celled organisms do not have organ systems and yet are able to carry out life processes because *human organ systems lack organelles found in single-celled organisms*. Cytology, the study of cell structure and function, has revealed that nearly all human cells contain all of the organelles found in typical single-celled organisms.

(2) It is *not* true that single-celled organisms do not have organ systems and yet are able to carry out life processes because *a human cell is more efficient than the cell of a single-celled organism*. Single-celled organisms are highly efficient in the operation of cellular processes essential to life. They are neither more efficient nor less efficient than human cells.

(3) It is *not* true that single-celled organisms do not have organ systems and yet are able to carry out life processes because *it is not necessary for single-celled organisms to maintain homeostasis*. Homeostasis is a state of dynamic equilibrium, or a "steady state," that is achieved in healthy living things and provides the optimum environment for the performance of life activities. All cells must maintain homeostasis in order to survive.

5. **3** The fact that certain poisons are toxic to organisms because they interfere with the function of enzymes in mitochondria results directly in the inability of the cell to *release energy from nutrients*. Mitochondrial enzymes catalyze the conversion of nutrients such as glucose into waste products and adenosine triphosphate (ATP) in the process of respiration. ATP is a molecule that stores the chemical bond energy released from the oxidation of glucose during respiration and is used in cells to provide the energy needed for certain cell processes that require activation energy.

WRONG CHOICES EXPLAINED

(1) The fact that certain poisons are toxic to organisms because they interfere with the function of enzymes does not result directly in the inability of the cell to *store information*. Information is stored in the nucleus of the cell in the form of DNA coding. Mitochondria are not involved directly in this process.

(2) The fact that certain poisons are toxic to organisms because they interfere with the function of enzymes does not result directly in the inability of the cell to *build proteins*. Proteins are built on the ribosomes of the cell. Mitochondria are not involved directly in this process.

(4) The fact that certain poisons are toxic to organisms because they interfere with the function of enzymes does not result directly in the inability of the cell to *dispose of metabolic wastes*. Metabolic wastes are removed from the cell through the action of the cell membrane and contractile vacuole. Mitochondria are not involved directly in this process.

6. **1** *Gene expression can be modified by interactions with the environment* is the statement that describes why this change in the color of the bread mold occurs. The mold's color change in warm versus cool temperatures indicates that the genes for the production of pigmentation are expressed differently under different environmental conditions. It is likely that the enzymes that catalyze the biochemical reactions for pigmentation are produced under one set of conditions but not under the other set of conditions.

WRONG CHOICES EXPLAINED

(2) *Every organism has a different set of coded instructions* is *not* the statement that describes why this change in the color of the bread mold occurs. Mold typically reproduces by a form of mitotic cell division in which cells are genetic duplicates of each other. For this reason, every organism has the same, not a different, set of coded instructions.

(3) *The DNA was altered in response to an environmental condition* is *not* the statement that describes why this change in the color of the bread mold occurs. DNA may be altered by random mutations, which are normally caused by chemicals or radiation, not temperature change, present in the environment. It is extremely unlikely that DNA alteration is responsible for this change.

(4) *There is no replication of genetic material in the cooler environment* is *not* the statement that describes why this change in the color of the bread mold occurs. Replication is a process by which chromosomes and the genes they carry are duplicated exactly during the process of mitotic cell division. Replication must occur each time a cell divides regardless of the temperature of the environment.

7. **1** Asexually reproducing organisms pass on hereditary information as *sequences of A, T, C, and G*. DNA, which controls hereditary information in the cell nucleus, is passed from generation to generation in asexually reproducing organisms by means of mitotic cell division. DNA molecules contain complex sequences of nitrogenous bases (A, T, G, and C) that code for the production of specific proteins. Each specific cell protein is capable of contributing to the development of an observable hereditary trait in the adult organism.

WRONG CHOICES EXPLAINED

(2), (3) Asexually reproducing organisms do *not* pass on hereditary information as *chains of complex amino acids* or *folded protein molecules*. Proteins are formed from specific sequences of amino acids linked in a chain, then folded as a result of electromagnetic attractions within the complex polypeptide molecule. DNA molecules, not protein molecules, are responsible for passing on hereditary information in asexually reproducing organisms.

(4) Asexually reproducing organisms do *not* pass on hereditary information as *simple organic sugars*. Simple organic sugars such as glucose and fructose serve as a source of nutrition and cell energy and have no role in heredity in living things. DNA molecules, not sugar molecules, are responsible for passing on hereditary information in asexually reproducing organisms.

8. **4** Species of bacteria can evolve more quickly than species of mammals because bacteria have *higher rates of reproduction*. Some strains of bacteria in an optimum growth environment can double their numbers every 20 minutes, theoretically allowing a single bacterium to produce as many as

4,722,366,482,869,645,213,696 (4.722×10^{21}) offspring within a 24-hour period! At such high reproductive rates, many opportunities for genetic mutation may occur, leading to large numbers of genetic variants and leading in turn to a relatively high rate of successful adaptation to changing environmental conditions.

WRONG CHOICES EXPLAINED

(1) It is *not* true that species of bacteria can evolve more quickly than species of mammals because bacteria have *less competition*. Bacteria face the same sorts of inter- and intraspecies competition as do mammals. Competition for finite quantities of food, water, space, and other resources may limit the actual growth rates of bacteria significantly, as they would for mammal species.

(2) It is *not* true that species of bacteria can evolve more quickly than species of mammals because bacteria have *more chromosomes*. Bacteria are extremely simple organisms that display relatively few genetic traits as compared to mammals. Because of this, the genomes of bacterial species are far simpler and the numbers of chromosomes far fewer than those of mammal species.

(3) It is *not* true that species of bacteria can evolve more quickly than species of mammals because bacteria have *lower mutation rates*. Mutations of genetic material are random events caused by mutagenic agents (such as chemicals or radiation) present in the environment. While the rate at which genetic mutation occurs is relatively constant in all species, the high reproductive rates of bacteria magnify the effects of these mutations greatly.

9. **2** Row 2 in the chart could be used to identify the building blocks and product in the diagram. The diagram illustrates the linkage of four different molecules into a single-chain molecule, so we must look for a row whose building blocks match a chain molecule product. Row 2 lists as building blocks amino acid molecules, which are known to link together chemically to produce a polypeptide chain, or protein molecule, as a product. Because proteins are constructed of hundreds of amino acid molecules linked in a specific sequence, the four-unit polypeptide shown is only "part" of a protein molecule.

WRONG CHOICES EXPLAINED

(1) Row *1* in the chart could *not* be used to identify the building blocks and product in the diagram. In this row, starch is incorrectly labeled as a building block and glucose is incorrectly labeled as a product. Starch molecules are complex polymers (product) made of linked glucose molecules (building block).

(3) Row *3* in the chart could *not* be used to identify the building blocks and product in the diagram. In this row, sugar molecules are incorrectly labeled as

being a building block of adenosine triphosphate (ATP). ATP is not a product constructed from linked sugar molecules.

(4) Row *4* in the chart could *not* be used to identify the building blocks and product in the diagram. In this row, DNA molecules are incorrectly labeled as being a building block of starch. Starch is a product constructed from linked sugar molecules, not DNA molecules.

10. **2** Diagram *2* best represents the relative locations of the structures in the list. A cell (*C*) is a living structure that normally contains a nucleus (*B*). The cell nucleus normally contains a number of chromosomes (*A*). Chromosomes are known to contain a large number of genes (*D*) in their structures. When these structures are arranged from smallest to largest, the sequence is *D, A, B, C*, as shown in diagram *2*.

WRONG CHOICES EXPLAINED

(1), (3), (4) Diagrams *1, 3*, and *4* do *not* best represent the relative locations of the structures in the list. Each of these diagrams contains one or more structures out of order by size and location. See correct answer above.

11. **3** *Replication* is the nuclear process represented. Replication is a process by which chromosomes and the genes they carry are duplicated exactly during the process of mitotic cell division. Replication occurs at the biochemical level when the double-helical structure of DNA untwists and separates, allowing free nucleotides (A, T, G, and C) in the cytoplasm to pair with complementary nucleotides bound in the DNA strand. When these nucleotides link together chemically, new DNA molecules result that are exact copies of the original "parent" DNA molecule.

WRONG CHOICES EXPLAINED

(1) *Recombination* is *not* the nuclear process represented. Recombination is the process by which the members of segregated allele pairs are randomly reunited in the zygote as a result of fertilization.

(2) *Fertilization* is *not* the nuclear process represented. Fertilization is the fusion of male and female gametic nuclei that occurs during sexual reproduction.

(4) *Mutation* is *not* the nuclear process represented. Mutation is the random alteration of genetic material in the cells of an organism.

12. **4** The technique described that results in the production of organisms with specific qualities is known as *selective breeding*. To begin this process, the breeder selects male and female organisms with the desired traits to be bred and crosses them with each other. The breeder then selects from among the offspring those displaying the desired trait, crossing them with each other to produce a third-generation offspring, and so on. After several generations of

such selective breeding, the desired trait becomes established and will appear in all offspring in the "pure breeding" line. Unfortunately, this technique may also establish undesirable traits in the line.

WRONG CHOICES EXPLAINED

(1) The technique described that results in the production of organisms with specific qualities is *not* known as *gene replication*. Replication is a process by which chromosomes and the genes they carry are duplicated exactly during the process of cell division so that they can be passed on to the next generation without alteration.

(2) The technique described that results in the production of organisms with specific qualities is *not* known as *natural selection*. Natural selection is a central concept in Darwin's theory of evolution that states that organisms best adapted to their natural environments tend to survive and pass their favorable traits on to the next generation.

(3) The technique described that results in the production of organisms with specific qualities is *not* known as *random mutation*. Mutations are random events that alter the sequence of genetic information in the cell. The effects of a particular mutation may be favorable or unfavorable and may or may not be passed on to the next generation.

13. **2** *Natural selection played a role in the development of this protective coloration* is the statement that provides a possible explanation for this resemblance. Natural selection is a central concept in Darwin's theory of evolution that states that organisms best adapted to their natural environments tend to survive and pass their favorable traits on to the next generation. In this case, the favorable trait is the camouflage coloration of the insect that may provide protection from predators.

WRONG CHOICES EXPLAINED

(1) *The insects needed camouflage so they developed protective coloration* is *not* the statement that provides a possible explanation for this resemblance. This statement would be consistent with Lamarck's discredited theory of evolution, which claimed that traits appeared in organisms as they were needed by those organisms. Lamarck's theory was disproved by credible scientific experimentation.

(3) *The lack of mutations resulted in the protective coloration* is *not* the statement that provides a possible explanation for this resemblance. Mutations are random events that alter the sequence of genetic information in the cell. Genetic mutation is thought to be a driving force of variation that is a central concept in the process of natural selection. The presence, not the lack, of mutations probably resulted in the protective coloration.

(4) *The trees caused mutations in the insects that resulted in protective coloration* is *not* the statement that provides a possible explanation for this resem-

blance. Mutagenic agents include chemicals and electromagnetic radiation. It is highly unlikely that the trees provided either chemicals or electromagnetic radiation that would produce specific genetic mutations leading to protective coloration in these insects.

14. **4** Extinction of a species is most likely to occur *when environmental conditions change and the members of a species lack adaptive traits to survive and reproduce*. A species lacking variations that provide adaptive advantages under various conditions may not be able to adapt readily to a changing environment. Environmental pressures may eliminate susceptible members of the species, reducing its numbers drastically over time. Eventually, such a nonadaptive species reaches a point at which it can no longer sustain a breeding population and may become extinct.

WRONG CHOICES EXPLAINED

(1), (2) Extinction of a species is *not* most likely to occur *when environmental conditions remain the same and the proportion of individuals within the species that lack adaptive traits increases* or *when environmental conditions remain the same and the proportion of individuals within the species that possess adaptive traits increases*. As long as environmental conditions remain stable, selection pressures will be relatively constant. This situation would tend to stabilize the species, neither increasing nor decreasing the likelihood of its extinction regardless of the presence or absence of adaptive traits.

(3) Extinction of a species is *not* most likely to occur *when environmental conditions change and the adaptive traits of the species favor the survival and reproduction of some of its members*. Species that face environmental change and whose adaptability is high tend to evolve new varieties that can respond to the changes and produce new generations of adaptable offspring. This situation would favor the species' evolution, not its extinction.

15. **3** Photosynthesis and cellular respiration are similar in that *they both involve organic and inorganic molecules*. Photosynthesis is a series of enzyme-controlled biochemical reactions in which inorganic carbon dioxide (CO_2) and water (H_2O) molecules are combined to produce organic glucose ($C_6H_{12}O_6$) molecules and inorganic oxygen (O_2) molecules. Cellular respiration is a series of enzyme-controlled biochemical reactions in which organic glucose ($C_6H_{12}O_6$) molecules and inorganic oxygen (O_2) molecules interact to produce inorganic carbon dioxide (CO_2) and water (H_2O) molecules.

WRONG CHOICES EXPLAINED

(1) It is *not* true that photosynthesis and cellular respiration are similar in that *they both occur in chloroplasts*. Photosynthesis occurs in chloroplasts, whereas cellular respiration occurs in mitochondria.

(2) It is *not* true that photosynthesis and cellular respiration are similar in that *they both require sunlight*. Photosynthesis directly requires sunlight, whereas cellular respiration does not.

(4) It is *not* true that photosynthesis and cellular respiration are similar in that *they both require oxygen and produce carbon dioxide*. Cellular respiration directly requires oxygen and produces carbon dioxide, whereas photosynthesis directly requires carbon dioxide and produces oxygen.

16. **3** *Recombination of genes* is the process that will increase variations that could be inherited. Recombination is the process by which the members of segregated allele pairs are randomly reunited in the zygote as a result of fertilization. One set of these alleles is contributed by the male, and the other set is contributed by the female. The new gene combinations that result provide the basis of genetic variations in the species that can be passed on to the next generation.

WRONG CHOICES EXPLAINED

(1) *Mitotic cell division* is *not* the process that will increase variations that could be inherited. Mitotic cell division is a process that results in the creation of genetically identical daughter cells. Mitotic cell division limits, not increases, the production of new variations in living things.

(2) *Active transport* is *not* the process that will increase variations that could be inherited. Active transport is a type of cross-membrane transport in which the energy of ATP molecules is used by a living cell to move molecules of a substance against the concentration gradient from a region of relatively low concentration to a region of relatively high concentration of that substance. Active transport is not directly associated with the production of new variations in living things.

(4) *Synthesis of proteins* is *not* the process that will increase variations that could be inherited. Protein synthesis occurs as a result of instructions provided by genes housed in the cell nucleus. These proteins are responsible for enabling the traits controlled by these genes. Protein synthesis is not directly associated with the production of new variations in living things.

17. **2** The process of meiosis formed *cells 1 and 2*. Cells 1 and 2 are a sperm cell and an egg cell, respectively. Both of these cells are monoploid (n) cells that formed in the gametic tissues of the human being and of many other sexually reproducing species. Monoploid gametes are formed as a result of the process of meiotic cell division.

WRONG CHOICES EXPLAINED

(1) The process of meiosis did *not* form *cell 1, only*. Cell 2 (egg cell) was also formed by meiosis. See correct answer above.

(3), (4) The process of meiosis did *not* form *cell 3, only* or *cells 2 and 3*. Cell 3 is a diploid (2*n*) zygote formed when cell 2 was fertilized by cell *1*.

18. **1** Kangaroos, as mammals that lack a placenta, must have an alternate way of supplying the developing embryo with *nutrients*. In placental mammals, the placenta supplies oxygen and nutrients to the embryo. In nonplacental mammals such as kangaroos, the embryo is born prematurely and develops in the mother's pouch, where mammary glands supply milk as the primary nutrient.

WRONG CHOICES EXPLAINED

(2) It is *not* true that kangaroos, as mammals that lack a placenta, must have an alternate way of supplying the developing embryo with *carbon dioxide*. Carbon dioxide is a waste material formed as a result of cellular respiration. It is excreted from the cells of the developing embryo, not supplied to it, during development.

(3) It is *not* true that kangaroos, as mammals that lack a placenta, must have an alternate way of supplying the developing embryo with *enzymes*. Enzymes are protein molecules that catalyze cellular reactions. Enzymes are synthesized inside the cells of the embryo and so do not have to be supplied during development.

(4) It is *not* true that kangaroos, as mammals that lack a placenta, must have an alternate way of supplying the developing embryo with *genetic information*. Genetic information is supplied to the embryo by the parents at the time of fertilization. Each cell in the embryo contains an exact copy of that genetic information.

19. **1** *ATP* is the substance that is the most direct source of the energy that an animal cell uses for the synthesis of materials. Adenosine triphosphate (ATP) is a chemical that receives energy that is released during the process of cellular respiration and stores it for use in operating many biochemical reactions in the cell, including synthesis reactions.

WRONG CHOICES EXPLAINED

(2), (4) *Glucose* and *starch* are not the substances that are the most direct source of the energy that an animal cell uses for the synthesis of materials. Glucose and starch are carbohydrates that are used as nutrients and sources of metabolic energy in cells. However, they are not the most direct source of this energy since they release their energy to molecules of ATP.

(3) *DNA* is *not* the substance that is the most direct source of the energy that an animal cell uses for the synthesis of materials. Deoxyribonucleic acid (DNA) is a molecule that contains and communicates the genetic code in the cell. DNA is not directly involved in energy transfer in cells.

20. **2** The purpose of these medications is to *decrease the immune response in the person receiving the transplant*. Because the transplanted tissues will probably not match the recipient's tissues in terms of blood type and other critical chemical factors, the recipient's body is likely to defend itself by employing the natural immune response mechanism to reject it. The special medications given to the recipient have the effect of suppressing the immune response so that the transplanted organ will be able to function in the recipient's body without being rejected.

WRONG CHOICES EXPLAINED

(1) The purpose of these medications is *not* to *increase the immune response in the person receiving the transplant*. A medication that increased the immune response would make it more likely that the recipient's body would reject the transplanted organ.

(3), (4) The purpose of these medications is *not* to *decrease mutations in the person receiving the transplant* or *increase mutations in the person receiving the transplant*. Mutations of genetic material are random events caused by mutagenic agents (such as chemicals or radiation) present in the environment. Increasing or decreasing mutations would have no effect on organ transplantation in humans.

21. **1** Compared to each cell of the original carrot plant, each cell of the new plant will have *the same number of chromosomes and the same types of genes*. Cloning is an asexual reproductive process in which all cloned cells retain the genetic information of the original parent plant. The new carrot plant will be a genetic duplicate (clone) of the original parent carrot plant and will share all its characteristics and capabilities, including the number of chromosomes and types of genes in its cells.

WRONG CHOICES EXPLAINED

(2), (3), (4) Compared to each cell of the original carrot plant, each cell of the new plant will *not* have *the same number of chromosomes, but different types of genes; half the number of chromosomes and different types of genes*; or *half the number of chromosomes, but different types of genes*. Each of these distracters describes a condition in which the cells of the new (clone) plant are different from the cells of the parent plant. Because the cloned cells are genetically identical to the parent cells, these conditions cannot exist as described. See correct answer above.

22. **2** These changes in the form of the embryo are the direct result of *differentiation and growth*. Growth is a process by which new cells are added to a developing embryo by means of mitotic cell division. Differentiation is a

process by which these new cells take on specialized characteristics in the developing embryo. The diagram illustrates the processes of growth and differentiation as they would occur in a developing human embryo.

WRONG CHOICES EXPLAINED

(1) These changes in the form of the embryo are *not* the direct result of *uncontrolled cell division and mutations*. Mutations are random events caused by mutagenic agents that alter the genetic information in a cell. Mutations are thought to be one possible cause of cancer, uncontrolled cell division that can crowd out and damage normal tissues. Mutation and cancer are not responsible for the changes illustrated in the diagram.

(3) These changes in the form of the embryo are *not* the direct result of *antibodies and antigens inherited from the father*. Antibodies and antigens are specific proteins that are produced within the bodies of developing embryos, not inherited from their fathers. Antibodies and antigens are not responsible for the changes illustrated in the diagram.

(4) These changes in the form of the embryo are *not* the direct result of *meiosis and fertilization*. Meiosis is a type of cell division in which monoploid (*n*) gametes are produced from diploid (2*n*) primary sex cells. Fertilization is the fusion of male and female gametic nuclei that occurs during sexual reproduction. Meiosis and fertilization are not responsible for the changes illustrated in the diagram.

23. **3** *Cell A is a white blood cell engulfing disease-causing organisms* is the statement that best describes the event illustrated in the diagram. The diagram illustrates a white blood cell surrounding a number of bacteria and enclosing them in a food vacuole, where they will be destroyed by digestive enzymes. White blood cells found in the blood tissue ingest and destroy bacteria that enter the bloodstream through breaks in the skin surface. This function is essential to the immune response in humans.

WRONG CHOICES EXPLAINED

(1) *Cell A is a white blood cell releasing antigens to destroy bacteria* is *not* the statement that best describes the event illustrated in the diagram. The sequence of drawings in this illustration clearly depicts a white blood cell taking in smaller structures (presumably bacteria) and enclosing them within a membrane-bound structure inside the cell, not releasing them into the blood.

(2) *Cell A is a cancer cell produced by the immune system and it is helping to prevent disease* is *not* the statement that best describes the event illustrated in the diagram. The immune system does not produce cancer cells but rather destroys them. Cancer cells in the blood do not prevent disease but rather are evidence of the disease of cancer.

(4) *Cell A is protecting bacteria so they can reproduce without being destroyed by predators* is *not* the statement that best describes the event illustrated in the diagram. This is a nonsense distracter. This process does not occur in human blood tissue or in any other known living system.

24. **1** Energy is available to organisms in the ecosystem because *producers have the ability to store energy from light in organic molecules*. Producers (green plants or autotrophs) are capable of using the energy of sunlight to convert carbon dioxide and water into glucose and oxygen in the process of photosynthesis. The Sun's energy is stored in the chemical bonds of glucose ($C_6H_{12}O_6$) molecules. Producers then become food for a large variety of plant-eating species, which serve in turn as food for meat-eating species. For this reason, producers are considered to be the basis of the food/energy pyramid in any balanced ecosystem.

WRONG CHOICES EXPLAINED

(2) It is *not* true that energy is available to organisms in the ecosystem because *consumers have the ability to transfer energy stored in bonds to plants*. Consumers (animals, bacteria, and fungi) do not have this ability. Producers (green plants) derive their energy from sunlight by the process of photosynthesis, not as a result of transfer from animals.

(3) It is *not* true that energy is available to organisms in the ecosystem because *all organisms in a food web have the ability to use light energy*. Only green plant producers have the ability to use light energy directly to manufacture glucose in their cells. All other organisms must consume organic plant or animal matter in order to receive a portion of this energy.

(4) It is *not* true that energy is available to organisms in the ecosystem because *all organisms in a food web feed on autotrophs*. Organisms that feed on autotrophs (green plants) are known as herbivores; those that feed on herbivores are known as carnivores. The category carnivore is represented by a large number of species that do not consume autotrophs in their normal nutritional activities.

25. **2** *Climate conditions* is the factor that has the greatest influence on the type of ecosystem that will form in a particular area. Climate conditions such as temperature, rainfall, and sunlight are important abiotic factors that determine the types of plants and animals that will survive in an environment. In any ecosystem, it is the delicate balance between biotic (living) and abiotic (nonliving) factors that promotes the stability of the environment and works to support the many species of living things contained within it.

WRONG CHOICES EXPLAINED

(1) *Genetic variations in the animals* is *not* the factor that has the greatest influence on the type of ecosystem that will form in a particular area. Genetic variation is common in all sexually reproducing plant and animal species. Although genetic variation in animals can influence the ecosystem to a degree, it is not the factor that has the greatest amount of influence.

(3) *Number of carnivores* is *not* the factor that has the greatest influence on the type of ecosystem that will form in a particular area. Carnivores are animals that consume other animals for food. Although the number of carnivores can influence the ecosystem to a degree, it is not the factor that has the greatest amount of influence.

(4) *Percentage of nitrogen gas in the atmosphere* is *not* the factor that has the greatest influence on the type of ecosystem that will form in a particular area. The atmosphere contains approximately 20% nitrogen gas at sea level. Although the percentage of nitrogen gas in the atmosphere can influence the ecosystem to a degree, it is not the factor that has the greatest amount of influence.

26. **2** This situation is an example of *a trade-off*. Although the term "trade-off" is not a scientific term, it is well understood in our day-to-day vocabulary to mean "making a choice between two options based on their relative merits." In this case, a choice is made between the merits of preserving natural biodiversity and the merits of increasing the food supply for human beings. In the short term, the decision to sacrifice natural biodiversity for the sake of human survival seems an easy trade-off to make. However, if this decision is made without a clear understanding of the importance of protecting natural ecosystems and the wealth of biodiversity they contain, then the trade-off will ultimately be a poor one. In many ways, intelligent preservation of the natural environment is more important to the long-term survival of our species than is the use of technology to support continued growth of the human population beyond the carrying capacity of the ecosystem.

WRONG CHOICES EXPLAINED

(1), (3), (4) This situation is *not* an example of *poor land use, conservation,* or *a technological fix*. Although these distracters are identified as incorrect according to the examination answer key, they may be considered correct responses depending on the context in which they are discussed. To a person concerned about preserving the natural environment, the idea of destroying natural diversity by farming on environmentally sensitive lands may represent "poor land use." To a person interested in preserving open space for future generations, the idea of farming a tract of land that might otherwise be developed may represent "conservation." To a person who wants to fight malnutrition in impoverished countries, the idea of farming land to maximize food production may represent a "technological fix" for a serious social problem.

[Note: There is no clearly correct scientific answer to this question. It is important to help students understand that their answers will depend to a large degree on their points of view, and that making good choices based on their knowledge of biology is an important outcome of this course of study.]

27. **4** This food chain contains *2 predators, 1 herbivore, and 1 producer*. Producers (such as grass) form the basis of any food chain because of their ability to capture light energy and convert it to chemical bond energy. Producers are consumed by herbivores (such as crickets) as a means of transferring food energy from the producers to higher levels of the food chain. Herbivores are consumed by predators (such as frogs) in order to continue the transfer of energy. Some of these predators may in turn be consumed by higher-order predators (such as owls) in order to continue the energy transfer still further. The arrows in the diagram indicate the direction of energy transfer in this food chain from grass, to cricket, to frog, to owl.

WRONG CHOICES EXPLAINED

(1) This food chain does *not* contain *4 consumers and no producers*. Grass is a producer organism, not a consumer organism, in this food chain.

(2) This food chain does *not* contain *1 predator, 1 parasite, and 2 producers*. Grass is the only producer represented in this food chain. None of the organisms shown is a parasite, which is a type of organism that lives in or on the body of another (host) organism and derives its nutrition from the host.

(3) This food chain does *not* contain *2 carnivores and 2 herbivores*. Grass is a producer organism, not a herbivore organism, in this food chain.

28. **3** These changes are an example of *ecological succession*. "Ecological succession" is a term that describes the replacement of one plant community by other, progressively more complex, plant communities until a stable climax community is established. In the example given, barren volcanic ash is gradually converted to rich soil by the action of pioneer species (small plants), which are then followed by deep-rooted shrubs and trees. As succession continues over many years, a self-sustaining climax community of mixed plants will eventually become established that is compatible with the climate and soil conditions of the area.

WRONG CHOICES EXPLAINED

(1) These changes are *not* an example of *manipulation of genes*. The manipulation of genes occurs as a result of the laboratory technique known as genetic engineering. The events described are natural, not laboratory-controlled, events.

(2) These changes are *not* an example of *evolution of a species*. "Evolution" is a term used to describe the way that new species are thought to arise from existing species by the process of natural selection. The events described

involve replacement of one plant community by another plant community, not the rise of new species.

(4) These changes are *not* an example of *equilibrium*. "Equilibrium" is a term used to describe a balanced steady state in a living or nonliving system. The event described demonstrates that this system is undergoing significant change and for that reason cannot be in equilibrium.

29. **1** A major reason that humans can have such a significant impact on an ecological community is that humans *can modify their environment through technology*. The fact that humans can design technologies to pump water to desert environments, transport heating fuels to cold regions, explore new environments hostile to human survival, and move easily to all parts of the Earth makes it possible for us to modify the environment to meet our needs as a species. In many cases, this ability has damaged the environment, overusing limited natural resources and making the environment less able to support native species.

WRONG CHOICES EXPLAINED

(2) It is *not* true that a major reason that humans can have such a significant impact on an ecological community is that humans *reproduce faster than most other species*. In fact, the human gestation period of 9 months represents an exceptionally long time compared to that for most other species. In addition, humans have an exceptionally long period of parental care, which further limits their reproductive rate.

(3) It is *not* true that a major reason that humans can have such a significant impact on an ecological community is that humans *are able to increase the amount of finite resources available*. Most natural resources such as water, air, and minerals are nonrenewable materials formed on Earth over millions of years. Once nonrenewable resources are used up, they cannot be replaced.

(4) It is *not* true that a major reason that humans can have such a significant impact on an ecological community is that humans *remove large amounts of carbon dioxide from the air*. In fact, humans do not remove carbon dioxide from the air but rather add tremendous quantities of this compound to the air every day, both through their metabolic processes and through their technological practices.

30. **3** One likely reason the rabbit population was able to grow so large is that the rabbits *successfully competed with native herbivores for food*. When humans import non-native species and release them into the native ecosystem, these species may become invasive, crowding out native species and altering the natural community substantially. Such an event would tend to upset the equilibrium of the ecosystem, making control of the Australian rabbit population difficult.

WRONG CHOICES EXPLAINED

(1) One likely reason the rabbit population was able to grow so large is *not* that the rabbits *were able to prey on native herbivores*. As herbivores, rabbits consume only plants for food and do not prey on other animals, including those native to Australia.

(2) One likely reason the rabbit population was able to grow so large is *not* that the rabbits *reproduced more slowly than the native animals*. In fact, rabbits reproduce far more quickly than many other animal species, including those native to Australia. This rapid reproductive rate made it even more inevitable that they would push out native species.

(4) One likely reason the rabbit population was able to grow so large is *not* that the rabbits *could interbreed with the native animals*. A defining characteristic of a distinct species is its ability to interbreed with other members of its own species and produce fertile offspring. Rabbits cannot naturally interbreed with other species, including those native to Australia.

PART B–1

31. **3** The laboratory procedure represented in the diagram is one in which a researcher is *adding stain to a slide without removing the coverslip*. In this procedure, a droplet of stain is placed on the slide at one edge of the coverslip. The stain is drawn under the coverslip and around the specimen by means of capillary action. A small piece of paper towel is used to absorb excess moisture from under the opposite edge of the coverslip, speeding the rate of capillary action.

WRONG CHOICES EXPLAINED

(1) The laboratory procedure represented in the diagram is *not* one in which a researcher is *placing a coverslip over a specimen*. Placing a coverslip involves holding it over the specimen and touching one edge to the slide, then lowering the coverslip in a manner that allows the majority of air bubbles to escape from under it.

(2) The laboratory procedure represented in the diagram is *not* one in which a researcher is *removing a coverslip from a slide*. Removing a coverslip involves moving it to the edge of the slide and using a fingertip or forceps to lift it off the specimen.

(4) The laboratory procedure represented in the diagram is *not* one in which a researcher is *reducing the size of air bubbles under a coverslip*. Reducing the size of air bubbles under the coverslip involves holding the slide at a slight angle and gently tapping the slide so that the bubbles escape from under the coverslip at its upper edge.

32. **2** One of the reasons for the lack of experiments on inserting other species' genes into human DNA is that *there are many ethical questions to be answered before inserting foreign genes into human DNA*. Although much has been learned about the process of genetic engineering over the past several decades, we are still not completely certain of its effects on the human species. This process, if not ethically conducted, could alter the human genome in unacceptable ways, potentially creating races of human clones engineered to perform menial tasks and others engineered to be a perpetual ruling class. There are also ethical questions of who will, and who will not, be the beneficiaries of gene therapies that reduce or remove the possibility of disease or extend the human lifespan. These and other ethical questions will have to be answered before this process receives universal approval to proceed.

WRONG CHOICES EXPLAINED

(1) One of the reasons for the lack of experiments on inserting other species' genes into human DNA is *not* that *the subunits of human DNA are different from the DNA subunits of other species*. The basic subunit of any DNA molecule is the nucleotide unit found in four types, A, T, C, and G, in all known species.

(3) One of the reasons for the lack of experiments on inserting other species' genes into human DNA is *not* that *inserting foreign DNA into human DNA would require using techniques completely different from those used to insert foreign DNA into the DNA of other mammals*. The basic gene insertion technique involves the use of restriction enzymes to snip out specific portions of a DNA molecule of one organism and insert it into the DNA of another organism.

(4) One of the reasons for the lack of experiments on inserting other species' genes into human DNA is *not* that *human DNA always promotes human survival, so there is no need to alter it*. By and large, human genes do promote human survival. However, some human genes make their owners susceptible to life-threatening genetic disorders that, if left untreated, will eventually kill these individuals.

33. **4** The development of an experimental research plan should *not* include a *conclusion based on data expected to be collected in the experiment*. A researcher should never draw conclusions about the anticipated results of an experiment but rather draw conclusions only after the experiment has been conducted and data objectively analyzed. To draw conclusions in advance about the outcome of a scientific experiment would be to introduce bias into the experiment and should never be done.

WRONG CHOICES EXPLAINED

(1), (2), (3) The development of an experimental research plan *should* include a *list of precautions for the experiment*, a *list of equipment needed for conducting the experiment*, and a *procedure for the use of technologies needed for the experiment*. These are prudent and necessary steps for the researcher to take in preparing to conduct an experiment that will help the researcher to gather and analyze data that will help to answer a scientific question.

34. **1** *Starch is synthesized from the glucose produced in the green areas of the leaf* is the statement that represents an assumption the student had to make in order to draw this conclusion. The direct product of the process of photosynthesis is not starch but glucose, a simple sugar. Starch is a polymeric carbohydrate molecule synthesized from glucose molecules and stored in plant cell vacuoles as a food source for the plant. The presence of starch in the green areas of the leaf can be linked to the process of photosynthesis only when this intervening step is known or assumed.

WRONG CHOICES EXPLAINED

(2) *Starch is converted to chlorophyll in the green areas of the leaf* is *not* the statement that represents an assumption the student had to make in order to draw this conclusion. Chlorophyll is a complex organic molecule containing carbon, hydrogen, oxygen, nitrogen, and magnesium. It is synthesized in the cells of the leaf from organic precursor molecules but not directly from starch.

(3) *The white areas of the leaf do not have cells* is *not* the statement that represents an assumption the student had to make in order to draw this conclusion. The physical structures of all plants are dependent on the presence of plant cells. The cells in the white areas of the leaf do not contain chlorophyll but are living cells nonetheless.

(4) *The green areas of the leaf are heterotrophic* is *not* the statement that represents an assumption the student had to make in order to draw this conclusion. Green plants are autotrophic (self-feeders) by virtue of their ability to carry on the process of photosynthesis. Only a few nongreen plants are adapted to carry on heterotrophic nutrition.

35. **3** The term *chemicals* in this diagram represents *hormone molecules*. Hormones are biochemicals (usually proteins) produced in special glands that regulate the metabolic activities of the body's tissues and organs. When the metabolic activity produces the desired effect, a feedback loop mechanism may cause the gland to stop secreting the hormone or may stimulate the secretion of other hormones.

WRONG CHOICES EXPLAINED

(1) The term *chemicals* in this diagram does *not* represent *starch molecules*. Starch is a complex carbohydrate composed of repeating glucose units. Starch has no regulatory influence over organs in the body.

(2) The term *chemicals* in this diagram does *not* represent *DNA molecules*. DNA is a complex nucleic acid composed of repeating nucleotide units. While DNA provides the underlying blueprint for hormone production, it does not directly participate in the feedback loop illustrated in the diagram.

(4) The term *chemicals* in this diagram does *not* represent *receptor molecules*. Receptors are proteins embedded in the cell membrane that are specialized to recognize and link to neurochemicals, hormones, or other regulatory biochemicals in the body.

36. **1** *It aids in the removal of metabolic wastes in both cell X and cell Y* is the statement that is correct concerning the structure labeled *A*. The structure labeled *A* represents the cell membrane. Cell *X* represents an animal cell and cell *Y* represents a plant cell. The cell membrane functions in both animal cells and plant cells to regulate the passage of materials into and out of the cell, including the removal of wastes from the cell interior.

WRONG CHOICES EXPLAINED

(2) *It is involved in cell communication in cell X, but not in cell Y* is *not* the statement that is correct concerning the structure labeled *A*. The cell membrane functions in the process of cell communication in both animal and plant cells.

(3) *It prevents the absorption of CO_2 in cell X and O_2 in cell Y* is *not* the statement that is correct concerning the structure labeled *A*. The cell membrane facilitates the free passage of both CO_2 and O_2 into or out of the cell.

(4) *It represents the cell wall in cell X and the cell membrane in cell Y* is *not* the statement that is correct concerning the structure labeled *A*. Animal cells do not have cell walls.

37. **4** *Individual species reproduce within a specific range of temperatures* is the statement that is a valid conclusion based on the information in the graph. The graphed data demonstrates that each of the four bacterial species studied has a different optimum temperature at which reproductive rates are highest. It also demonstrates that, for each species, there is a specific range of temperatures only within which reproduction is even possible.

WRONG CHOICES EXPLAINED

(1) *Changes in temperature cause bacteria to adapt to form new species* is *not* the statement that is a valid conclusion based on the information in the graph. The formation of new species (speciation) is thought to occur as a result of the natural selection of organisms displaying favorable variations caused by

mutations within their genome. It is not correct to assume that the temperature of the environment is responsible for speciation.

(2) *Increasing temperatures speed up bacterial reproduction* is *not* the statement that is a valid conclusion based on the information in the graph. The data illustrates that the rates of bacterial reproduction increase with temperature to an optimal level, then decline rapidly as temperatures rise above optimum. This is often due to the fact that denaturation of enzymes found in living cells occurs at high temperatures, changing their shape permanently and rendering them unable to support life functions.

(3) *Bacteria can survive only at temperatures between 0°C and 100°C* is *not* the statement that is a valid conclusion based on the information in the graph. In this experiment, only four bacterial species were studied out of hundreds of thousands of species that may exist in nature. This conclusion is too broad to be made from the information presented.

38. **2** The section of DNA being used to make the strand of mRNA is known as a *gene*. DNA provides the genetic code for the synthesis of complex proteins in the cell. This code is "read" and carried to the ribosomes by molecules of mRNA. At the ribosome, amino acid molecules are linked in a specific sequence guided by the genetic code on mRNA.

WRONG CHOICES EXPLAINED

(1) The section of DNA being used to make the strand of mRNA is *not* known as a *carbohydrate*. Carbohydrates are a class of organic compounds composed of carbon, hydrogen, and oxygen in a 1:2:1 ratio. Carbohydrate molecules are not directly involved in protein synthesis.

(3) The section of DNA being used to make the strand of mRNA is *not* known as a *ribosome*. The ribosome is a cell organelle that serves as the site of protein synthesis. Ribosomes do not "read" genetic code directly from DNA molecules.

(4) The section of DNA being used to make the strand of mRNA is *not* known as a *chromosome*. A chromosome is a complex structure that contains genes in specific locations along the chromosome strand. A single chromosome may contain hundreds of individual genes.

39. **4** Graph *4* best represents the relative energy content of this pyramid. The base of an energy pyramid contains the producers (plants) and must have the greatest amount of energy in a balanced ecosystem. In this energy pyramid, solar energy captured by producer organisms on level *4* is inefficiently passed on to the herbivores on level *3*, with as much as 90% lost to the environment as heat in the exchange. The energy exchange from level *3* (herbivores) to level *2* (predators) is similarly inefficient, as is the exchange from level *2* (predators) to level 1 (top predators). In a balanced ecosystem, an

energy pyramid must have the most energy in its lowest level and progressively less in each higher level.

WRONG CHOICES EXPLAINED

(1), (2), (3) Graphs *1, 2,* and *3* do not best represent the relative energy content of this pyramid. Each of these graphs shows energy amounts in one or more levels of the energy pyramid to be higher than in the level(s) below it. This inverse energy relationship cannot exist in a balanced ecosystem.

40. **1** *Species A has the best chance of survival because it has the most genetic diversity* is the statement that is correct concerning the chances of survival for these species if there is a change in the environment. In general, the more genetic variation that is present in a species, the more likely the species will be to adapt to long-term changes in the environment. Species *A* has three variants illustrated and so is expected to be more adaptable than species *B* and *D* (with two variants) or species *C* (with one variant). [Note: In order to answer this question correctly, the student must infer that the different patterns shown for the members of each bacterial species represent significant genetic variations that will give these variants an adaptive advantage over the other variants of that species in the event of environmental change.]

WRONG CHOICES EXPLAINED

(2) *Species C has the best chance of survival because it has no gene mutations* is *not* the statement that is correct concerning the chances of survival for these species if there is a change in the environment. Species *C* is illustrated as having only one variant and so is assumed to have no gene mutations. For this reason, of these four hypothetical species, species *C* is the least adaptable and has the least, not the best, chance for survival.

(3) *Neither species B nor species D will survive because they compete for resources* is *not* the statement that is correct concerning the chances of survival for these species if there is a change in the environment. There is no information given in the item concerning the relative levels of competition that exist among these four hypothetical species. If they did compete for limited resources, one species would be likely to be better than another, allowing it to survive at the expense of the other.

(4) *None of the species will survive because bacteria reproduce asexually* is *not* the statement that is correct concerning the chances of survival for these species if there is a change in the environment. Asexual reproduction tends to limit the degree of variation in species that employ it for their survival because it is based on mitotic cell division. However, because of the rapid mutation and reproduction rate of bacteria, it is likely that at least some will survive this hypothetical environmental change.

41. **3** *Earthworms and sea stars have a common ancestor* is the statement that is a valid conclusion that can be drawn from the diagram. In fact, all the evolutionary lines shown in the diagram share a single common ancestor identified as "Ancestral Protists." [Note: In order to answer this question correctly, students must infer that the bottom of the diagram is ancient time and that the top of the diagram is present time in Earth's geological history.]

WRONG CHOICES EXPLAINED

(1) *Snails appeared on Earth before corals* is *not* the statement that is a valid conclusion that can be drawn from the diagram. The diagram shows that corals branched off the ancestral root at an earlier time than snails, so snails appeared on Earth after, not before, corals.

(2) *Sponges were the last new species to appear on Earth* is *not* the statement that is a valid conclusion that can be drawn from the diagram. The diagram shows that sponges branched off the ancestral root before any other evolutionary line, indicating that they were the first, not the last, new species to appear on Earth.

(4) *Insects are more complex than mammals* is *not* the statement that is a valid conclusion that can be drawn from the diagram. There is no information given in the diagram concerning the relative complexity of the organisms shown.

42. **2** The population represented in the graph reached the carrying capacity of the ecosystem on *day 8*. At this point in the experiment, the population stopped growing and leveled off at about 70 individuals, a population sustainable given the environment's available resources. It is likely that this population will be at, slightly above, or slightly below 70 individuals as long as the environmental conditions, availability of resources, and level of interspecies competition remain relatively constant.

WRONG CHOICES EXPLAINED

(1) The population represented in the graph did *not* reach the carrying capacity of the ecosystem on *day 11*. The graph shows that on day 11 the population was reasonably stable at just under 70 individuals.

(3), (4) The population represented in the graph did *not* reach the carrying capacity of the ecosystem on *day 3* or *day 5*. The graph shows that on day 3 and day 5 the population was growing at a steady pace, but that it had not yet reached the carrying capacity of the environment.

PART B–2

43. One credit is allowed for correctly marking an appropriate scale on the axis labeled "Percentage of Electricity Generated." [1]

44. One credit is allowed for correctly constructing and shading vertical bars to represent the data in the table. [1]

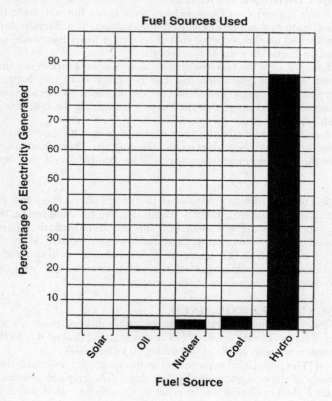

Fuel Sources Used

45. One credit is allowed for correctly identifying *one* fuel source in the table that is considered a fossil fuel. Acceptable responses include: [1]

- *Oil*
- *Coal*

46. One credit is allowed for correctly identifying *one* fuel source in the table that is classified as a renewable resource. Acceptable responses include: [1]

- *Hydro (water)*
- *Solar (Sun)*

47. One credit is allowed for correctly stating *one* specific environmental problem that can result from burning coal to generate electricity. Acceptable responses include, but are not limited to: [1]

- *Burning coal can produce air pollution.*
- *Acid rain can result when sulfur gases from coal burning mix with water in the atmosphere.*
- *Carbon dioxide and other greenhouse gases from coal burning can add to the global warming problem.*
- *Mining coal can be destructive to ecological communities around the mines.*

48. One credit is allowed for correctly stating *one* way the hawk population may change and supporting your answer. Acceptable responses include, but are not limited to: [1]

- *The hawk population will decrease because there will be fewer snakes since there are fewer frogs for them to eat.*
- *The hawk population will increase because there will be more grasshoppers for the shrews to eat and more shrews for the hawks to eat.*
- *The hawk population will decrease because the increased number of grasshoppers will eat more grass, leaving less grass for rabbits to eat and fewer rabbits for hawks to eat.*

49. One credit is allowed for correctly identifying *one* cell structure found in a producer in this meadow ecosystem that is *not* found in carnivores. Acceptable responses include, but are not limited to: [1]

- *Chloroplast (grana, stroma)*
- *Cell wall*
- *Cell plate*
- *Food (starch) vacuole*

50. One credit is allowed for correctly stating *one* genetic advantage of an earthworm mating with another earthworm for the production of offspring. Acceptable responses include, but are not limited to: [1]

- *Mating with another earthworm allows for variety in the species.*
- *The species has a better chance for survival because of variation.*

- *Genetic recombination allows for new genetic combinations that can increase variation and promote species survival in the event of environmental change.*

51. One credit is allowed for correctly completing the missing information for sections 5.a. and 5.b. so that the key is complete for all six species. Acceptable responses include, but are not limited to: [1]

- *5.a. has white (or clear or light) wings*
- *5.b. has shaded (or black or dark) wings*

52. One credit is allowed for correctly identifying the species, as shown below. [1]

Species _D_   Species _E_   Species _C_   Species _F_   Species _B_   Species _A_

53. One credit is allowed for correctly drawing one or more shapes on the virus that will fit with the receptor molecules on the human cell. Acceptable responses include, but are not limited to: [1]

54. One credit is allowed for correctly explaining *one* way this change in the virus could come about. Acceptable responses include, but are not limited to: [1]

- *Mutation*
- *Genetic change leading to a new protein*

55. One credit is allowed for correctly identifying *one* relationship that exists between a virus and a human when the virus infects the human. Acceptable responses include, but are not limited to: [1]

- *Parasite/host*
- *Pathogen/host*

- *The virus is a pathogenic (disease-causing) organism whose activities can disrupt the homeostatic balance of the human host.*
- *The virus commandeers the metabolic apparatus of the infected human cells, producing viral DNA from the host's DNA and releasing new viruses to infect the host's healthy tissues.*

PART C

56. One credit is allowed for correctly identifying the structure in the human body that is the usual source of insulin. Acceptable responses include: [1]

- *Pancreas*
- *Islets of Langerhans*

57. One credit is allowed for correctly identifying *one* substance in the blood, other than insulin, that could change in concentration and indicate that a person is not secreting insulin in normal amounts. Acceptable responses include, but are not limited to: [1]

- *Sugar*
- *Glucose*
- *Ketones*
- *Glucagon*

58. One credit is allowed for correctly stating *one negative* effect of importing exotic species into the United States. Acceptable responses include, but are not limited to: [1]

- *They can transfer pathogens to human and domestic animals.*
- *Imported species may displace native species.*
- *Imported species may increase competition with native species for food.*
- *The imported species may compete with native species for available habitats.*

59. One credit is allowed for correctly stating *one* way the human immune system might respond to an invading pathogen associated with handling a hedgehog. Acceptable responses include, but are not limited to: [1]

- *The body can make antibodies to fight invading antigens.*
- *White blood cells will engulf and destroy pathogens.*

60. One credit is allowed for correctly stating *one* effect the reintroduction of the wolf may have on the coyote population in the Adirondacks and explaining why it would have this effect. Acceptable responses include, but are not limited to: [1]

- *The coyote population will decrease because the wolf will be a competitor for the same prey as the coyote.*
- *The coyote population will be unaffected because there is sufficient prey for both the wolf and the coyote.*
- *The coyote population will be unaffected because the coyote is primarily a scavenger and will utilize this feeding habit if prey becomes scarce.*

61. One credit is allowed for correctly explaining why the coyote is considered a limiting factor in the Adirondack Mountains. Acceptable responses include, but are not limited to: [1]

- *The coyotes control growth of certain prey populations.*
- *When populations of a prey species increase their numbers, the coyote population increases in response and will reduce the prey population through normal predatory activities.*

62. One credit is allowed for correctly stating *one* ecological reason why some individuals might support the reintroduction of wolves into the Adirondacks. Acceptable responses include, but are not limited to: [1]

- *The wolf was once a natural part of this ecosystem.*
- *The wolf can help to control the deer population.*
- *There is adequate prey to support the wolf population.*
- *Reintroducing the wolf would increase biodiversity.*

63. Three credits are allowed for designing an experiment to determine the effects of light on the growth of tomato plants. In your experimental design, be sure to:

- State *one* hypothesis to be tested. *[1]*
- Identify the independent variable in the experiment. [1]
- Describe the type of data to be collected. [1]

Examples of acceptable responses are: [3]

My hypothesis is that tomato plants exposed to 16 hours of light will grow faster than tomato plants exposed to 8 hours of light. [1] The independent variable in this experiment will be the duration of light to which tomato plants are exposed. [1] In order to determine the effects of light on the growth of tomato plants, I will measure the lengths of their stems every day for 30 days. [1]

I think that tomato plants grown under bright light will grow better than tomato plants grown under dim light. [1] The intensity of light will be the independent variable in the experiment. [1] At the end of the 30-day experiment, I will remove the plants from their pots and measure their mass in grams. [1]

64. Three credits are allowed for explaining how guard cells in both land and water plants help maintain homeostasis. In your answer, be sure to:

- Identify *one* function regulated by the guard cells in leaves. [1]
- Explain how guard cells carry out this function. [1]
- Give one possible evolutionary advantage of the position of the guard cells on the leaves of land plants. [1]

Examples of acceptable responses are: [3]

The primary function regulated by guard cells is the exchange of gases between the leaf interior and the atmosphere. [1] In regulating this function, the guard cells change shape when they produce glucose during photosynthesis and absorb water to balance their diffusion gradients, becoming turgid. In this condition, the guard cells open stomates on the leaf's surface, allowing carbon dioxide to enter and oxygen to escape. [1] The evolutionary advantage of this regulation in land plants is to facilitate the process of photosynthesis by allowing carbon dioxide to enter the leaf at times when light is most intense and photosynthesis is occurring most rapidly. [1]

Guard cells regulate the function of water vapor exchange with the atmosphere. [1] When guard cells lose water during dry periods and become flaccid, they close the stomates they surround. [1] The advantage of this process to land plants is the reduction of water vapor loss from the plant during dry periods. [1]

65. One credit is allowed for correctly stating whether or not mutations that caused skin cancers will be passed on to offspring and supporting your answer. Acceptable responses include, but are not limited to: [1]

- *No, mutations in body cells cannot be transmitted to offspring.*
- *No, only mutations in gametes may be transmitted to offspring.*

66. Two credits are allowed, one credit each, for correctly stating *two* specific ways in which an ocean ecosystem will change (other than having fewer photosynthetic organisms) if populations of photosynthetic organisms die off as a result of damage to the ozone layer. Acceptable responses include, but are not limited to: [2]

- *Decrease in consumers*
- *Decrease in biodiversity*
- *Decrease in oxygen*
- *Decrease in available food energy*
- *Increase in carbon dioxide*
- *Interruption of the food chain*

67. One credit is allowed for correctly identifying *one* way that the practice of dumping leaves and grass clippings into landfills was harmful to the environment. Acceptable responses include, but are not limited to: [1]

- *Does not allow for recycling of nutrients in the lawn.*
- *Takes up landfill space.*
- *Encourages overuse of lawn fertilizers.*

PART D

68. **2** In preparation for an electrophoresis procedure, enzymes are added to DNA in order to *cut the DNA into fragments*. This technique involves the use of restriction enzymes to snip DNA molecules at specific points corresponding to the enzyme's active site. By this technique, each type of DNA is cut into pieces of different sizes that can then be separated in the charged electrophoresis gel.

WRONG CHOICES EXPLAINED
(1), (3), (4) It is not true that, in preparation for an electrophoresis procedure, enzymes are added to DNA in order to *convert the DNA into a gel*, *change the color of the DNA*, or *produce larger sections of DNA*. None of these is a technique for which restriction enzymes are used in the process of electrophoresis.

69. **1** Paper chromatography is a laboratory technique that is used to *separate different molecules from one another*. This technique involves dissolving the chemicals to be separated (such as leaf pigments) in a small amount of a solvent such as acetone, then placing a concentrated spot of the solution on a strip of chromatography paper. When the spot is dry, the strip is dipped into clean acetone, which wicks up the paper and separates the different chemicals into distinct bands based on their relative molecular sizes.

WRONG CHOICES EXPLAINED

(2) It is *not* true that paper chromatography is a laboratory technique that is used to *stain cell organelles*. In order to accomplish this, a cell stain such as bromthymol blue or iodine should be used.

(3) It is *not* true that paper chromatography is a laboratory technique that is used to *indicate the pH of a substance*. Litmus paper would be used to determine the relative acidity of a substance.

(4) It is *not* true that paper chromatography is a laboratory technique that is used to *compare relative cell sizes*. A compound microscope might profitably be used to make this determination.

70. **1** The cramping of the muscles most likely results from *lack of adequate oxygen supply to the muscle*. During periods of oxygen deprivation, muscle cells revert to an anaerobic respiratory process known as lactic acid fermentation. During this process, lactic acid builds up in the muscle cells, causing them to cramp. The lactic acid is also detected by nerves in the muscle, sending a sensation of pain to the brain and thus alerting it to the need to reestablish a proper oxygen level in the muscles.

WRONG CHOICES EXPLAINED

(2) The cramping of the muscles does *not* most likely result from *the runner running too slowly*. Usually running too fast for the body's capabilities, not too slowly, results in muscle cramping.

(3) The cramping of the muscles does *not* most likely result from *the runner warming up before running*. Warm-up techniques, in which the muscles are stretched and blood flow increased, generally reduce the incidence of muscle cramping.

(4) The cramping of the muscles does *not* most likely result from *increased glucose production in the muscle*. Glucose is produced in the muscles from the breakdown of glycogen stored in the muscle cells and in the liver. Its increased production would have little or no effect on the incidence of muscle cramping.

71. One credit is allowed for correctly stating which *two* plant species appear to be the most closely related and supporting your answer. Acceptable responses include, but are not limited to: [1]

- *A and C—They have the most characteristics in common.*
- *A and C—The same type of chlorophyll is present in both.*

72. One credit is allowed for correctly stating what additional information could be gathered to support your answer to question 71. Acceptable responses include, but are not limited to: [1]

- *Structure of protein molecules*
- *Types of enzymes present*
- *DNA sequences*
- *Other physical characteristics*

73. One credit is allowed for correctly stating *one* reason why scientists might want to know if two plant species are closely related. Acceptable responses include, but are not limited to: [1]

- *Two related plants may produce similar substances that could be used for medicines.*
- *A related plant may provide a cheaper source of a needed substance.*
- *If a plant becomes extinct, a related plant may provide an alternative source of a substance.*
- *A related plant may fill an environmental niche vacated by a less adaptive relative.*

74. **3** Based on its preferred food, species *B* would be classified as a *carnivore*. Carnivores are meat-eating animals. The diet of species *B* includes the bodies of other animals; therefore species *B* is carnivorous.

WRONG CHOICES EXPLAINED
(1) Based on its preferred food, species *B* would *not* be classified as a *decomposer*. Decomposers are organisms, such as bacteria and fungi, that consume the bodies of dead plants and animals.

(2) Based on its preferred food, species *B* would *not* be classified as a *producer*. Producers are organisms, such as green plants, that produce their own food through the process of photosynthesis.

(4) Based on its preferred food, species *B* would *not* be classified as a *parasite*. Parasites are organisms, such as mosquitoes and tapeworms, that live in or on the bodies of other organisms and harm their hosts while deriving a nutritional benefit from the association.

75. **2** Species *B and C* are the two species that would most likely be able to live in the same habitat without competing with each other for food. Species *B* depends on worms and insects for its food supply, whereas species *C* consumes fruits and seeds. Because they utilize different foods, these species are less likely to compete with each other in their nutritional activities.

WRONG CHOICES EXPLAINED
(1), (4) Species *A and C* and species *C and E* are *not* the two species that would most likely be able to live in the same habitat without competing with each other for food. All of these species consume plant matter for food and so will be more likely to compete with each other in their nutritional activities than species *B* and *C*.

(3) Species *B and D* are *not* the two species that would most likely be able to live in the same habitat without competing with each other for food. Both of these species include animal matter in their diets and so will be more likely to compete with each other in their nutritional activities than species *B* and *C*.

76. One credit is allowed for correctly drawing in the expected locations of the molecules after a period of 1 hour. [1]

77. One credit is allowed for correctly stating, when starch indicator is used, what observation would indicate the presence of starch. Acceptable responses include, but are not limited to: [1]

- *A blue-black color would indicate the presence of starch.*
- *A color change would occur in the presence of the starch indicator.*

78. One credit is allowed for correctly stating *one* reason why some molecules can pass through a certain membrane, but other molecules *cannot*. Acceptable responses include, but are not limited to: [1]

- *Some molecules are too large to pass through the membrane.*
- *Some molecules are not soluble.*
- *The size of the pores in the membrane allows small molecules to pass through but blocks large ones.*

79. **2** The width (w) of this plant cell is closest to *800 μm*. The width (w) of the cell illustrated in the microscope field can be placed along the 4,000-μm diameter of the microscope field approximately five (5) times. Dividing 4,000 μm by 5 yields a value for w of 800 μm.

WRONG CHOICES EXPLAINED
(1), (3), (4) The width (w) of this plant cell is *not* closest to *200 μm, 1,200 μm,* or *1,600 μm*. None of these values can be derived using the method provided above.

80. **3** Diagram 3 best represents the appearance of the cell after it has been placed in a 15% salt solution for 2 minutes. A plant cell in tap water has a water concentration roughly equal to that of tap water. If this cell is moved into a 15% saltwater environment, a concentration gradient is established that causes water to move by osmosis from a region of relatively high water concentration (the cell interior) to a region of relatively low water concentration (the saltwater environment). As a result, the cytoplasm of the cell shrinks and the cell membrane pulls away from the cell wall, concentrating the cell contents in the center of the cell.

WRONG CHOICES EXPLAINED
(1) Diagram *1* does *not* best represent the appearance of the cell after it has been placed in a 15% salt solution for 2 minutes. The cell in diagram *1* is virtually identical to the cell in tap water.

(2) Diagram *2* does *not* best represent the appearance of the cell after it has been placed in a 15% salt solution for 2 minutes. The cell in diagram *2* contains an enlarged vacuole and has a more rounded appearance that might occur if the cell had been placed in distilled water for a short time.

(4) Diagram *4* does *not* best represent the appearance of the cell after it has been placed in a 15% salt solution for 2 minutes. The cell in diagram *4* has burst open, an event that might occur if the cell had been placed in distilled water for a long time. In this case, water moves from the distilled water container through the cell membranes and into the cell cytoplasm by osmosis. This water movement continues until the cytoplasm bursts the membrane and wall of the cell.

STANDARDS/KEY IDEAS	JUNE 2008 QUESTION NUMBERS	NUMBER OF CORRECT RESPONSES
STANDARD 1		
Key Idea 1: The central purpose of scientific inquiry is to develop explanations of natural phenomena in a continuing and creative process.	32	
Key Idea 2: Beyond the use of reasoning and consensus, scientific inquiry involves the testing of proposed explanations involving the use of conventional techniques and procedures and usually requiring considerable ingenuity.	33, 34, 63	
Key Idea 3: The observations made while testing proposed explanations, when analyzed using conventional and invented methods, provide new insights into natural phenomena.	39, 43, 44	
Laboratory Checklist	31, 51, 52	
STANDARD 4		
Key Idea 1: Living things are both similar to and different from each other and from nonliving things.	1, 2, 4, 5, 9, 25, 35, 36, 42, 48, 53	
Key Idea 2: Organisms inherit genetic information in a variety of ways that result in continuity of structure and function between parents and offspring.	3, 6, 7, 10, 11, 12, 38, 50	
Key Idea 3: Individual organisms and species change over time.	8, 13, 14, 16, 40, 41, 54, 65	
Key Idea 4: The continuity of life is sustained through reproduction and development.	17, 18, 21, 22, 37	
Key Idea 5: Organisms maintain a dynamic equilibrium that sustains life.	15, 19, 20, 23, 49, 56, 57, 59, 64	
Key Idea 6: Plants and animals depend on each other and their physical environment.	24, 27, 28, 55, 60, 61, 62, 66	

STANDARDS/KEY IDEAS	JUNE 2008 QUESTION NUMBERS	NUMBER OF CORRECT RESPONSES
STANDARD 4		
Key Idea 7: Human decisions and activities have a profound impact on the physical and living environment.	26, 29, 30, 45, 46, 47, 58, 67	
REQUIRED LABORATORIES		
Lab 1: "Relationships and Biodiversity"	68, 69, 71, 72, 73	
Lab 2: "Making Connections"	70	
Lab 3: "The Beaks of Finches"	74, 75	
Lab 5: "Diffusion Through a Membrane"	76, 77, 78, 79, 80	

Examination
August 2008
Living Environment

PART A

Answer all questions in this part. [30]

Directions (1–30): For *each* statement or question, select the word or expression that best completes the statement or answers the question. Record your answers in the spaces provided.

1 Scientists studying ocean organisms are discovering new and unusual species. Which observation could be used to determine that an ocean organism carries out autotrophic nutrition?

1 Chloroplasts are visible inside the cells.
2 Digestive organs are visible upon dissection.
3 The organism lives close to the surface.
4 The organism synthesizes enzymes to digest food.   1____

2 Abiotic factors that characterize a forest ecosystem include

1 light and biodiversity
2 temperature and amount of available water
3 types of producers and decomposers
4 pH and number of heterotrophs   2____

3 One season, there was a shortage of producers in a food web. As a result, the number of deer and wolves decreased. The reason that both the deer and wolf populations declined is that

1 producers are not as important as consumers in a food web
2 more consumers than producers are needed to support the food web
3 organisms in this food web are interdependent
4 populations tend to stay constant in a food web   3____

4 Which statement best describes a population of organisms if cloning is the only method used to reproduce this population?

1 The population would be more likely to adapt to a changing environment.
2 There would be little chance for variation within the population.
3 The population would evolve rapidly.
4 The mutation rate in the population would be rapid.   4____

5 An organelle that releases energy for metabolic activity in a nerve cell is the

1 chloroplast   3 mitochondrion
2 ribosome   4 vacuole   5____

6 A student notices that fruit flies with the curly-wing trait develop straight wings if kept at a temperature of 16°C, but develop curly wings if kept at 25°C. The best explanation for this observation is that

1 wing shape is controlled by behavior
2 wing shape is influenced by light intensity
3 gene expression can be modified by interactions with the environment
4 gene mutations for wing shape can occur at high temperatures   6____

7 In all organisms, the coded instructions for specifying the characteristics of the organism are directly determined by the arrangement of the

1 twenty kinds of amino acids in each protein
2 twenty-three pairs of genes on each chromosome
3 strands of simple sugars in certain carbohydrate molecules
4 four types of molecular bases in the genes   7____

8 Which sequence shows a *decreasing* level of complexity?

1 organs → organism → cells → tissues
2 organism → cells → organs → tissues
3 cells → tissues → organs → organism
4 organism → organs → tissues → cells   8____

9 Which row in the chart below contains an event that is paired with an appropriate response in the human body?

Row	Event	Response
(1)	a virus enters the bloodstream	increased production of antibodies
(2)	fertilization of an egg	increased levels of testosterone
(3)	dehydration due to increased sweating	increased urine output
(4)	a drop in the rate of digestion	increased respiration rate

9 _____

10 The diagram below represents a genetic procedure.

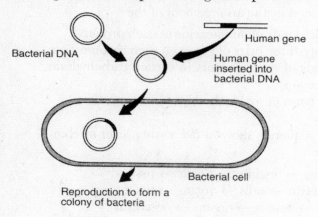

Bacterial DNA

Human gene

Human gene inserted into bacterial DNA

Bacterial cell

Reproduction to form a colony of bacteria

Which statement best describes the outcome of this procedure?

1 Bacterial cells will destroy defective human genetic material.

2 Bacterial cells may form a multicellular embryo.

3 The inserted human DNA will change harmful bacteria to harmless ones.

4 The inserted human DNA may direct the synthesis of human proteins.

10 _____

11 The diagram below represents the genetic contents of cells before and after a specific reproductive process.

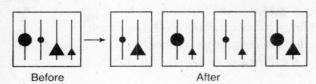

Before   After

This process is considered a mechanism of evolution because it

1 decreases the chance for new combinations of inheritable traits in a species
2 decreases the probability that genes can be passed on to other body cells
3 increases the chance for variations in offspring
4 increases the number of offspring an organism can produce   11____

12 One *disadvantage* of a genetic mutation in a human skin cell is that it

1 may result in the production of a defective protein
2 may alter the sequence of simple sugars in insulin molecules
3 can lead to a lower mutation rate in the offspring of the human
4 can alter the rate of all the metabolic processes in the human   12____

13 The DNA of a human cell can be cut and rearranged by using

1 a scalpel   3 hormones
2 electrophoresis   4 enzymes   13____

14 Much of the carbon dioxide produced by green plants is *not* excreted as a metabolic waste because it

    1 can be used for photosynthesis
    2 is too large to pass through cell membranes
    3 is needed for cellular respiration
    4 can be used for the synthesis of proteins   14_____

15 In several species of birds, the males show off their bright colors and long feathers. The dull-colored females usually pick the brightest colored males for mates. Male offspring inherit their father's bright colors and long feathers. Compared to earlier generations, future generations of these birds will be expected to have a greater proportion of

    1 bright-colored females
    2 dull-colored females
    3 dull-colored males
    4 bright-colored males   15_____

16 To determine evolutionary relationships between organisms, a comparison would most likely be made between all of the characteristics below *except*

    1 methods of reproduction
    2 number of their ATP molecules
    3 sequences in their DNA molecules
    4 structure of protein molecules present   16_____

17 The females of certain species of turtles will sneak into a nest of alligator eggs to lay their own eggs and then leave, never to return. When the baby turtles hatch, they automatically hide from the mother alligator guarding the nest and to the nearest body of water when it is safe to do so. Which statement best explains the behavior of these baby turtles?

1 More of the turtles' ancestors who acted in this way survived to reproduce, passing this behavioral trait to their offspring.
2 The baby turtles are genetically identical, so they behave in the same way.
3 Turtles are not capable of evolving, so they repeat the same behaviors generation after generation.
4 The baby turtles' ancestors who learned to behave this way taught the behaviors to their offspring.   17____

18 A pattern of reproduction and growth in a one-celled organism is shown below.

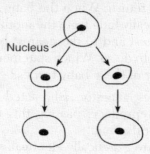

Which statement best describes this pattern of reproduction?

1 All genetic material comes from one parent.
2 Only some of the genetic material comes from one parent.
3 The size of the parent determines the amount of genetic material.
4 The size of the parent determines the source of the genetic material.   18____

19 A technique used to reproduce plants is shown in the diagram below.

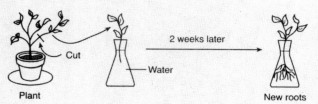

This technique is a form of

1 sexual reproduction   3 gamete production
2 asexual reproduction   4 gene manipulation   19____

20 The diagram below represents a biochemical process.

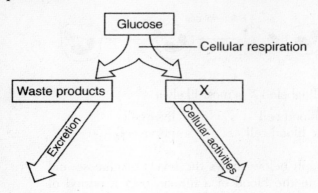

Which molecule is represented by X?

1 DNA   3 protein

2 starch   4 ATP   20____

21 The diagram below represents early stages of embryo development.

The greatest amount of differentiation for organ formation most likely occurs at arrow

(1) A   (3) C

(2) B   (4) D   21____

22 The diagram below shows a cell in the human body engulfing a bacterial cell.

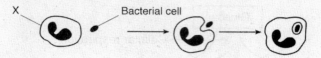

The cell labeled X is most likely a

1 red blood cell   3 liver cell
2 white blood cell   4 nerve cell   22____

23 The graph below shows the levels of glucose and insulin in the blood of a human over a period of time.

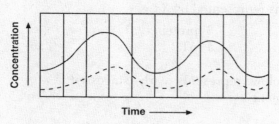

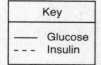

This graph represents

1 an allergic reaction
2 an antigen-antibody reaction
3 maintenance of homeostasis
4 autotrophic nutrition   23____

24 Which concept is best represented in the diagram shown below?

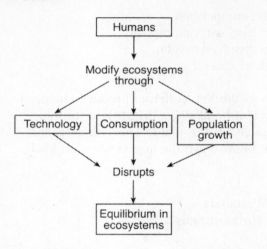

1 Human actions are a threat to equilibrium in ecosystems.
2 Equilibrium in ecosystems requires that humans modify ecosystems.
3 Equilibrium in ecosystems directly affects how humans modify ecosystems.
4 Human population growth is the primary reason for equilibrium in ecosystems. 24____

25 One possible reason for the rise in the average air temperature at Earth's surface is that

1 decomposers are being destroyed
2 deforestation has increased the levels of oxygen in the atmosphere
3 industrialization has increased the amount of carbon dioxide in the air
4 growing crops is depleting the ozone shield 25____

26 The size of a frog population in a pond remains fairly constant over a period of several years because of

  1 decreasing competition
  2 environmental carrying capacity
  3 excessive dissolved oxygen
  4 the depth of water   26____

27 Plants such as the Venus flytrap produce chemical compounds that break down insects into substances that are usable by the plant. The chemical compounds that break down the insects are most likely

  1 fats
  2 minerals
  3 biological catalysts
  4 complex carbohydrates   27____

28 In December 2004, a tsunami (giant wave) destroyed many of the marine organisms along the coast of the Indian Ocean. What can be expected to happen to the ecosystem that was most severely hit by the tsunami?

  1 The ecosystem will change until a new stable community is established.
  2 Succession will continue in the ecosystem until one species of marine organism is established.
  3 Ecological succession will no longer occur in this marine ecosystem.
  4 The organisms in the ecosystem will become extinct.   28____

29 Many homeowners who used to collect, bag, and discard grass clippings are now using mulching lawnmowers, which cut up the clippings into very fine pieces and deposit them on the soil. The use of mulching lawnmowers contributes most directly to

1 increasing the diversity of life
2 recycling of nutrients
3 the control of pathogens
4 the production of new species   29____

30 Deforestation of areas considered to be rich sources of genetic material could limit future agricultural and medical advances due to

1 the improved quality of the atmosphere
2 the maintenance of dynamic equilibrium
3 an increase in the rate of evolutionary change
4 the loss of biodiversity   30____

Part B–1

Answer all questions in this part. [12]

Directions (31–42): For *each* statement or question, select the word or expression that best completes the statement or answers the question. Record your answers in the spaces provided.

31 In 1883, Thomas Engelmann, a German botanist, exposed a strand of algae to different wavelengths of light. Engelmann used bacteria that concentrate near an oxygen source to determine which sections of the algae were releasing the most O_2. The results are shown below.

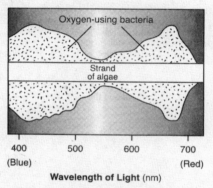

Oxygen-using bacteria

Strand of algae

400 (Blue)   500   600   700 (Red)

Wavelength of Light (nm)

Which statement is a valid inference based on this information?

1 Oxygen production decreases as the wavelength of light increases from 550 to 650 nm.
2 Respiration rate in the bacteria is greatest at 550 nm.
3 Photosynthetic rate in the algae is greatest in blue light.
4 The algae absorb the greatest amount of oxygen in red light.

31 _____

32 The graph below represents a predator-prey relationship.

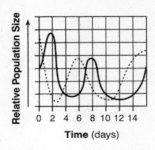

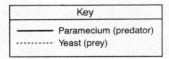

What is the most probable reason for the increasing predator population from day 5 to day 7?

1 an increasing food supply from day 5 to day 6
2 a predator population equal in size to the prey population from day 5 to day 6
3 the decreasing prey population from day 1 to day 2
4 the extinction of the yeast on day 3

32_____

33 A single cell and a multicellular organism are represented below.

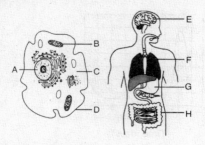

Which structures are correctly paired with their primary function?

(1) *A* and *G*—transmission of nerve impulses
(2) *B* and *E*—photosynthesis
(3) *C* and *H*—digestion of food
(4) *D* and *F*—gas exchange   33____

34 The graph below indicates the size of a fish population over a period of time.

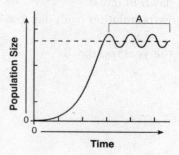

The section of the graph labeled *A* represents

1 biodiversity within the species
2 nutritional relationships of the species
3 a population becoming extinct
4 a population at equilibrium   34____

35 The data table below shows the presence or absence of DNA in four different cell organelles.

Data Table

Organelle	DNA
cell membrane	absent
cell wall	absent
mitochondrion	present
nucleus	present

Information in the table suggests that DNA functions

1 within cytoplasm and outside of the cell membrane
2 both inside and outside of the nucleus
3 only within energy-releasing structures
4 within cell vacuoles   35_____

Base your answers to questions 36 and 37 on the diagram below, which represents stages in the digestion of a starch, and on your knowledge of biology.

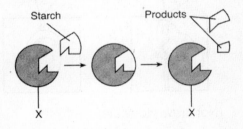

36 The products would most likely contain

  1 simple sugars   3 amino acids
  2 fats   4 minerals   36_____

37 The structure labeled X most likely represents

  1 an antibody   3 an enzyme
  2 a receptor molecule   4 a hormone   37_____

Base your answers to questions 38 through 40 on the diagram below and on your knowledge of biology. Each arrow in the diagram represents a different hormone released by the pituitary gland that stimulates the gland indicated in the diagram. All structures are present in the same organism.

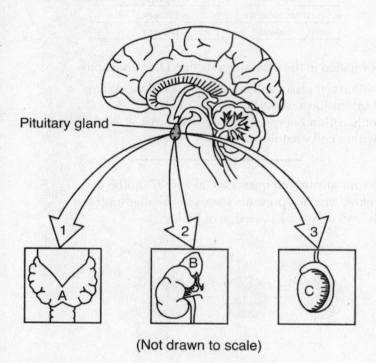

(Not drawn to scale)

38 The pituitary gland may release hormone 2 when blood pressure drops. Hormone 2 causes gland *B* to release a different hormone that raises blood pressure which, in turn, stops the secretion of hormone 2. The interaction of those hormones is an example of

1 DNA base substitution
2 manipulation of genetic instructions
3 a feedback mechanism
4 an antigen-antibody reaction   38_____

39 What would most likely occur if the interaction is blocked between the pituitary and gland *C*, the site of meiosis in males?

1 The level of progesterone would start to increase.
2 The pituitary would produce another hormone to replace hormone 3.
3 Gland *A* would begin to interact with hormone 3 to maintain homeostasis.
4 The level of testosterone may start to decrease.   39_____

40 Why does hormone 1 influence the action of gland *A* but *not* gland *B* or *C*?

1 Every activity in gland *A* is different from the activities in glands *B* and *C*.
2 The cells of glands *B* and *C* contain different receptors than the cells of gland *A*.
3 Each gland contains cells that have different base sequences in their DNA.
4 The distance a chemical can travel is influenced by both pH and temperature.   40_____

Base your answers to questions 41 and 42 on the information and diagram below and on your knowledge of biology.

A small water plant (elodea) was placed in bright sunlight for five hours as indicated below. Bubbles of oxygen gas were observed being released from the plant.

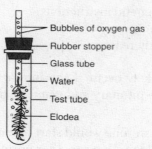

Bubbles of oxygen gas
Rubber stopper
Glass tube
Water
Test tube
Elodea

41 Since oxygen gas is being released, it can be inferred that the plant is

1 producing glucose
2 making protein
3 releasing energy from water
4 carrying on active transport   41____

42 What substance did the plant most likely absorb from the water for the process that produces the oxygen gas?

1 dissolved nitrogen   3 an enzyme
2 carbon dioxide   4 a hormone   42____

PART B-2

Answer all questions in this part.   [13]

Directions (43–55): **For those questions that are followed by four choices, record your answers in the spaces provided. For all other questions in this part, record your answers in accordance with the directions given in the question.**

Base your answers to questions 43 through 45 on the information below and on your knowledge of biology.

Human reproduction is influenced by many different factors.

43 Identify *one* reproductive hormone and state the role it plays in reproduction.   [1]

_____

_____

44 Identify the structure in the uterus where the exchange of material between the mother and the developing fetus takes place.   [1]

_____

45 Identify *one* harmful substance that can pass through this structure and describe the *negative* effect it can have on the fetus.   [1]

_____

_____

46 The flow of materials through ecosystems involves the interactions of many processes and organisms. State how decomposers aid in the flow of materials in an ecosystem.   [1]

_____

_____

Base your answers to questions 47 through 49 on the information below and on your knowledge of biology.

Honeybees have a very cooperative way of living. Scout bees find food, return to the hive, and do the "waggle dance" to communicate the location of the food source to other bees in the hive. The waggle, represented by the wavy line in the diagram below, indicates the direction of the food source, while the speed of the dance indicates the distance to the food. Different species of honeybees use the same basic dance pattern in slightly different ways as shown in the table below.

Number of Waggle Runs in 15 Seconds		Distance to Food (feet)
Giant Honeybee	Indian Honeybee	
10.6	10.5	50
9.6	8.3	200
6.7	4.4	1000
4.8	2.8	2000

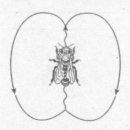

47 State the relationship between the distance to the food source and the number of waggle runs in 15 seconds.   [1]

_____

_____

48 Explain how waggle-dance behavior increases the reproductive success of the bees.   [1]

_____

_____

49 The number of waggle runs in 15 seconds for each of these species is most likely due to

1 behavioral adaptation as a result of natural selection

2 replacement of one species by another as a result of succession

3 alterations in gene structure as a result of diet

4 learned behaviors inherited as a result of asexual reproduction   [1]

49_____

Base your answers to questions 50 through 54 on the information and data table below and on your knowledge of biology.

The table shows data collected on the pH level of an Adirondack lake from 1980 to 1996.

Lake pH Level

Year	pH Level
1980	6.7
1984	6.3
1986	6.4
1988	6.2
1990	5.9
1992	5.6
1994	5.4
1996	5.1

Directions (50–54): Using the information in the data table, construct a line graph on the grid *on the next page,* following the directions below.

50 Label the axes.   [1]

51 Mark an appropriate scale on the *y*-axis. The scale has been started for you.   [1]

52 Plot the data from the data table. Surround each
point with a small circle and connect the points.
[1]

Example: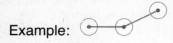

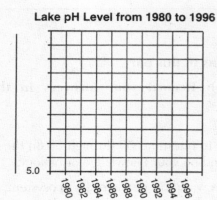

Lake pH Level from 1980 to 1996

53 Describe the trend in pH level in the lake over this
16-year period. [1]

_____

_____

54 Identify *one* factor that should have been kept constant each time water samples were collected from
the lake. [1]

_____

_____

55 Two cultures, each containing a different species of bacteria, were exposed to the same antibiotic. Explain how, after exposure to this antibiotic, the population of one species of bacteria could increase while the population of the other species of bacteria decreased or was eliminated. [1]

_____

_____

PART C

Answer all questions in this part. [17]

Directions (56–71): **Record your answers in the spaces provided.**

Base your answer to questions 56 through 58 on the information below and on your knowledge of biology.

Throughout the world, in nearly every ecosystem, there are animal and plant species present that were introduced into the ecosystem by humans or transported to the ecosystem as a result of human activities. Some examples are listed in the chart below.

Examples of Introduced Species

Organism	New Location
purple loosestrife (plant)	wetlands in New York State
zebra mussel	Great Lakes
brown tree snake	Guam

56 State *one* reason why an introduced species might be very successful in a new environment. [1]

_____

_____

57 Identify *one* action the government could take to prevent the introduction of additional new species. [1]

_____

_____

58 Identify *one* introduced organism and write its name in the space below. Describe *one* way in which this organism has altered an ecosystem in the new location. [1]

Organism: _____

_____

_____

Base your answers to questions 59 and 60 on the information and food web below and on your knowledge of biology.

The organisms in the food web below live near large cattle ranches. Over many years, mountain lions occasionally killed a few cattle. One year, a few ranchers hunted and killed many mountain lions to prevent future loss of their cattle. Later, ranchers noticed that animals from this food web were eating large amounts of grain from their fields.

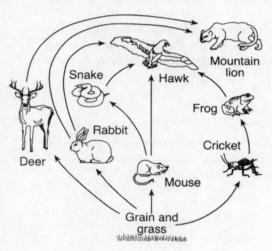

59 Identify *two* specific populations that most likely increased in number after the mountain lion population *decreased*. Support your answer. [2]

_____

_____

_____

_____

60 Explain how killing many mountain lions affected other ranchers in the community. [1]

_____

_____

_____

Base your answers to questions 61 through 64 on the passage below and on your knowledge of biology. The letters indicate paragraphs.

Yellow Fever

Paragraph *A*

A team of doctors was sent to Havana, Cuba, to study a yellow fever epidemic. The doctors wanted to find out how the pathogenic microbe that causes yellow fever is transferred from those who are sick to those who are well. Some people thought that the disease was spread by having contact with a person who had the disease or even through contact with clothing or bedding that they had used.

Paragraph *B*

It was known that yellow fever occurred more frequently in swampy environments than in environments that were dry. Consequently, some people thought that the disease was due to contact with the atmosphere of the swamps. A respected doctor in Havana was convinced that a particular species of mosquito, *Aedes calopus*, spread the disease.

Paragraph *C*

The team of doctors carried out several experiments and collected data. They built poorly ventilated houses in which American soldiers volunteered to sleep on bedding used by individuals who had recently died of yellow fever in local hospitals. The soldiers also wore the nightshirts of those who had died. The houses were fumigated to kill all mosquitoes and the doors and windows of the houses were screened. None of the soldiers living in these houses contracted the disease, though the experiment was tried repeatedly.

Paragraph *D*

In another experiment, the team built houses that were tightly sealed. The doors and windows were screened. The insides of the houses were divided into two parts by mosquito netting. One part of the house contained a species of mosquito, *Aedes calopus*, that had been allowed to bite yellow fever patients in the hospital. There were no mosquitoes in the other part of the house. A group of volunteers lived in each part of the house. A number of those who lived in the part of the house with the mosquitoes became infected; none of those in the other part of the house did.

Paragraph *E*

Putting these facts together with the other evidence, the team concluded that *Aedes calopus* spread the disease. The validity of this conclusion then had to be tested. All newly reported cases of yellow fever were promptly taken to well-screened hospitals and their houses were fumigated to kill any mosquitoes. The breeding places of the mosquitoes in and around Havana were drained or covered with a film of oil to kill mosquito larvae. Native fish species known to feed on mosquito larvae were introduced into streams and ponds. The number of yellow fever cases steadily declined until Havana was essentially free of the epidemic.

61 State the problem the team of doctors was trying to solve. [1]

_____

_____

62 State *one* hypothesis from paragraph *A* that was tested by one of the experiments. [1]

_____

_____

63 Describe the control that should have been set up for the experiment described in paragraph *C*. [1]

_____

_____

64 Explain why the use of native fish (described in paragraph *E*), rather than the use of pesticides, is less likely to have a *negative* impact on the environment. [1]

_____

_____

Base your answers to questions 65 through 67 on the information below and on your knowledge of biology.

Vaccines play an important role in the ability of the body to resist certain diseases.

65 Describe the contents of a vaccine. [1]

_____

_____

66 Identify the system in the body that is most directly affected by a vaccination.   [1]

_____

67 Explain how a vaccination results in the long-term ability of the body to resist disease.   [1]

_____

_____

_____

_____

Base your answers to questions 68 and 69 on the information below and on your knowledge of biology.

A factory in Florida had dumped toxic waste into the soil for 40 years. Since the company is no longer in business, government officials removed the toxic soil and piled it up into large mounds until they can finish evaluating how to treat the waste.

68 State _one_ way these toxins could move from the soil into local ecosystems, such as nearby lakes and ponds.   [1]

_____

_____

69 State _one_ way these toxins might affect local ecosystems.   [1]

_____

_____

Base your answers to questions 70 and 71 on the diagram below and on your knowledge of biology. The diagram shows some of the gases that, along with their sources, contribute to four major problems associated with air pollution.

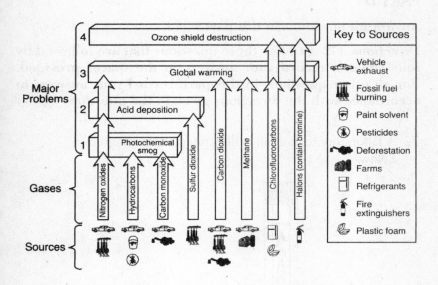

70 Select *one* of the four major problems from the diagram and record the number of the problem on the line below. Identify a gas that contributes to the problem you selected and state *one* way in which the amount of this gas can be reduced.   [1]

Problem number:_____

Gas:_____

_____

_____

71 Explain why damage to the ozone shield is considered a threat to many organisms.　[1]

_____

_____

PART D

Answer all questions in this part.　[13]

Directions: **(72–84): For those questions that are followed by four choices, record your answers in the spaces provided. For all other questions in this part, record your answers in accordance with the directions given in the question.**

72 A laboratory technique is illustrated in the diagram below.

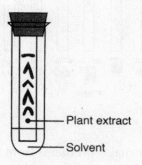

Plant extract

Solvent

This technique is used to

1　determine volume
2　separate molecules in a mixture
3　measure length
4　analyze data from an experiment　　　　72____

73 As part of an experiment, the heart rate of a person at rest was measured every hour for 7 hours. The data are shown in the table below.

Data Table

Hour	Heart Rate (beats/min)
1	72
2	63
3	61
4	61
5	60
6	63
7	68

Which graphed line best represents this data?

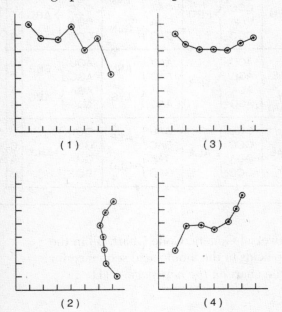

(1)   (3)

(2)   (4)

73_____

Base your answers to questions 74 through 77 on the Universal Genetic Code Chart below and on your knowledge of biology. Some DNA, RNA, and amino acid information from the analysis of a gene present in five different species is shown in the chart on the next page.

Universal Genetic Code Chart
Messenger RNA Codons and Amino Acids for Which They Code

		Second base				
		U	C	A	G	
First base	**U**	UUU } PHE UUC } UUA } LEU UUG }	UCU ⌉ UCC } UCA } SER UCG ⌋	UAU } TYR UAC } UAA } STOP UAG } STOP	UGU } CYS UGC } UGA } STOP UGG } TRP	U C A G
	C	CUU ⌉ CUC } CUA } LEU CUG ⌋	CCU ⌉ CCC } CCA } PRO CCG ⌋	CAU } HIS CAC } CAA } GLN CAG }	CGU ⌉ CGC } CGA } ARG CGG ⌋	U C A G
	A	AUU ⌉ AUC } ILE AUA } AUG } MET or START	ACU ⌉ ACC } ACA } THR ACG ⌋	AAU } ASN AAC } AAA } LYS AAG }	AGU } SER AGC } AGA } ARG AGG }	U C A G
	G	GUU ⌉ GUC } GUA } VAL GUG ⌋	GCU ⌉ GCC } GCA } ALA GCG ⌋	GAU } ASP GAC } GAA } GLU GAG }	GGU ⌉ GGC } GGA } GLY GGG ⌋	U C A G

Third base

74 Using the Universal Genetic Code Chart, fill in the missing amino acids in the amino acid sequence for species *A* in the chart *on the next page.* [1]

75 Using the information given, fill in the missing mRNA bases in the mRNA strand for species *B* in the chart below.   [1]

76 Using the information given, fill in the missing DNA bases in the DNA strand for species *C* in the chart below.   [1]

Species A	DNA strand:	TAC	CGA	CCT	TCA
	mRNA strand:	AUG	GCU	GGA	AGU
	Amino acid sequence:	___	___	___	___
Species B	DNA strand:	TAC	TTT	GCA	GGA
	mRNA strand:	___	___	___	___
	Amino acid sequence:	MET	LYS	ARG	PRO
Species C	DNA strand:	___	___	___	___
	mRNA strand:	AUG	UUU	UGU	CCC
	Amino acid sequence:	MET	PHE	CYS	PRO
Species D	DNA strand:	TAC	GTA	GTT	GCA
	mRNA strand:	AUG	CAU	CAA	CGU
	Amino acid sequence:	MET	HIS	GLN	ARG
Species E	DNA strand:	TAC	TTC	GCG	GGT
	mRNA strand:	AUG	AAG	CGC	CCA
	Amino acid sequence	MET	LYS	ARG	PRO

77 According to the information, which *two* species are most closely related? Support your answer.   [1]

Species: _____ and _____

_____

_____

_____

Base your answers to questions 78 and 79 on the information below and on your knowledge of biology. The diagram below represents the relationship between beak structure and food in several species of finches in the Galapagos Islands.

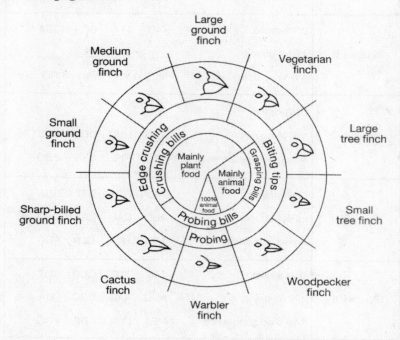

From: *Galapagos: A Natural History Guide*

Variations in Beaks of Galapagos Islands Finches

78 Which factor most directly influenced the evolution of the diverse types of beaks of these finches?

  1 predation by humans
  2 available food sources
  3 oceanic storms
  4 lack of available niches   78_____

79 State *one* reason why the large tree finch and the large ground finch are able to coexist on the same island.   [1]

_____

_____

80 Phenolphthalein is a chemical that turns pink in the presence of a base. A student set up the demonstration shown in the diagram below.

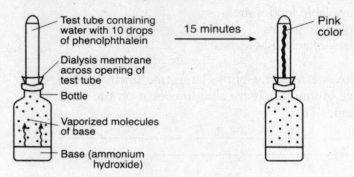

The appearance of the pink color was due to the movement of

1 phenolphthalein molecules from low concentration to high concentration
2 base molecules from high concentration through the membrane to low concentration
3 water molecules through the membrane from high concentration to low concentration
4 phenolphthalein molecules in the water from high concentration to low concentration

80____

Base your answers to questions 81 and 82 on the information and data table below and on your knowledge of biology.

A student cut three identical slices from a potato. She determined the mass of each slice. She then placed them in labeled beakers and added a different solution to each beaker. After 30 minutes, she removed each potato slice from its solution, removed the excess liquid with a paper towel, and determined the mass of each slice. The change in mass was calculated and the results are shown in the data table below.

Change in Mass of Potato in Different Solutions

Beaker	Solution	Change in Mass
1	distilled water	gained 4.0 grams
2	6% salt solution	lost 0.4 gram
3	16% salt solution	lost 4.7 grams

81 Identify the process that is responsible for the change in mass of each of the three slices.   [1]

_____

82 Explain why the potato slice in beaker 1 increased in mass.   [1]

_____

_____

83 Which activity might lead to damage of a microscope and specimen?

   1 cleaning the ocular and objectives with lens paper

   2 focusing with low power first before moving the high power into position

   3 using the coarse adjustment to focus the specimen under high power

   4 adjusting the diaphragm to obtain more light under high power   83_____

84 A solution containing both starch and glucose was placed inside the model cell represented below. The model cell was then placed in a beaker containing distilled water.

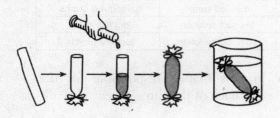

Identify *one* specific substance that should have been added to the distilled water so that observations regarding movement of starch could be made. [1]

_____

Answers
August 2008
Living Environment

Answer Key

PART A

1. 1	6. 3	11. 3	16. 2	21. 4	26. 2
2. 2	7. 4	12. 1	17. 1	22. 2	27. 3
3. 3	8. 4	13. 4	18. 1	23. 3	28. 1
4. 2	9. 1	14. 1	19. 2	24. 1	29. 2
5. 3	10. 4	15. 4	20. 4	25. 3	30. 4

PART B–1

31. 3	33. 4	35. 2	37. 3	39. 4	41. 1
32. 1	34. 4	36. 1	38. 3	40. 2	42. 2

PART B–2

43. *See* Answers Explained.
44. *See* Answers Explained.
45. *See* Answers Explained.
46. *See* Answers Explained.
47. *See* Answers Explained.
48. *See* Answers Explained.
49. 1

50. *See* Answers Explained.
51. *See* Answers Explained.
52. *See* Answers Explained.
53. *See* Answers Explained.
54. *See* Answers Explained.
55. *See* Answers Explained.

PART C

56. *See* Answers Explained.
57. *See* Answers Explained.
58. *See* Answers Explained.
59. *See* Answers Explained.
60. *See* Answers Explained.
61. *See* Answers Explained.
62. *See* Answers Explained.
63. *See* Answers Explained.

64. *See* Answers Explained.
65. *See* Answers Explained.
66. *See* Answers Explained.
67. *See* Answers Explained.
68. *See* Answers Explained.
69. *See* Answers Explained.
70. *See* Answers Explained.
71. *See* Answers Explained.

PART D

72. 2
73. 3
74. *See* Answers Explained.
75. *See* Answers Explained.
76. *See* Answers Explained.
77. *See* Answers Explained.
78. 2
79. *See* Answers Explained.
80. 2
81. *See* Answers Explained.
82. *See* Answers Explained.
83. 3
84. *See* Answers Explained.

Answers Explained

PART A

1. **1** *Chloroplasts are visible inside the cells* is the observation that could be used to determine that an ocean organism carries out autotrophic nutrition. Autotrophic (self-feeding) organisms derive their nutritional requirements from the process of photosynthesis. Chloroplasts are organelles in which the chemical reactions of photosynthesis occur. If chloroplasts are observed in the cells of this organism, it is logical to conclude that the organism is autotrophic.

WRONG CHOICES EXPLAINED

(2) *Digestive organs are visible upon dissection* is *not* the observation that could be used to determine that an ocean organism carries out autotrophic nutrition. Digestive organs that are used to break down and absorb foods are commonly found in the bodies of heterotrophic (other-feeding) organisms, not in the bodies of autotrophic organisms that are able to manufacture their own food.

(3) The *organism lives close to the surface* is *not* the observation that could be used to determine that an ocean organism carries out autotrophic nutrition. Many autotrophic organisms commonly inhabit the top layers of the ocean where they can absorb light energy more easily. However, because many heterotrophic organisms also inhabit this layer, it is not logical to classify an organism as autotrophic or heterotrophic based solely on this observation.

(4) *The organism synthesizes enzymes to digest food* is *not* the observation that could be used to determine that an ocean organism carries out autotrophic nutrition. Enzymes that are used to break down foods are commonly found in the bodies of both heterotrophic and autotrophic organisms. Because both types of organisms need to break down complex nutrient molecules into simpler subunits that are used in the processes of respiration and synthesis, it is not logical to classify an organism as autotrophic or heterotrophic based solely on this observation.

2. **2** Abiotic factors that characterize a forest ecosystem include *temperature and the amount of available water*. Abiotic factors are the nonliving conditions present in an ecosystem. Both temperature and water are nonliving conditions in the environment and therefore are abiotic factors.

ANSWERS August 2008

WRONG CHOICES EXPLAINED

(1) Abiotic factors that characterize a forest ecosystem do *not* include both *light and biodiversity*. While light is a nonliving condition, biodiversity (the variety of species that inhabit an ecosystem) is a biotic (living) factor in the environment.

(3) Abiotic factors that characterize a forest ecosystem do *not* include *types of producers and decomposers*. The types of producers (e.g., green plants, algae) and decomposers (e.g., bacteria, fungi) are both biotic (living) factors in the environment.

(4) Abiotic factors that characterize a forest ecosystem do *not* include both *pH and number of heterotrophs*. While pH (relative acidity) is a nonliving condition, the number of heterotrophs (e.g., animals, decomposers) is a biotic (living) factor in the environment.

3. **3** The reason that both the deer and wolf populations declined is that *organisms in this food web are interdependent*. The shortage of producers will leave the deer (primary consumers) without an adequate food supply, causing many to die of starvation. As the deer population declines, the wolves (secondary consumers) will eventually die of starvation as well. This situation will quickly destabilize the ecosystem until the producer populations increase and an appropriate balance can be reestablished.

WRONG CHOICES EXPLAINED

(1) The reason that both the deer and wolf populations declined is *not* that *producers are not as important as consumers in a food web*. Every species population in an ecosystem is important to the establishment and maintenance of the homeostatic balance of an ecosystem. Because producers form the basis of the food pyramid, they are essential to the survival of the consumer species in that ecosystem.

(2) The reason that both the deer and wolf populations declined is *not* that *more consumers than producers are needed to support the food web*. There must always be more producers than consumers in a stable ecosystem because plants form the basis of the food pyramid and inefficiently pass on the energy they have captured to consumer organisms.

(4) The reason that both the deer and wolf populations declined is *not* that *populations tend to stay constant in a food web*. Species populations in a food web are in a constant state of flux as they adjust to changing abiotic (nonliving) and biotic (living) conditions in the environment. The fact that the deer and wolf populations declined is an example of such an adjustment.

4. **2** *There would be little chance for variation within the population* is the statement that best describes a population of organisms if cloning is the only method used to reproduce this population. The chief sources of variation in living organisms are mutation of genetic material and gene recombination that occurs through sexual reproduction. A population that depends on cloning (a form of asexual reproduction) can vary only to the extent that its genetic material is exposed to mutagenic agents (e.g., chemicals, radiation).

WRONG CHOICES EXPLAINED
(1) *The population would be more likely to adapt to a changing environment* is *not* the statement that best describes a population of organisms if cloning is the only method used to reproduce this population. Because this population would have little chance for variation, its members would be less likely, not more likely, to adapt to environmental changes.

(3) *The population would evolve rapidly* is *not* the statement that best describes a population of organisms if cloning is the only method used to reproduce this population. Evolution is the gradual change in the characteristics of a species population that results from genetic variation, natural selection of favorable variations, and reproduction of the members of the population displaying those favorable variations. A population that depends solely on cloning for reproduction would evolve slowly, not rapidly.

(4) *The mutation rate in the population would be rapid* is *not* the statement that best describes a population of organisms if cloning is the only method used to reproduce this population. Mutation of genetic material is a random event that speeds up only if their exposure to mutagenic agents increases. All things being equal, a cloning population experiences a rate of mutation equivalent to that of noncloning species in a particular environment.

5. **3** An organelle that releases energy for metabolic activity in a nerve cell is the *mitochondrion*. Mitochondrial enzymes catalyze the conversion of nutrients such as glucose to waste products and adenosine triphosphate (ATP) in the process of respiration. ATP is a molecule that stores the chemical bond energy released from the oxidation of glucose during respiration and is used in cells to provide the energy needed for certain cell processes.

WRONG CHOICES EXPLAINED
(1) An organelle that releases energy for metabolic activity in a nerve cell is *not* the *chloroplast*. The chloroplast is an organelle that absorbs light energy and, through the action of enzymatically controlled reactions, converts it to the chemical bond energy found in molecules of glucose.

(2) An organelle that releases energy for metabolic activity in a nerve cell is *not* the *ribosome*. The ribosome is an organelle that serves as the site of enzymatically controlled reactions that link free amino acid subunits into polypeptide chains (proteins).

(4) An organelle that releases energy for metabolic activity in a nerve cell is *not* the *vacuole*. The vacuole is an organelle that functions to store materials (e.g., food, wastes) in cells.

6. **3** The best explanation for this observation is that *gene expression can be modified by interactions with the environment*. The fruit flies' wing shape in warm versus cool temperatures indicates that the genes for the production of wing shape are expressed differently under different environmental conditions for this species. It is likely that the enzymes that catalyze the biochemical reactions for the curly-wing trait are produced under one set of conditions but not under the other set of conditions.

WRONG CHOICES EXPLAINED
(1) The best explanation for this observation is *not* that *wing shape is controlled by behavior*. The fruit flies' behavior was not measured or reported for this experiment. There are no data provided that support this as an explanation for the observation.

(2) The best explanation for this observation is *not* that *wing shape is influenced by light intensity*. Light intensity is not an environmental condition that was measured or reported for this experiment. There are no data provided that support this as an explanation for the observation.

(4) The best explanation for this observation is *not* that *gene mutations for wing shape can occur at high temperatures*. DNA may be altered by random mutations, which are normally caused by chemicals or radiation, not temperature change, present in the environment. It is extremely unlikely that DNA alteration is responsible for this observation.

7. **4** In all organisms, the coded instructions for specifying the characteristics of the organism are directly determined by the arrangement of the *four types of molecular bases in the genes*. DNA, which controls hereditary information in the cell nucleus, is passed from generation to generation through the reproductive process. DNA molecules contain complex sequences of nitrogenous bases (A, T, G, and C) that code for the production of specific proteins. Each specific cell protein is capable of contributing to the development of an observable hereditary trait in the adult organism.

WRONG CHOICES EXPLAINED
(1) It is *not* true that, in all organisms, the coded instructions for specifying the characteristics of the organism are directly determined by the arrangement of the *twenty kinds of amino acids in each protein*. Proteins are formed from specific sequences of amino acids linked in a chain, then folded as a result of electromagnetic attractions within the complex polypeptide molecule. DNA molecules, not protein molecules, are responsible for passing on hereditary information in asexually reproducing organisms.

(2) It is *not* true that, in all organisms, the coded instructions for specifying the characteristics of the organism are directly determined by the arrangement of the *twenty-three pairs of genes on each chromosome*. Chromosomes are complex cell structures that contain hundreds or thousands of genes. The number 23 is significant in that it represents the monoploid (n) number of chromosomes found in human gametes, but it does not relate to the number of genes on a chromosome.

(3) It is *not* true that, in all organisms, the coded instructions for specifying the characteristics of the organism are directly determined by the arrangement of the *strands of simple sugars in certain carbohydrate molecules*. Simple organic sugars such as glucose and fructose serve as a source of nutrition and cell energy and have no direct role in heredity in living things. DNA molecules, not strands of simple sugars, are responsible for passing on hereditary information in asexually reproducing organisms.

8. **4** The sequence *organism* → *organs* → *tissues* → *cells* shows a *decreasing* level of complexity. An organism such as the human being is a complex living thing made up of a number of organ systems (e.g., circulatory system). Organ systems are composed of organs of similar function (e.g., heart, artery). Organs are made up of tissues (e.g., heart muscle tissue, blood tissue) that contain specific types of cells (e.g., cardiac muscle cells, red blood cells). In this sequence, the organism is the most complex structure; the organ is less complex than the organism but more complex than the tissue; the tissue is less complex than the organ but more complex than the cell; the cell is the least complex structure of those listed.

WRONG CHOICES EXPLAINED

(1), (2), (3) It is *not* true that the sequences *organs* → *organism* → *cells* → *tissues, organism* → *cells* → *organs* → *tissues*, and *cells* → *tissues* → *organs* → *organism* show a *decreasing* level of complexity. Each of these sequences contains one or more items out of order of decreasing size. See correct answer above.

9. **1** Row *1* in the chart contains an event that is paired with an appropriate response in the human. Row *1* lists "a virus enters the bloodstream" as the event, and "increased production of antibodies" as the response. In humans, the immune system responds to the presence of foreign proteins (antigens) embedded in the membranes of viruses and other pathogens by producing specific antibodies designed to neutralize and destroy the foreign invaders, thus keeping the body free of disease.

WRONG CHOICES EXPLAINED

(2) Row *2* in the chart does *not* contain an event that is paired with an appropriate response in the human. In this row, "fertilization of an egg" is incorrectly paired with "increased levels of testosterone." In humans, eggs are fertilized inside the reproductive tract of the female, whereas testosterone is a hormone produced in males.

(3) Row *3* in the chart does *not* contain an event that is paired with an appropriate response in the human. In this row, "dehydration due to increased sweating" is incorrectly paired with "increased urine output." In humans, dehydration normally results in decreased, not increased, urine production.

(4) Row *4* in the chart does *not* contain an event that is paired with an appropriate response in the human. In this row, "a drop in the rate of digestion" is incorrectly paired with "increased respiration rate." In humans, the rate of digestion is not directly linked to the rate of respiration.

10. **4** *The inserted human DNA may direct the synthesis of human proteins* is the statement that best describes the outcome of this procedure. The genetic procedure illustrated is known as "gene splicing" and is characterized by the insertion of genes from one species (typically human) into the genome of another species (typically bacteria). The spliced gene controls the synthesis of a specific desired protein (e.g., insulin), which can be produced abundantly and cheaply by colonies of bacteria containing the human gene.

WRONG CHOICES EXPLAINED

(1) *Bacterial cells will destroy defective human genetic material* is *not* the statement that best describes the outcome of this procedure. Bacterial cells receiving the human genes are independent organisms and are incapable of destroying defective genetic material.

(2) *Bacterial cells may form a multicellular embryo* is *not* the statement that best describes the outcome of this procedure. Bacteria are simple unicellular organisms that reproduce by mitotic cell division and are therefore incapable of forming a multicellular embryo.

(3) *The inserted human DNA will change harmful bacteria to harmless ones* is *not* the statement that best describes the outcome of this procedure. The bacterial cells receiving the spliced human genes retain all of their original genetic characteristics and functions, including any "harmful" effects.

11. **3** This process is considered a mechanism of evolution because it *increases the chance for variations in offspring*. The process illustrated is meiotic cell division, during which homologous chromosome pairs segregate into gametes. This process enhances variation by providing new genetic combinations that result in offspring with increased variation. This assists evolution by creating new varieties of species that may display enhanced survival rates during periods of environmental change.

WRONG CHOICES EXPLAINED

(1) This process is *not* considered a mechanism of evolution because it *decreases the chance for new combinations of inheritable traits in a species.* In fact, this process is considered by biologists to be the single most important factor at work to drive organic evolution. The process shown increases, not decreases, the probability for new combinations of inherititable traits in a species.

(2) This process is *not* considered a mechanism of evolution because it *decreases the probability that genes can be passed on to other body cells.* Genes are passed on to body cells through the process of mitotic cell division of the zygote. The process shown increases, not decreases, the probability that genes will be passed on to body cells.

(4) This process is *not* considered a mechanism of evolution because it *increases the number of offspring an organism can produce.* In general, sexually reproducing species produce fewer offspring in a given amount of time than do asexually reproducing species.

12. **1** One disadvantage of a genetic mutation in a human skin cell is that it *may result in the production of a defective protein.* Defective proteins may be either nonfunctional or harmful to the health of the cell in which they are produced. Harmful proteins may lead to a cancerous condition in the cell or may produce other deleterious results.

WRONG CHOICES EXPLAINED

(2) One disadvantage of a genetic mutation in a human skin cell is *not* that it *may alter the sequence of simple sugars in insulin molecules.* Insulin molecules are proteins, which are composed of chains of amino acids, not simple sugars.

(3) One disadvantage of a genetic mutation in a human skin cell is *not* that it *can lead to a lower mutation rate in the offspring of the human.* Mutations are random events that occur because of the presence of mutagenic factors (e.g., chemicals, radiation) in the environment. The presence of one mutation does not increase or decrease the likelihood of future mutations.

(4) One disadvantage of a genetic mutation in a human skin cell is *not* that it *can alter the rate of all the metabolic processes in the human.* The rate of metabolic activity in humans is regulated by physiological factors and hormone secretions, not by the presence of mutations in skin cells.

13. **4** The DNA of a human cell can be cut and rearranged by using *enzymes.* The basic DNA manipulation technique involves the use of restriction enzymes to snip out specific portions of a human DNA molecule for scientific analysis.

WRONG CHOICES EXPLAINED

(1) The DNA of a human cell can *not* be cut and rearranged by using *a scalpel*. A scalpel is a tool used in gross dissection; it is much too large to be used to cut and rearrange a DNA molecule.

(2) The DNA of a human cell can *not* be cut and rearranged by using *electrophoresis*. Electrophoresis is a technique used to separate charged molecular fragments. This process may be employed to separate fragments of human DNA but is not used to cut and rearrange those DNA molecules.

(3) The DNA of a human cell can *not* be cut and rearranged by using *hormones*. Hormones are specialized proteins produced as cell secretions within the endocrine glands. Hormones regulate specific metabolic processes within the human body but do not function in the cutting and rearrangement of DNA molecules.

14. **1** Much of the carbon dioxide produced by green plants is *not* excreted as a metabolic waste because it *can be used for photosynthesis*. Carbon dioxide produced as a waste product of aerobic respiration in the living cells of the leaf is released into the air spaces in the leaf interior. Carbon dioxide is an inorganic raw material used in the process of photosynthesis, during which its atoms are linked with those of water to produce glucose and oxygen gas. Much of the carbon dioxide produced during daylight hours is used in photosynthetic reactions rather than escaping into the atmosphere as a waste product of respiration.

WRONG CHOICES EXPLAINED

(2) It is *not* true that much of the carbon dioxide produced by green plants is *not* excreted as a metabolic waste because it *is too large to pass through cell membranes*. Carbon dioxide is a very small molecule that can pass readily though the cell membrane.

(3) It is *not* true that much of the carbon dioxide produced by green plants is *not* excreted as a metabolic waste because it *is needed for cellular respiration*. Carbon dioxide is a waste product, not a raw material, of respiration.

(4) It is *not* true that much of the carbon dioxide produced by green plants is *not* excreted as a metabolic waste because it *can be used for the synthesis of proteins*. Proteins are synthesized from units of amino acid, not from molecules of carbon dioxide.

15. **4** Compared to earlier generations, future generations of these birds will be expected to have a greater proportion of *bright-colored males*. In this context, the males' brightly colored, long feathers represent a favorable variation that increases the chance that males displaying this characteristic will produce male offspring that will in turn display the characteristic of brightly colored, long feathers. If in each generation the males with brightly colored, long feathers are more successful at mating than males without this trait, then this trait will become more and more common in the males of these species.

WRONG CHOICES EXPLAINED

(1), (2) Compared to earlier generations, future generations of these birds will *not* be expected to have a greater proportion of *bright-colored females* or *dull-colored females*. The observation given indicates that the mating selection is made by the females, so the color of the females is evidently not important in the selection of favorable variations in these species. For this reason, other factors (e.g., camouflage) will help determine the feather color in females.

(3) Compared to earlier generations, future generations of these birds will *not* be expected to have a greater proportion of *dull-colored males*. Given the fact that males with brightly colored, long feathers are selected as breeding partners by females in preference to males with dull-colored feathers, it is likely that future generations of these species will have a smaller, not a greater, proportion of dull-colored males.

16. **2** To determine evolutionary relationships between organisms, a comparison would most likely be made between all of the characteristics below *except* the *number of their ATP molecules*. Adenosine triphosphate (ATP) is a chemical that receives energy released during the process of cellular respiration and stores it for use in operating many biochemical reactions in the cell. Because ATP is commonly found in the cells of all organisms that carry on respiration, it is not possible to use ATP concentration as a measure of evolutionary relationships.

WRONG CHOICES EXPLAINED

(1), (3), (4) It is *not* true that, in order to determine evolutionary relationships between organisms, a comparison would most likely be made between all of the characteristics below *except* the *methods of reproduction*, the *sequences of their DNA molecules*, or the *structure of protein molecules present*. Each of these characteristics prove to be more similar in closely related species than in remotely related species. A study of the similarities of reproductive methods among different species is known as "comparative physiology," while studies of the similarities in DNA or protein structure among different species are known as "comparative biochemistry."

17. **1** *More of the turtles' ancestors who acted in this way survived to reproduce, passing this behavioral trait to their offspring* is the statement that best explains the behavior of the baby turtles. The baby turtles' behavior is an example of a favorable adaptation to an environmental condition. Presumably, ancestral turtles varied in their reproductive behaviors and the variation described proved to be a successful one within the species. As a result, the behavior became established in the majority of the species.

WRONG CHOICES EXPLAINED

(2) *The baby turtles are genetically identical, so they behave the same way* is *not* the statement that best explains the behavior of the baby turtles. Turtles reproduce sexually and for that reason cannot be genetically identical. In any sexually reproducing species, the members vary in numerous ways.

(3) *Turtles are not capable of evolving, so they repeat the same behaviors generation after generation* is *not* the statement that best explains the behavior of the baby turtles. All species evolve over time because of the natural selection of favorable variations that exist within the members of the species.

(4) *The baby turtles' ancestors who learned to behave this way taught the behaviors to their offspring* is *not* the statement that best explains the behavior of the baby turtles. Baby turtles, newly hatched from their eggs, cannot have learned these behaviors from other members of the species. This behavior is most likely genetically programmed.

18. **1** *All genetic material comes from one parent* is the statement that best describes this pattern of reproduction. The method of reproduction illustrated is an asexual form known as mitotic cell division. In mitotic cell division a single diploid ($2n$) parental nucleus undergoes mitosis, creating two genetically identical daughter nuclei. These nuclei are then separated by cytoplasmic division into two genetically identical daughter cells.

WRONG CHOICES EXPLAINED

(2) *Only some of the genetic material comes from one parent* is *not* the statement that best describes this pattern of reproduction. In asexual forms of reproduction, all genetic material comes from a single parent. In sexual reproduction, each parent provides half of the genetic information needed to produce an offspring.

(3) *The size of the parent determines the amount of genetic material* is *not* the statement that best describes this pattern of reproduction. The amount of genetic material in the nucleus is determined by the amount of genetic material in the prior generation. The amount of genetic material does not change with a change in cell size.

(4) *The size of the parent determines the source of the genetic material* is *not* the statement that best describes this pattern of reproduction. The source of genetic material is the nucleus. The source of the genetic material does not change with a change in cell size.

19. **2** The technique illustrated in the diagram is a form of *asexual reproduction*. The technique illustrated is known as reproduction by cuttings. When the cut stem is placed in water, some differentiated stem cells are stimulated to modify themselves into root cells. When the new roots that result from these root cells are sufficiently developed, the cut stem can be planted in soil where it can grow into a self-sustaining organism with genetic traits identical to those of the original parent plant.

WRONG CHOICES EXPLAINED

(1) The technique illustrated in the diagram is *not* a form of *sexual reproduction*. A diagram illustrating sexual reproduction would most likely show the fusion of monoploid (*n*) gametes from two different parents to produce a diploid (2*n*) zygote.

(3) The technique illustrated in the diagram is *not* a form of *gamete production*. A diagram illustrating gamete production would most likely show monoploid (*n*) sperm or egg cells being produced as a result of meiotic cell division of a diploid (2*n*) primary sex cell.

(4) The technique illustrated in the diagram is *not* a form of *gene manipulation*. A diagram illustrating gene manipulation would most likely show DNA molecules being cut by restriction enzymes and then rearranged in a manner that will result in the production of specific proteins.

20. **4** *ATP* is the molecule represented by *X*. The biochemical process illustrated is respiration. In this process, enzymes catalyze the conversion of nutrients such as glucose to waste products and adenosine triphosphate (ATP). ATP is a molecule that stores the chemical bond energy released from the oxidation of glucose during respiration and is used in cells to provide the energy needed for certain cell processes.

WRONG CHOICES EXPLAINED

(1), (2), (3) *DNA*, *starch*, and *protein* are *not* the molecules represented by *X*. Each of these molecules plays a specific role in the metabolic processes of the cell, but none is directly involved in the biochemical process of respiration.

21. **4** The greatest amount of differentiation for organ formation most likely occurs at arrow *D*. Differentiation is a process by which new cells take on specialized characteristics in the developing embryo. Differentiation begins following the blastula stage of embryonic development. Arrow *D* in the diagram illustrates the point at which the greatest amount of differentiation occurs in the stages of human embryo development illustrated.

WRONG CHOICES EXPLAINED

(1) The greatest amount of differentiation for organ formation does *not* most likely occur at arrow *A*. This arrow illustrates the process of fertilization, at which the least amount of differentiation has occurred.

(2) The greatest amount of differentiation for organ formation does *not* most likely occur at arrow *B*. This arrow illustrates the process of cleavage, at which no differentiation has yet occurred.

(3) The greatest amount of differentiation for organ formation does *not* most likely occur at arrow *C*. This arrow illustrates the process of blastula formation, at which only rudimentary differentiation has occurred.

22. **2** The cell labeled X is most likely a *white blood cell*. In humans, white blood cells found in the blood tissue ingest and destroy bacteria that enter the bloodstream through breaks in the skin surface. This function is essential to the immune response in humans.

WRONG CHOICES EXPLAINED

(1) The cell labeled X is *not* most likely a *red blood cell*. In humans, red blood cells are specialized for carrying oxygen to the body's tissues, not for engulfing bacterial cells.

(3), (4) The cell labeled X is *not* most likely a *liver cell* or a *nerve cell*. In humans, liver cells and nerve cells carry on specialized tasks related to the functions of the systems of which they are a part. These cells do not engulf bacterial cells.

23. **3** This graph represents *maintenance of homeostasis*. Insulin is a hormone that is secreted from specialized endocrine tissues in the human pancreas. Insulin is secreted in response to elevated glucose concentrations in the blood and acts to convert soluble glucose to insoluble glycogen, which is then stored in liver tissues. As the blood glucose concentration drops, the secretion of insulin slows. This feedback mechanism contributes to the maintenance of homeostasis, or a "steady state," in the human body.

WRONG CHOICES EXPLAINED

(1) This graph does *not* represent *an allergic reaction*. An allergic reaction sets off processes that result in the production of histamines. Histamines irritate mucous membranes and result in swelling of the affected tissues.

(2) This graph does *not* represent *an antigen-antibody reaction*. An antigen-antibody reaction involves the body's reaction to invasion by foreign organisms. This reaction includes the production of antibodies that are specifically designed to link to and neutralize the invading antigens.

(4) This graph does *not* represent *autotrophic nutrition*. Autotrophic (self-feeding) nutrition involves processes by which certain organisms (particularly green plants) derive their nutritional requirements from biochemical reactions (such as photosynthesis) that result in the capture of energy in the chemical bonds of organic molecules (such as glucose).

24. **1** *Human actions are a threat to equilibrium in ecosystems* is the concept that is best represented in the diagram. Human population growth and consumption of natural resources have put tremendous pressure on natural ecosystems throughout the biosphere. In addition, humans' use of technology to alter the environment has destroyed the habitats of countless species, further threatening the balance of nature.

WRONG CHOICES EXPLAINED

   (2) *Equilibrium in ecosystems requires that humans modify ecosystems* is *not* the concept that is best represented in the diagram. Equilibrium is established in ecosystems through the interactions of the species that inhabit those ecosystems. This balance is normally struck without the intervention of human technology.

   (3) *Equilibrium in ecosystems directly affects how humans modify ecosystems* is *not* the concept that is best represented in the diagram. It is unclear how the equilibrium in ecosystems could affect the manner in which humans alter their environment. In fact, it is because humans have traditionally ignored the balance of nature that so much damage has been done to the environment through their actions.

   (4) *Human population growth is the primary reason for equilibrium in ecosystems* is *not* the concept that is best represented in the diagram. In fact, human population growth is the single most significant factor responsible for *de*stabilizing ecosystems throughout the world. The growth of the human population has exceeded the natural carrying capacity of the environment, requiring ever more aggressive technological interventions to sustain it.

   25. **3** One possible reason for the rise in average air temperature at Earth's surface is that *industrialization has increased the amount of carbon dioxide in the air*. The burning of fossil fuels to produce energy for industrial processes is a prime contributor to the increasing concentration of various greenhouse gases, especially carbon dioxide, in the atmosphere. As carbon dioxide concentration increases in the atmosphere, solar radiation is trapped and warms Earth's surface.

WRONG CHOICES EXPLAINED

   (1) One possible reason for the rise in average air temperature at Earth's surface is *not* that *decomposers are being destroyed*. Decomposers are organisms such as bacteria and fungi that consume the bodies of dead plants and animals. Although the decomposition process releases carbon dioxide into the atmosphere, it does so much more slowly than do industrial processes. Destroying decomposers (not a good idea) would tend to decrease, not increase, surface temperatures on Earth.

   (2) One possible reason for the rise in average air temperature at Earth's surface is *not* that *deforestation has increased the levels of oxygen in the atmosphere*. Deforestation, the wholesale removal of trees from large tracts of land, decreases the biomass of plants available to absorb carbon dioxide from, and release oxygen gas into, the atmosphere during photosynthesis. Deforestation contributes to decreased, not increased, levels of oxygen in the atmosphere,

   (4) One possible reason for the rise in average air temperature at Earth's surface is *not* that *growing crops is depleting the ozone shield*. The growing of crops does not have any known negative effect on the ozone shield. Rather, the ozone shield is depleted by the presence of chlorofluorohydrocarbons (CFCs), chemicals once commonly used to pressurize certain aerosol cans.

26. **2** The size of a frog population in a pond remains fairly stable over a period of several years because of the *environmental carrying capacity*. The carrying capacity is the number of individuals in a species population that can be sustained in a particular environment. The population's size is controlled by limiting factors (e.g., amount of available food, amount of sunlight, average temperature, available oxygen) within the environment. When the population grows too large and uses up the available resources, its numbers are reduced until they once again fall within the carrying capacity of the environment.

WRONG CHOICES EXPLAINED
(1) It is *not* true that the size of a frog population in a pond remains fairly stable over a period of several years because of *decreasing competition*. Competition within species and competition between species is an ever-present biotic factor in any natural environment. Decreasing competition would tend to increase the frog's population size, not keep it stable.

(3) It is *not* true that the size of a frog population in a pond remains fairly stable over a period of several years because of *excessive dissolved oxygen*. Excessive oxygen dissolved in the frogs' water environment would be unlikely to affect the adult frog population because they are air breathers. Excessive dissolved oxygen may, however, affect the tadpole stage either positively or negatively, which would be unlikely to result in a stable population.

(4) It is *not* true that the size of a frog population in a pond remains fairly stable over a period of several years because of *the depth of water*. Frogs tend to inhabit and lay their eggs in the warmer shallow levels of their pond habitats. For this reason, it is unlikely that the depth of the water in the pond would have an effect on the frog population one way or the other.

27. **3** The chemical compounds that break down the insects are most likely *biological catalysts*. Such catalysts, known as enzymes, speed up or slow down chemical reactions in living cells without themselves being changed by the reaction. Enzymes are specialized protein molecules produced from coded messages held in DNA molecules and carried to ribosomes by RNA molecules.

WRONG CHOICES EXPLAINED
(1), (2), (4) The chemical compounds that break down the insects are *not* most likely *fats*, *minerals*, or *complex carbohydrates*. Enzyme molecules are protein in nature and so do not fall into any of these nonprotein categories.

28. **1** *The ecosystem will change until a new stable community is established* is the statement that best describes what will happen to the ecosystem that was most severely hit by the tsunami. "Ecological succession" is a term that describes the replacement of one plant community by other, progressively more complex, plant communities until a stable climax community is established. In the example given, a tsunami-ravaged shoreline is gradually inhabited by pioneer species (small plants), which are then followed by deep-rooted shrubs and trees. As succession continues over many years, a self-sustaining climax community of mixed plants will eventually become established that is compatible with the climate and soil conditions of the area.

WRONG CHOICES EXPLAINED

(2) *Succession will continue in the ecosystem until one species of marine organism is established* is *not* the statement that best describes what will happen to the ecosystem that was most severely hit by the tsunami. Healthy biological communities always contain numerous species of plants, animals, and decomposers. Only by establishing and maintaining biodiversity in an ecosystem will the ecosystem flourish.

(3) *Ecological succession will no longer occur in this marine ecosystem* is *not* the statement that best describes what will happen to the ecosystem that was most severely hit by the tsunami. Ecological succession cannot be "turned off" as a result of a tsunami or other natural phenomenon. This process will continue to restore ravaged ecosystems to a stable state despite such disasters.

(4) *The organisms in the ecosystem will become extinct* is *not* the statement that best describes what will happen to the ecosystem that was most severely hit by the tsunami. Species extinction is the total elimination of all members of a species from Earth. It is unlikely that a tsunami, no matter how powerful, would cause the extinction of all species inhabiting the affected area.

29. **2** The use of mulching lawnmowers contributes most directly to *recycling of nutrients*. Nutrients (such as nitrogen, phosphorus, magnesium, and potassium) held in the chemical makeup of the grass clippings are released through the action of decomposers in the soil. Once converted to a soluble form by the decomposers, these nutrients can be absorbed into the roots of the grass plants to promote their continued healthy growth without the need for artificial fertilizers.

WRONG CHOICES EXPLAINED

(1), (4) The use of mulching lawnmowers does *not* contribute most directly to *increasing the diversity of life* or to *the production of new species*. Increasing the diversity of life and producing new species is the role of biological evolution, not mulching lawnmowers.

(3) The use of mulching lawnmowers does *not* contribute most directly to *the control of pathogens*. Controlling pathogens is the role of immune systems operating within organisms, not mulching lawnmowers.

30. **4** Deforestation of areas considered to be rich sources of genetic material could limit future agricultural and medical advances because of *the loss of biodiversity*. Biodiversity is the variety of species that inhabit an ecosystem. When any forest ecosystem is destroyed by human activities such as logging or farming, the species that have inhabited that ecosystem are also destroyed. In some cases, these species carry unique genetic characteristics from which scientists might extract new medicines or develop new crops. However, if these species are driven to extinction, their unique genetic characteristics may be lost to science forever.

WRONG CHOICES EXPLAINED

(1) It is *not* true that deforestation of areas considered to be rich sources of genetic material could limit future agricultural and medical advances because of *the improved quality of the atmosphere*. Deforestation removes photosynthesizing plants that could have absorbed Earth-warming carbon dioxide from, and added life-giving oxygen to, the atmosphere. Deforestation does not improve, but degrades, the quality of the atmosphere.

(2) It is *not* true that deforestation of areas considered to be rich sources of genetic material could limit future agricultural and medical advances because of *the maintenance of dynamic equilibrium*. Deforestation upsets the natural balance of the ecosystem and interferes with the maintenance of dynamic equilibrium.

(3) It is *not* true that deforestation of areas considered to be rich sources of genetic material could limit future agricultural and medical advances because of *an increase in the rate of evolutionary change*. Deforestation reduces genetic variation by removing species from an area and decreases the opportunity for evolutionary change within that area.

PART B–1

31. **3** *Photosynthetic rate in the algae is greatest in blue light* is the statement that is a valid inference based on this information. This inference is based on the observation that oxygen-loving bacteria were most concentrated in the 400–500 nm (blue) region of the light spectrum. Except for another, smaller, bacterial concentration in the 675–700 nm (red) region, the blue region appears to be the best for photosynthetic activity, during which oxygen is released from the algae.

WRONG CHOICES EXPLAINED

(1) *Oxygen production decreases as the wavelength of light increases from 550 to 650 nm* is *not* the statement that is a valid inference based on this information. The data indicate that oxygen production increases, not decreases, in this region of the spectrum.

(2) *Respiration rate in the bacteria is greatest at 550 nm* is *not* the statement that is a valid inference based on this information. There are no data provided that would either support or refute this inference. It is likely that respiratory activity is unaffected by the wavelength of light present in the environment.

(4) *The algae absorb the greatest amount of oxygen in red light* is *not* the statement that is a valid inference based on this information. When they are photosynthesizing, algae release more oxygen than they absorb.

32. **1** The most probable reason for the increasing predator population from day 5 to day 7 is *an increasing food supply from day 5 to day 6*. As the yeast population increases, they provide more food opportunities for the paramecia. As the paramecia consume the yeast, they reproduce more rapidly and the newly formed paramecia consume still more yeast. As a result of this heavy predation, the number of yeast begins to decline on day 6, leaving less food available to the paramecia, whose population begins to level off on day 7 and to decline on day 8.

WRONG CHOICES EXPLAINED

(2) It is *not* true that the most probable reason for the increasing predator population from day 5 to day 7 is *a predator population equal in size to the prey population from day 5 to day 6*. The graph indicates that the population of yeast far exceeded that of the paramecia on days 5 and 6 and so could not have equaled it at the same time.

(3) It is *not* true that the most probable reason for the increasing predator population from day 5 to day 7 is *the decreasing prey population from day 1 to day 2*. Although the graph does indicate a significant decline in the yeast population during this time, it is unrelated to the increase in the paramecium population from day 5 to day 7.

(4) It is *not* true that the most probable reason for the increasing predator population from day 5 to day 7 is *the extinction of the yeast on day 3*. The graph indicates that the population of yeast continued to flourish through day 16 of the experiment. Therefore, the yeast cannot be extinct.

33. **4** *D and F—Gas exchange* is the correct pairing of structures and function. Structure *D* is the cell membrane, which facilitates the exchange of dissolved gases such as oxygen and carbon dioxide into and out of the cell. Structure *F* is the lung of a human, which contains small, moist alveoli specialized for the exchange of these same gases into and out of the blood.

WRONG CHOICES EXPLAINED

(1) *A and G—Transmission of nerve impulses* is *not* the correct pairing of structures and function. Structure *A* is the cell nucleus, which contains the genetic information needed to operate the cell's biochemical mechanism and to produce new cells. Structure *G* is the stomach of a human, which is specialized to aid in the digestion of foods, especially proteins, for the body's needs.

(2) *B and E—Photosynthesis* is *not* the correct pairing of structures and function. Structure *B* is the mitochondrion of an animal cell, an organelle that contains the enzymes needed to carry on aerobic respiration. Structure *E* is the brain of a human, which is specialized to control the organism's conscious and unconscious behaviors.

(3) *C and H—Digestion of food* is *not* the correct pairing of structures and function. Structure *C* is the cell cytoplasm, which is the watery medium that suspends the cell's organelles and provides an environment for operation of the cell's metabolic activities. Structure *H* is the large intestine of a human, which is specialized to aid in the digestion of foods by consolidating waste materials and absorbing excess water back into the bloodstream.

34. **4** The section of the graph labeled *A* represents *a population at equilibrium*. Equilibrium is reached when a species population reaches the carrying capacity of the environment, which is the number of individuals of that species that can be sustained in a particular environment. The population's size is controlled by limiting factors (e.g., amount of available food, amount of sunlight, average temperature, available oxygen) within the environment. When the population grows too large and uses up available resources, its numbers are reduced until they once again fall within the carrying capacity of the environment.

WRONG CHOICES EXPLAINED

(1) It is *not* true that the section of the graph labeled *A* represents *biodiversity within the species*. Biodiversity refers to the variety of different species that inhabit a particular environment. The term cannot apply to a single species.

(2) It is *not* true that the section of the graph labeled *A* represents *nutritional relationships of the species*. A graphic representation of nutritional relationships resembles a web connecting organisms that either serve as food or consume other organisms.

(3) It is *not* true that the section of the graph labeled *A* represents *a population becoming extinct*. Extinction is the complete elimination of a species from Earth. The species population illustrated has grown from near zero to the carrying capacity of its environment and so is far from becoming extinct.

35. **2** Information in the table suggests that DNA functions *both inside and outside of the nucleus*. DNA that exists within the nucleus provides the instructions for the production of enzymes that operate the biochemical mechanisms of the cell. The fact that DNA is also located in mitochondria suggests that mitochondria may have been independent organisms at some point in time but became incorporated into living cells in a symbiotic relationship, thereby losing their independence in exchange for a secure living environment inside the cell.

WRONG CHOICES EXPLAINED

(1) Information in the table does *not* suggest that DNA functions *within cytoplasm and outside of the cell membrane*. No information is given in the table concerning the presence or absence of DNA in cytoplasm.

(3) Information in the table does *not* suggest that DNA functions *only within energy-releasing structures*. Information given in the table identifies both the mitochondrion and the nucleus as structures that contain DNA. The mitochondrion is an energy-releasing organelle, but the nucleus is not.

(4) Information in the table does *not* suggest that DNA functions *within cell vacuoles*. No information is given in the table concerning the presence or absence of DNA in cell vacuoles.

36. **1** The products would most likely contain *simple sugars*. Starch is a complex carbohydrate made up of many repeating units of glucose. When starch is broken down (digested) by the action of hydrolytic enzymes, the products that result are simple sugars, including glucose.

WRONG CHOICES EXPLAINED

(2), (3), (4) The products would *not* most likely contain *fats, amino acids*, or *minerals*. None of these materials make up the structure of a molecule of starch, so none of them will result as products when starch is broken down.

37. **3** The structure labeled X most likely represents *an enzyme*. Structure X is shown with a specialized section, known as the active site, that has a chemical shape that is complementary to the chemical shape of the starch molecule (substrate). In the second diagram, the enzyme and starch are shown as being linked at the active site in an enzyme-substrate complex. In the third diagram, the starch has been split into products through the catalyzing action of the enzyme, which remains unchanged by the reaction.

WRONG CHOICES EXPLAINED

(1) The structure labeled X does *not* most likely represent *an antibody*. A diagram illustrating an antibody would show structure X linking to an antigen molecule, not to a starch molecule.

(2) The structure labeled X does *not* most likely represent *a receptor molecule*. A diagram illustrating a receptor molecule would show structure X linking to a protein molecule, not to a starch molecule.

(4) The structure labeled X does *not* most likely represent *a hormone*. A diagram illustrating a hormone would show structure X linking to a receptor molecule, not to a starch molecule.

38. **3** The interaction of these hormones is an example of *a feedback mechanism*. The description given in the question is a classic example of the feedback loop in a living organism. By utilizing a series of feedback mechanisms, the body is kept in a state of dynamic equilibrium that promotes survival of the individual.

WRONG CHOICES EXPLAINED

(1) The interaction of these hormones is *not* an example of *DNA base substitution*. An illustration of this phenomenon would show the effects of a mutagenic agent as it alters the sequence of nitrogenous bases on a strand of DNA.

(2) The interaction of these hormones is *not* an example of *manipulation of genetic instructions*. An illustration of this phenomenon would show the technique of gene splicing, including the use of restriction enzymes to snip the DNA of the gene at specific points in order to insert it into the genome of another organism.

(4) The interaction of these hormones is *not* an example of *an antigen-antibody reaction*. An illustration of this phenomenon would show the body's reaction to the introduction of a foreign invader such as a bacterium or virus in which specific antibodies are produced to seek out and neutralize the antigens of these invaders.

39. **4** *The level of testosterone may start to decrease* is the statement that best describes what would most likely occur if the interaction is blocked between the pituitary and gland C, the site of meiosis in males. Gland C is the testis, the male organ responsible for the production and storage of sperm as well as for the production of the male sex hormone testosterone. The secretion of testosterone is stimulated by hormonal secretions from the pituitary gland.

WRONG CHOICES EXPLAINED

(1) *The level of progesterone would start to increase* is *not* the statement that best describes what would most likely occur if the interaction is blocked between the pituitary and gland C, the site of meiosis in males. Progesterone is a female hormone secreted by the cells of the corpus luteum, whose effect is to maintain the lining of the uterus intact through pregnancy. A female hormone would not be produced in the body of a male under these circumstances.

(2) *The pituitary would produce another hormone to replace hormone 3* is *not* the statement that best describes what would most likely occur if the interaction is blocked between the pituitary and gland *C*, the site of meiosis in males. The action of hormones in the body is highly specific. It is unlikely that the pituitary would produce an alternate hormone under these circumstances.

(3) *Gland A would begin to interact with hormone 3 to maintain homeostasis* is *not* the statement that best describes what would most likely occur if the interaction is blocked between the pituitary and gland *C*, the site of meiosis in males. Gland *A* is the thyroid gland, which produces the hormone thyroxin. Thyroxin has no known effect on the testis and would be unlikely to interact with it under these circumstances.

40. **2** *The cells of glands B and C contain different receptors than the cells of gland A* is the statement that best explains why hormone 1 influences the action of gland *A* but not that of gland *B* or *C*. Hormones are protein molecules with specific chemical shapes that are complementary to the chemical shapes of specific receptor molecules embedded in the cell membranes of the tissues targeted by those hormones. Hormone 1 is specifically designed to link to receptors on, and influence the actions of, cells in gland *A* (thyroid gland) but not those of cells in gland *B* (adrenal gland) or gland *C* (testis).

WRONG CHOICES EXPLAINED

(1) *Every activity in gland A is different from the activities in glands B and C* is *not* the statement that best explains why hormone 1 influences the action of gland *A* but not gland *B* or *C*. The living cells of every tissue must carry on certain functions in common, such as absorption, circulation, respiration, excretion, and regulation, among others.

(3) *Each gland contains cells that have different base sequences in their DNA* is *not* the statement that best explains why hormone 1 influences the action of gland *A* but not that of gland *B* or *C*. All somatic (body) cells of a human being contain identical complements of DNA in their nuclei. This is true for the cells in endocrine glands as well as for the cells in other body tissues.

(4) *The distance a chemical can travel is influenced by both pH and temperature* is *not* the statement that best explains why hormone 1 influences the action of gland *A* but not that of gland *B* or *C*. Hormones are carried throughout the body via the bloodstream. The distance hormones travel in the body is not influenced by factors such as pH or temperature.

41. **1** Since oxygen gas is being released, it can be inferred that the plant is *producing glucose*. Elodea is a green water plant commonly found in freshwater ponds and lakes throughout New York State. When elodea is exposed to sunlight, it carries on photosynthetic reactions that produce glucose and oxygen gas from carbon dioxide and water. It is logical to infer that, when oxygen bubbles are observed being secreted from elodea, glucose production is occurring in its leaves.

WRONG CHOICES EXPLAINED

(2), (4) It is *not* true that, since oxygen gas is being released, it can be inferred that the plant is *making protein* or *carrying on active transport*. While it is probable that the elodea is carrying on these functions as part of its normal metabolic activities, they are not directly associated with the release of oxygen gas, so this inference cannot be made from the observation.

(3) It is *not* true that, since oxygen gas is being released, it can be inferred that the plant is *releasing energy from water*. In photosynthetic reactions, water is split through the use of light energy during photolysis. For this reason, water cannot serve as a source of released energy in this process.

42. **2** *Carbon dioxide* is the substance that the plant most likely absorbed from the water for the process that produces oxygen gas. The process referenced is photosynthesis. When elodea is exposed to sunlight, it carries on photosynthetic reactions that produce glucose and oxygen gas from carbon dioxide and water according to the formula

$$6 \ CO_2 + 6 \ H_2O \rightarrow C_6H_{12}O_6 + 6 \ O_2$$

Both water and carbon dioxide are absorbed into the leaves of the elodea from the watery environment.

WRONG CHOICES EXPLAINED

(1) *Dissolved nitrogen* is *not* the substance that the plant most likely absorbed from the water for the process that produces oxygen gas. While dissolved nitrogen gas (N_2) may be absorbed from the water by the elodea, nitrogen does not enter into the chemical process of photosynthesis as described above.

(3), (4) *An enzyme* or *a hormone* is *not* a substance that the plant most likely absorbed from the water for the process that produces oxygen gas. These substances are manufactured within the cells of the elodea through the process of protein synthesis, not absorbed from the water.

PART B–2

43. One credit is allowed for correctly identifying *one* reproductive hormone and stating the role it plays in reproduction. Acceptable responses include, but are not limited to: [1]

- *Testosterone influences the formation of sperm cells.*
- *Testosterone influences the formation of gametes.*
- *Estrogen regulates female reproductive cycles.*

- *Estrogen builds the uterine lining for implantation of the developing embryo.*
- *Progesterone maintains the uterine lining during pregnancy.*
- *Follicle-stimulating hormone (FSH) controls the maturation of the ovarian follicle leading to ovulation.*
- *Luteinizing hormone (LH) stimulates the development of the corpus luteum from the cells of the ovarian follicle.*

44. One credit is allowed for correctly identifying the structure in the uterus where exchange of material between the mother and the developing fetus takes place. Acceptable responses include: [1]

- *placenta*

45. One credit is allowed for correctly identifying *one* harmful substance that can pass through this structure and describing the *negative* effect it can have on the fetus. Acceptable responses include, but are not limited to: [1]

- *drugs—fetal addiction*
- *alcohol—low birth weight*
- *alcohol—premature birth*
- *alcohol—brain damage*
- *alcohol—fetal alcohol syndrome*
- *nicotine—brain damage*
- *nicotine—low birth weight*
- *viruses—disease or birth defects*

46. One credit is allowed for correctly stating how decomposers aid in the flow of materials in an ecosystem. Acceptable responses include, but are not limited to: [1]

- *Decomposers recycle nutrients.*
- *Decomposers convert organic molecules to inorganic molecules.*
- *Decomposers recycle carbon, hydrogen, oxygen, nitrogen, and other elements for reuse by other organisms.*
- *Decomposers break down the bodies of dead plants and animals and convert their complex structures to simple inorganic materials that can be absorbed into the roots of plants and incorporated into new organic molecules in living plants and animals.*

47. One credit is allowed for correctly stating the relationship between the distance to the food source and the number of waggle runs in 15 seconds. Acceptable responses include, but are not limited to: [1]

- *The closer the food source, the more waggle runs in 15 seconds.*
- *Fewer waggles mean the food is farther away.*

- *As one variable increases, the other decreases.*
- *There is an inverse relationship between the distance to the food source and the number of waggle runs per 15 seconds.*

48. One credit is allowed for correctly explaining how waggle-dance behavior increases the reproductive success of bees. Acceptable responses include, but are not limited to: [1]

- *If bees have better access to food, they can produce more offspring.*
- *Finding food is easier; thus more bees can exist.*
- *Bees convert some of the food they eat into beeswax used to build the comb cells in which eggs are deposited, allowing more bees to develop successfully.*
- *Honey produced by bees from flower nectar is fed to developing bee larvae, helping to ensure their growth into adults.*

49. **1**   The number of waggle runs in 15 seconds for each of these species is most likely due to *behavioral adaptation as a result of natural selection*. The difference in waggle runs recorded for the giant honeybee and the Indian honeybee demonstrates that this trait varies even between these closely related honeybee species. It is likely that this behavior varied even more widely in ancestral honeybees. It is also likely that the waggle-dance behavior was selected for its positive effect on honeybee survival, with those bee colonies displaying the trait surviving periods of food scarcity and reproducing more successfully than those colonies lacking the trait.

WRONG CHOICES EXPLAINED

(2) The number of waggle runs in 15 seconds for each of these species is *not* most likely due to *replacement of one species by another as a result of succession*. "Succession" is a term that describes the replacement of one plant community by other, progressively more complex, plant communities until a stable climax community is established. An individual species replacing another in a habitat is known as interspecies competition, not succession.

(3) The number of waggle runs in 15 seconds for each of these species is *not* most likely due to *alterations in gene structure as a result of diet*. Gene alteration occurs by mutation of DNA that results from the effects of mutagenic agents such as chemicals and radiation. Diet is not considered to be a mutagenic agent.

(4) The number of waggle runs in 15 seconds for each of these species is *not* most likely due to *learned behaviors as a result of asexual reproduction*. Learned behaviors are those obtained by an animal during its lifetime. Learned behaviors cannot be passed on to the next generation by any reproductive method but must be relearned by each succeeding generation.

50. One credit is allowed for correctly labeling the axes with "Year" on the horizontal (x) axis and "pH level" on the vertical (y) axis.   [1]

51. One credit is allowed for marking an appropriate scale on the y-axis. [1]

52. One credit is allowed for plotting the data and circling and connecting the points.   [1]

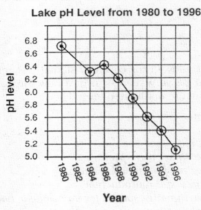

Lake pH Level from 1980 to 1996

53. One credit is allowed for correctly describing the trend in pH level in the lake over this 16-year period. Acceptable responses include, but are not limited to: [1]

- *The pH level decreased over this time period.*
- *The relative acidity increased over this time period.*

54. One credit is allowed for correctly identifying *one* factor that should have been kept constant each time water samples were collected from the lake. Acceptable responses include, but are not limited to: [1]

- *The samples should be taken at the same water depth.*
- *The samples should be taken at the same time of year.*
- *The samples should be taken under similar weather conditions.*
- *The samples should be taken from the same site.*
- *The technique used to measure pH should remain constant.*
- *The sampling containers should be free of all chemical contaminants for each sampling period.*

55. One credit is allowed for correctly explaining how, after exposure to this antibiotic, the population of one species of bacteria could increase while the population of the other species of bacteria decreased or was eliminated. Acceptable responses include, but are not limited to: [1]

- *Organisms with antibiotic resistance would survive and reproduce. Others without resistance would decrease in number.*
- *Antibiotics target specific bacteria species.*
- *Different bacterial species have different genetic characteristics, so one species may be more resistant to a particular antibiotic than another species.*

PART C

56. One credit is allowed for correctly stating *one* reason why an introduced species might be very successful in a new environment. Acceptable responses include, but are not limited to: [1]

- *No competition*
- *No predators*
- *Large food supply*
- *Bigger/stronger than other species*
- *Better at competing for limited resources*
- *Faster/more successful reproduction*
- *May compete with native species for an available habitat*
- *May increase competition for food with native species*

57. One credit is allowed for correctly identifying *one* action the government could take to prevent the introduction of additional new species. Acceptable responses include, but are not limited to: [1]

- *Pass laws to ban importation of invasive species.*
- *Enforce existing laws that prohibit dumping of water ballast containing non-native species.*
- *Conduct inspections of cargo ships/trucks to ensure that they contain no imported species.*
- *Increase public knowledge about the problem.*
- *Fund projects that will reduce/eliminate invasive species in affected areas.*

58. One credit is allowed for identifying *one* introduced organism, writing its name in the space provided, and describing one way in which this organism has altered an ecosystem in the new location. Acceptable responses include, but are not limited to: [1]

- *Purple loosestrife—crowds out native plants such as the cattail*

- *Zebra mussel—outcompetes native species*
- *Brown tree snake—eats birds' eggs and reduces bird populations*
- *Gypsy moth—eats oak leaves and can kill oak trees*
- *Elm bark beetle—infects elm trees with Dutch elm disease; kills elm trees*
- *English sparrow—competes with native bird species for nesting space*
- *Starling—lays eggs in the nests of native bird species; chicks outcompete native offspring for food/nest space*
- *Sea lamprey—preys on native fish species in the Great Lakes and feeder streams*
- *Killer bee—competes for native honeybee habitats; extremely aggressive when disturbed*

59. Two credits are allowed for correctly identifying *two* specific populations that most likely increased in number after the mountain lion population *decreased* and for supporting your answer. Acceptable responses include, but are not limited to: [2]

- *Rabbit and deer—they have fewer predators.*
- *Two species shown in the food web as prey for mountain lions are the deer and the rabbit. If the mountain lion population were decreased, it is likely that the deer and rabbit populations would increase because less predation would occur.*
- *Rabbit and hawk populations would increase. There would be more rabbits because there would be fewer mountain lions to hunt them. There would be more hawks because there would be more rabbits for them to eat and their reproductive rate would increase.*

60. One credit is allowed for correctly explaining how killing many mountain lions affected other ranchers in the community. Acceptable responses include, but are not limited to: [1]

- *Their grain crops decreased.*
- *Fewer cattle were killed.*
- *Rabbit and deer populations became nuisance species.*

61. One credit is allowed for correctly stating the problem the team of doctors was trying to solve. Acceptable responses include, but are not limited to: [1]

- *How is the yellow fever microbe transferred from person to person?*
- *How is yellow fever spread?*

62. One credit is allowed for correctly stating *one* hypothesis from paragraph *A* that was tested by one of the experiments. Acceptable responses include, but are not limited to: [1]

- *Yellow fever is spread by contact with the clothing of people who have yellow fever.*

63. One credit is allowed for describing the control that should have been set up for the experiment described in paragraph *C*. Acceptable responses include, but are not limited to: [1]

- *The control should have been a group of people sleeping in nightshirts.*
- *The control should have been the use of bedding that had not been used by yellow fever patients.*

64. One credit is allowed for correctly explaining why the use of native fish (described in paragraph *E*), rather than the use of pesticides, is less likely to have a *negative* impact on the environment. Acceptable responses include, but are not limited to: [1]

- *Pesticides can harm other parts of the environment (other species), but native fish will not.*
- *Native species will target the larvae with less disruption of food chains.*
- *Pesticides may disrupt the food chains in the area, but native fish will not.*
- *Pesticides may cause human illness.*

65. One credit is allowed for correctly describing the contents of a vaccine. Acceptable responses include, but are not limited to: [1]

- *A vaccine contains dead pathogens.*
- *A vaccine contains weakened pathogens.*
- *A vaccine contains the products (membranes or antigens) of pathogens.*

66. One credit is allowed for identifying the system in the body that is most directly affected by a vaccination. Acceptable responses include: [1]

- *Immune system*

67. One credit is allowed for correctly explaining how vaccination results in the long-term ability of the body to resist disease. Acceptable responses include, but are not limited to: [1]

- *White blood cells produce antibodies for a particular pathogen.*
- *White blood cells are prepared to recognize a particular pathogen in the future.*
- *Vaccination causes the immune system to produce antibodies.*
- *Vaccination stimulates an immune response.*

68. One credit is allowed for correctly stating *one* way these toxins could move from the soil into local ecosystems, such as nearby lakes and ponds. Acceptable responses include, but are not limited to: [1]

- *Rain may wash the toxins into lakes.*
- *Toxins may seep into groundwater.*

69. One credit is allowed for correctly stating *one* way these toxins might affect local ecosystems. Acceptable responses include, but are not limited to: [1]

- *They may move through the food web.*
- *They may change the pH of pond water.*
- *They may kill organisms.*

70. One credit is allowed for selecting *one* of the four major problems from the diagram, recording the number of the problem on the line provided, identifying a gas that contributes to the problem, and stating *one* way in which the amount of this gas can be reduced. Acceptable responses include, but are not limited to: [1]

1	2	3	4
hydrocarbons	sulfur dioxide	carbon dioxide	chlorofluorocarbons
reduce pesticide use reduce auto exhaust	reduce burning of fossil fuel	reduce car use reduce deforestation	use alternatives to chlorofluorocarbons

71. One credit is allowed for correctly explaining why damage to the ozone shield is considered a threat to many organisms. Acceptable responses include, but are not limited to: [1]

- *It exposes organisms to UV rays.*
- *It increases the chance of mutations in cells.*

PART D

72. **2** This technique is used to *separate molecules in a mixture.* The technique illustrated is known as paper chromatography. A concentrated spot of plant extract containing different plant pigments is placed on the lower portion of the chromatography strip. The bottom of the strip is then placed in the solvent, which wicks up the strip, carrying the different pigments up the strip at different rates depending on their molecular size and other characteristics.

WRONG CHOICES EXPLAINED

(1) This technique is *not* used to *determine volume*. Equipment used for this task might include a graduated cylinder, beaker, or flask.

(3) This technique is *not* used to *measure length*. Equipment used for this task might include a meterstick.

(4) This technique is *not* used to *analyze data from an experiment*. Equipment used for this task might include a computer or a graphing calculator.

73. **3** Graph 3 has the graphed line that best represents these data. The graphed line, like the data in the table, gradually decreases, then increases nearly to the starting point. To properly graph this data set, the horizontal (x) axis should be labeled "Hour" and numbered 1 through 7; the vertical (y) axis should be labeled "Heart Rate" and numbered 10 through 80 (increments of 10).

WRONG CHOICES EXPLAINED

(1), (4) Graphs *1* and *4* do *not* have the graphed lines that best represent these data. Neither of these graphs has points that have been placed on the grid correctly.

(2) Graph 2 does *not* have the graphed line that best represents these data. In this graph, the labels and numbering of the axes have been reversed from the method described above. By convention, time is always marked on the horizontal (x) axis of a line graph.

74. One credit is allowed for correctly filling in the missing amino acid sequence for species A in the chart. Acceptable responses include: [1]

- *MET-ALA-GLY-SER*
- *START-ALA-GLY-SER*

75. One credit is allowed for correctly filling in the missing mRNA bases in the mRNA strand for species B in the chart. Acceptable responses include: [1]

- *AUG-AAA-CGU-CCU*

76. One credit is allowed for correctly filling in the missing DNA bases in the DNA strand for species C in the chart. Acceptable responses include: [1]

- *TAC-AAA-ACA-GGG*

77. One credit is allowed for correctly stating which two species are most closely related and supporting your answer. Acceptable responses include, but are not limited to: [1]

- *Species B and E—Their amino acid sequences are the same.*
- *Species B and E—Their DNA codes result in the same amino acid sequence in the protein strand.*
- *Species B and E—Their mRNA codons attract and link the same amino acids in the same order.*

78. **2**   *Available food sources* is the factor that most directly influenced the evolution of the diverse types of beaks of these finches. When the ancestral finches arrived at the remote Galapagos Islands, it is likely that they were adapted to abundant food sources on the mainland of South America. The variations in beak size and shape present in these ancestral finches allowed the species to diversify and take advantage of available feeding niches on the islands. Over time, these variants separated themselves into distinct finch species that differ in beak shape and size, among other characteristics.

WRONG CHOICES EXPLAINED

(1) *Predation by humans* is *not* the factor that most directly influenced the evolution of the diverse types of beaks of these finches. Humans were not present during the period of time when the finches were diversifying in the Galapagos. In the early 19th century, Charles Darwin was among the first European naturalists to visit and document the species inhabiting the Galapagos Islands, including the finches.

(3) *Oceanic storms* is *not* the factor that most directly influenced the evolution of the diverse types of beaks of these finches. While it is likely that oceanic storms were responsible for allowing the finches to reach the Galapagos from the South American mainland, it is unlikely that these storms had any influence on the evolutionary history of the finch species in the Galapagos.

(4) *Lack of available niches* is *not* the factor that most directly influenced the evolution of the diverse types of beaks of these finches. In fact, the finches moved readily into a number of available feeding niches on the Galapagos Islands, as described in the chart. It is unknown whether these niches had been occupied by native species prior to the finches' arrival.

79. One credit is allowed for correctly stating *one* reason why the large tree finch and the large ground finch are able to coexist on the same island. Acceptable responses include, but are not limited to: [1]

- *The large tree finch eats mainly animal food, while the large ground finch eats mainly plant food.*
- *They occupy different environmental niches.*

- *They eat different kinds of food.*
- *They have different-shaped bills that are adapted to deal with different types of food.*

80. **2** The appearance of the pink color was due to the movement of *base molecules from high concentration through a membrane to low concentration.* The diagram illustrates an experimental setup that allows a concentrated base (ammonium hydroxide) to vaporize and fill the bottle with molecules mixed with air. As these molecules collide with the dialysis membrane, some pass through the membrane by diffusion (from high relative concentration to low relative concentration). As these molecules cross the membrane boundary, they react with the phenolphthalein and cause it to turn pink, indicating that they have dissolved in the phenolphthalein solution.

WRONG CHOICES EXPLAINED
(1) The appearance of the pink color was *not* due to the movement of *phenolphthalein molecules from low concentration to high concentration.* Movement from low to high concentration requires active transport, which is not readily achievable in a laboratory experiment involving nonliving components.
(3) The appearance of the pink color was *not* due to the movement of *water molecules through the membrane from high concentration to low concentration.* The movement of water through membranes would not produce a reaction in the phenolphthalein.
(4) The appearance of the pink color was *not* due to the movement of *phenolphthalein molecules in the water from high concentration to low concentration.* Phenolphthalein molecules are too large to pass through a dialysis tubing membrane.

81. One credit is allowed for correctly identifying the process that is responsible for the change in mass of each of the three slices. Acceptable responses include, but are not limited to: [1]

- *Diffusion*
- *Osmosis*
- *Passive transport*

82. One credit is allowed for correctly explaining why the potato slice in beaker 1 increased in mass. Acceptable responses include, but are not limited to: [1]

- *Water diffused into the cells of the potato because there was a higher concentration of water outside than inside the slice.*
- *The potato slice increased in water content.*

83. **3** *Using the coarse adjustment to focus the specimen under high power* is the activity that might lead to damage of a microscope and specimen. This action can easily crush a coverslip and specimen and even break a slide. Only the fine adjustment should be used when the high-power objective lens is in place.

WRONG CHOICES EXPLAINED
(1), (2), (4) *Cleaning the ocular and objectives with lens paper, focusing with low power first before moving the high power into position,* and *adjusting the diaphragm to obtain more light under high power* are *not* the activities that might lead to damage of a microscope and specimen. These are all proper techniques to employ when using a microscope to view a specimen on a slide.

84. One credit is allowed for correctly identifying *one* specific substance that should have been added to the distilled water so that observations regarding the movement of starch could be made. Acceptable responses include, but are not limited to: [1]

- *Starch indicator*
- *Iodine solution*
- *Lugol's solution*

STANDARDS/KEY IDEAS	AUGUST 2008 QUESTION NUMBERS	NUMBER OF CORRECT RESPONSES
STANDARD 1		
Key Idea 1: The central purpose of scientific inquiry is to develop explanations of natural phenomena in a continuing and creative process.		
Key Idea 2: Beyond the use of reasoning and consensus, scientific inquiry involves the testing of proposed explanations involving the use of conventional techniques and procedures and usually requiring considerable ingenuity.	54, 61, 62, 63	
Key Idea 3: The observations made while testing proposed explanations, when analyzed using conventional and invented methods, provide new insights into natural phenomena.	31, 47	
Laboratory Checklist	50, 51, 52	
STANDARD 4		
Key Idea 1: Living things are both similar to and different from each other and from nonliving things.	2, 3, 5, 8, 9, 33, 36, 40, 59, 60	
Key Idea 2: Organisms inherit genetic information in a variety of ways that result in continuity of structure and function between parents and offspring.	6, 7, 10, 12, 13, 18, 19, 35	
Key Idea 3: Individual organisms and species change over time.	11, 15, 16, 17, 48, 49, 55	
Key Idea 4: The continuity of life is sustained through reproduction and development.	4, 21, 39, 43, 44, 45	
Key Idea 5: Organisms maintain a dynamic equilibrium that sustains life.	20, 22, 23, 27, 37, 38, 41, 42, 65, 66, 67	
Key Idea 6: Plants and animals depend on each other and their physical environment.	1, 14, 26, 28, 30, 32, 34, 46	

STANDARDS/KEY IDEAS	AUGUST 2008 QUESTION NUMBERS	NUMBER OF CORRECT RESPONSES
STANDARD 4		
Key Idea 7: Human decisions and activities have a profound impact on the physical and living environment.	24, 25, 29, 56, 57, 58, 64, 68, 69, 70, 71	
REQUIRED LABORATORIES		
Lab 1: "Relationships and Biodiversity"	72, 74, 75, 76, 77	
Lab 2: "Making Connections"	73	
Lab 3: "The Beaks of Finches"	78, 79	
Lab 5: "Diffusion Through a Membrane"	80, 81, 82, 83, 84	

Examination
June 2009
Living Environment

PART A

Answer all 30 questions in this part. [30]

Directions (1–30): For *each* statement or question, select the word or expression that, of those given, best completes the statement or answers the question. Record your answers in the spaces provided.

1 Which statement best describes one of the stages represented in the diagram below?

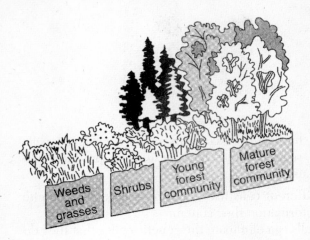

1 The mature forest will most likely be stable over a long period of time.
2 If all the weeds and grasses are destroyed, the number of carnivores will increase.
3 As the population of the shrubs increases, it will be held in check by the mature forest community.
4 The young forest community will invade and take over the mature forest community.   1_____

2 Which organ system in humans is most directly involved in the transport of oxygen?

1 digestive   3 excretory
2 nervous   4 circulatory   2_____

3 Which cell structure contains information needed for protein synthesis?

(1)
(2)
(3)
(4)

4 The human liver contains many specialized cells that secrete bile. Only these cells produce bile because

 1 different cells use different parts of the genetic information they contain
 2 cells can eliminate the genetic codes that they do not need
 3 all other cells in the body lack the genes needed for the production of bile
 4 these cells mutated during embryonic development

5 Although identical twins inherit exact copies of the same genes, the twins may look and act differently from each other because

 1 a mutation took place in the gametes that produced the twins.
 2 the expression of genes may be modified by environmental factors
 3 the expression of genes may be different in males and females
 4 a mutation took place in the zygote that produced the twins

6 Which hormone does *not* directly regulate human reproductive cycles?

   1 testosterone   3 insulin

   2 estrogen   4 progesterone   6_____

7 Owls periodically expel a mass of undigested material known as a pellet. A student obtained several owl pellets from the same location and examined the animal remains in the pellets. He then recorded the number of different prey animal remains in the pellets. The student was most likely studying the

   1 evolution of the owl

   2 social structure of the local owl population

   3 role of the owl in the local ecosystem

   4 life cycle of the owl   7_____

8 Which sequence best represents the relationship between DNA and the traits of an organism?

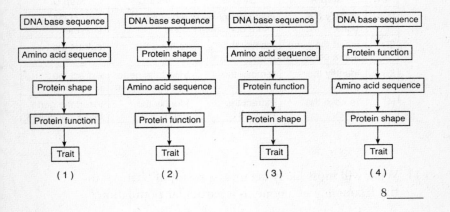

                                                                 8_____

9 A sequence of events associated with ecosystem stability is represented below.

sexual reproduction → genetic variation →
biodiversity → ecosystem stability

The arrows in this sequence should be read as

1 leads to   3 prevents
2 reduces   4 simplifies   9_____

10 In some people, the lack of a particular enzyme causes a disease. Scientists are attempting to use bacteria to produce this enzyme for the treatment of people with the disease. Which row in the chart below best describes the sequence of steps the scientists would most likely follow?

Row	Step A	Step B	Step C	Step D
(1)	identify the gene	insert the gene into a bacterium	remove the gene	extract the enzyme
(2)	insert the gene into a bacterium	identify the gene	remove the gene	extract the enzyme
(3)	identify the gene	remove the gene	insert the gene into a bacterium	extract the enzyme
(4)	remove the gene	extract the enzyme	identify the gene	insert the gene into a bacterium

10_____

11 What will most likely occur as a result of changes in the frequency of a gene in a particular population?

1 ecological succession
2 biological evolution
3 global warming
4 resource depletion

11_____

12 The puppies shown in the photograph below are all from the same litter.

The differences seen within this group of puppies are most likely due to

1 overproduction and selective breeding
2 mutations and elimination of genes
3 evolution and asexual reproduction
4 sorting and recombination of genes   12_____

13 Carbon dioxide makes up less than 1 percent of Earth's atmosphere, and oxygen makes up about 20 percent. These percentages are maintained most directly by

  1 respiration and photosynthesis
  2 the ozone shield
  3 synthesis and digestion
  4 energy recycling in ecosystems   13_____

14 Which sequence represents the order of some events in human development?

  1 zygote → sperm → tissues → egg
  2 fetus → tissues → zygote → egg
  3 zygote → tissues → organs → fetus
  4 sperm → zygote → organs → tissues   14_____

15 A variety of plant produces small white fruit. A stem was removed from this organism and planted in a garden. If this stem grows into a new plant, it would most likely produce

  1 large red fruit, only
  2 large pink fruit, only
  3 small white fruit, only
  4 small red and small white fruit on the same plant   15_____

16 A mutation that can be inherited by offspring would result from

  1 random breakage of chromosomes in the nucleus of liver cells
  2 a base substitution in gametes during meiosis
  3 abnormal lung cells produced by toxins in smoke
  4 ultraviolet radiation damage to skin cells   16_____

17 The diagram below represents a process that occurs in organisms.

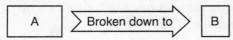

Which row in the chart indicates what A and B in the boxes could represent?

Row	A	B
(1)	starch	proteins
(2)	starch	amino acids
(3)	protein	amino acids
(4)	protein	simple sugars

17_____

18 Some organs of the human body are represented in the diagram below.

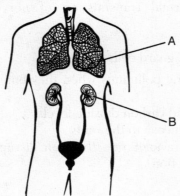

Which statement best describes the functions of these organs?

1 *B* pumps blood to *A* for gas exchange.
2 *A* and *B* both produce carbon dioxide, which provides nutrients for other body parts.
3 *A* releases antibodies in response to an infection in *B*.
4 The removal of wastes from both *A* and *B* involves the use of energy from ATP.

18_____

19 *Salmonella* bacteria can cause humans to have stomach cramps, vomiting, diarrhea, and fever. The effect these bacteria have on humans indicates that *Salmonella* bacteria are

  1 predators
  2 pathogenic organisms
  3 parasitic fungi
  4 decomposers   19_____

20 The virus that causes AIDS is damaging to the body because it

  1 targets cells that fight invading microbes
  2 attacks specific red blood cells
  3 causes an abnormally high insulin level
  4 prevents the normal transmission of nerve impulses   20_____

21 In the leaf of a plant, guard cells help to

  1 destroy atmospheric pollutants when they enter the plant
  2 regulate oxygen and carbon dioxide levels
  3 transport excess glucose to the roots
  4 block harmful ultraviolet rays that can disrupt chlorophyll production   21_____

22 An antibiotic is effective in killing 95% of a population of bacteria that reproduce by the process shown below.

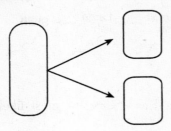

Which statement best describes future generations of these bacteria?

1 They will be produced by asexual reproduction and will be more resistant to the antibiotic.
2 They will be produced by sexual reproduction and will be more resistant to the antibiotic.
3 They will be produced by asexual reproduction and will be just as susceptible to the antibiotic.
4 They will be produced by sexual reproduction and will be just as susceptible to the antibiotic.   22_____

23 The size of plant populations can be influenced by the

1 molecular structure of available oxygen
2 size of the cells of decomposers
3 number of chemical bonds in a glucose molecule
4 type of minerals present in the soil   23_____

24 Competition between two species occurs when

1 mold grows on a tree that has fallen in the forest
2 chipmunks and squirrels eat sunflower seeds in a garden
3 a crow feeds on the remains of a rabbit killed on the road
4 a lion stalks, kills, and eats an antelope   24_____

25 A food chain is illustrated below.

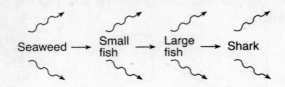

The arrows represented as ∼∼∼→ most likely indicate

1 energy released into the environment as heat
2 oxygen produced by respiration
3 the absorption of energy that has been synthesized
4 the transport of glucose away from the organism   25_____

26 If several species of carnivores are removed from an ecosystem, the most likely effect on the ecosystem will be

1 an increase in the kinds of autotrophs
2 a decrease in the number of abiotic factors
3 a decrease in stability among populations
4 an increase in the rate of succession   26_____

27 Some people make compost piles consisting of weeds and other plant materials. When the compost has decomposed, it can be used as fertilizer. The production and use of compost is an example of

1 the introduction of natural predators
2 the use of fossil fuels
3 the deforestation of an area
4 the recycling of nutrients   27_____

28 Which statement best describes a chromosome?

  1 It is a gene that has thousands of different forms.
  2 It has genetic information contained in DNA.
  3 It is a reproductive cell that influences more than one trait.
  4 It contains hundreds of genetically identical DNA molecules.   28_____

29 The graph below shows how the level of carbon dioxide in the atmosphere has changed over the last 150,000 years.

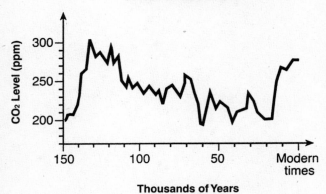

Carbon Dioxide Level

Thousands of Years

Which environmental factor has been most recently affected by these changes in carbon dioxide level?

  1 light intensity
  2 types of decomposers
  3 size of consumers
  4 atmospheric temperature   29_____

30 One reason why people should be aware of the impact of their actions on the environment is that

1 ecosystems are never able to recover once they have been adversely affected
2 the depletion of finite resources cannot be reversed
3 there is a decreased need for new technology
4 there is a decreased need for substances produced by natural processes

30 _____

PART B-1

Answer all question in this part.   [11]

Directions (31–41): For *each* statement or question, select the word or expression that, of those given, best completes the statement or answers the question. Record your answers in the spaces provided.

31 The diagram below represents the process used in 1996 to clone the first mammal, a sheep named Dolly.

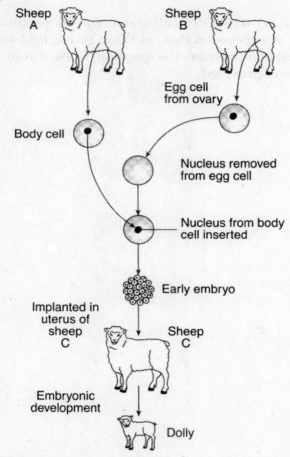

Which statement concerning Dolly is correct?

1 Gametes from sheep *A* and sheep *B* were united to produce Dolly.

2 The chromosome makeup of Dolly is identical to that of sheep *A*.

3 Both Dolly and sheep *C* have identical DNA.

4 Dolly contains genes from sheep *B* and sheep *C*.

31_____

32 The diagram below represents a cell.

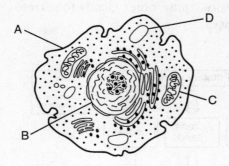

Which statement concerning ATP and activity within the cell is correct?

1 The absorption of ATP occurs at structure A.
2 The synthesis of ATP occurs within structure B.
3 ATP is produced most efficiently by structure C.
4 The template for ATP is found in structure D.   32_____

33 The diagram below illustrates some functions of the pituitary gland. The pituitary gland secretes substances that, in turn, cause other glands to secrete different substances.

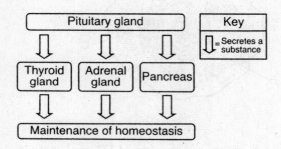

Which statement best describes events shown in the diagram?

1 Secretions provide the energy needed for metabolism.
2 The raw materials for the synthesis of secretions come from nitrogen.
3 The secretions of all glands speed blood circulation in the body.
4 Secretions help the body to respond to changes from the normal state.

33_____

34 A pond ecosystem is shown in the diagram below.

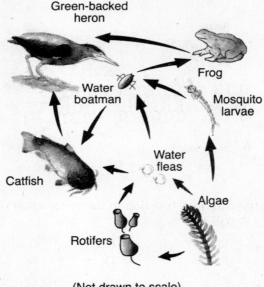

(Not drawn to scale)

Which statement describes an interaction that helps maintain the dynamic equilibrium of this ecosystem?

1 The frogs make energy available to this ecosystem through the process of photosynthesis.
2 The algae directly provide food for both the rotifers and the catfish.
3 The green-backed heron provides energy for the mosquito larvae.
4 The catfish population helps control the populations of water boatman and water fleas.

34_____

35 The diagram below represents a portion of a cell membrane.

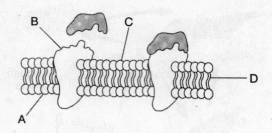

Which structure may function in the recognition of chemical signals?

(1) *A*   (3) *C*

(2) *B*   (4) *D*   35_____

36 Some evolutionary pathways are represented in the diagram below.

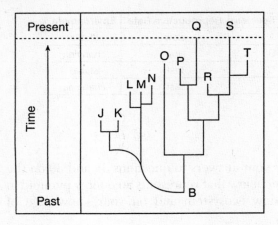

An inference that can be made from information in the diagram is that

1 many of the descendants of organism *B* became extinct

2 organism *B* was probably much larger than any of the other organisms represented

3 most of the descendants of organism *B* successfully adapted to their environment and have survived to the present time

4 the letters above organism *B* represent members of a single large population with much biodiversity

36_____

37 Which species in the chart below is most likely to have the fastest rate of evolution?

Species	Reproductive Rate	Environment
A	slow	stable
B	slow	changing
C	fast	stable
D	fast	changing

(1) A
(2) B
(3) C
(4) D

37_____

Base your answers to questions 38 and 39 on the diagram below that represents an energy pyramid in a meadow ecosystem and on your knowledge of biology.

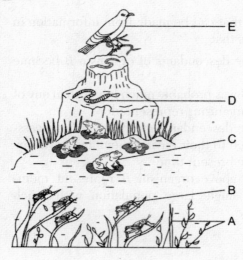

38 Which species would have the largest amount of available energy in this ecosystem?

(1) A
(2) B
(3) C
(4) D

38_____

39 Which two organisms are carnivores?

(1) *A* and *B*   (3) *B* and *D*

(2) *A* and *E*   (4) *C* and *E*   39_____

40 The kit fox and red fox species are closely related. The kit fox lives in the desert, while the red fox inhabits forests. Ear size and fur color are two differences that can be observed between the species. An illustration of these two species is shown below.

Kit Fox   Red Fox

Which statement best explains how the differences between these two species came about?

1 Different adaptations developed because the kit fox preferred hotter environments than the red fox.

2 As the foxes adapted to different environments, differences in appearance evolved.

3 The foxes evolved differently to prevent overpopulation of the forest habitat.

4 The foxes evolved differently because their ancestors were trying to avoid competition.   40_____

41 An ecosystem is represented below.

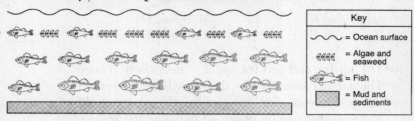

The organisms represented as ꙮꙮ are found in the area shown due to which factor?

1 pH

2 sediment

3 light intensity

4 colder temperature   41_____

PART B–2

Answer all questions in this part.   [14]

Directions **(42–51): For those questions that are followed by four choices, record your answers in the spaces provided. For all other questions in this part, record your answers in accordance with the directions given in the question.**

42 The graphs below show dissolved oxygen content, sewage waste content, and fish populations in a lake between 1950 and 1970.

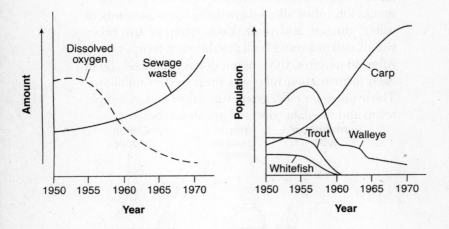

State what happened to the amount of dissolved oxygen and the number of fish species as the amount of sewage waste increased.   [1]

_____

_____

Base your answers to questions 43 through 46 on the information below and on your knowledge of biology.

Yeast cells carry out the process of cellular respiration as shown in the equation below.

$$C_6H_{12}O_6 \rightarrow 2C_2H_5OH + 2CO_2$$

glucose   ethyl   carbon
          alcohol   dioxide

An investigation was carried out to determine the effect of temperature on the rate of cellular respiration in yeast. Five experimental groups, each containing five fermentation tubes, were set up. The fermentation tubes all contained the same amounts of water, glucose, and yeast. Each group of five tubes was placed in a water bath at a different temperature. After 30 minutes, the amount of gas produced (D) in each fermentation tube was measured in milliliters. The average for each group was calculated. A sample setup and the data collected are shown below.

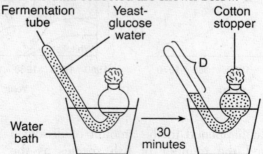

Average Amount of Gas Produced (D)
After 30 Minutes at Various Temperatures

Group	Temperature (°C)	D (mL)
1	5	0
2	20	5
3	40	12
4	60	6
5	80	3

Directions (43 and 44): Using the information in the data table, construct a line graph on the grid below, following the directions below.

43 Mark an appropriate scale on each labeled axis.   [1]

44 Plot the data from the data table. Surround each point with a small circle, and connect the points. [1]

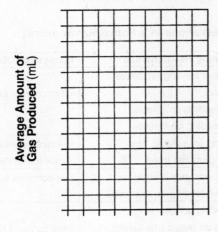

Example:

**Average Amount of Gas Produced
at Various Temperatures**

Average Amount of
Gas Produced (mL)

Temperature (°C)

45 The maximum rate of cellular respiration in yeast occurred at which temperature?

(1) 5°C
(2) 20°C
(3) 40°C
(4) 60°C

45 _____

46 Compared to the other tubes at the end of 30 minutes, the tubes in group 3 contained the

1 smallest amount of CO_2
2 smallest amount of glucose
3 smallest amount of ethyl alcohol
4 same amounts of glucose, ethyl alcohol, and CO_2 46_____

Base your answers to questions 47 through 49 on the information below and on your knowledge of biology.

An ecologist made some observations in a forest ecosystem over a period of several days. Some of the data collected are shown in the table below.

Observations in a Forest Environment

Date	Observed Feeding Relationships	Ecosystem Observations
6/2	• white-tailed deer feeding on maple tree leaves • woodpecker feeding on insects • salamander feeding on insects	• 2 cm of rain in 24 hours
6/5	• fungus growing on a maple tree • insects feeding on oak trees	• several types of sedimentary rock are in the forest
6/8	• woodpecker feeding on insects • red-tailed hawk feeding on chipmunk	• air contains 20.9% oxygen
6/11	• chipmunk feeding on insects • insect feeding on maple tree leaves • chipmunk feeding on a small salamander	• soil contains phosphorous

47 On the diagram below, complete the food web by placing the names of *all* the organisms in the correct locations.   [1]

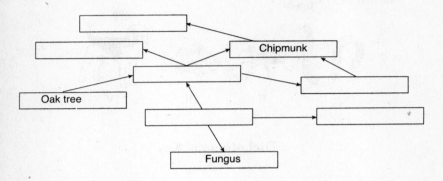

48 Identify *one* producer recorded by the ecologist in the data table.   [1]

_____

49 Which statement describes how one biotic factor of the forest uses one of the abiotic factors listed in the data table?

1 Trees absorb water as a raw material for photosynthesis.
2 Insects eat and digest the leaves of trees.
3 Erosion of sedimentary rock adds phosphorous to the soil.
4 Fungi release oxygen from the trees back into the air.   49_____

50 Fill in all of the blanks in parts 2 and 3 of the dichotomous key below, so that it contains information that could be used to identify the four animals shown below.   [2]

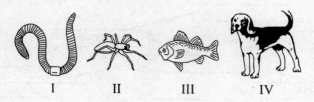

I   II   III   IV

Dichotomous Key

1. a. Legs present ....Go to 2
   b. Legs not present...............................Go to 3

   Characteristic   **Organism**

2. a. _____ ...._____

   b. _____ ...._____

3. a. _____ ...._____

   b. _____ ...._____

51 The human female reproductive system is represented in the diagram below.

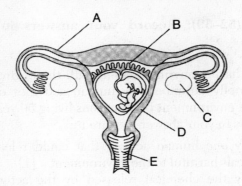

Complete boxes 1 through 4 in the chart below using the information from the diagram. [4]

Name of Structure	Letter on Diagram	Function of Structure
1 _____	2 _____	produces gametes
uterus	D	3 _____
4 _____	B	transports oxygen directly to the embryo

PART C

Answer all questions in this part. [17]

Directions (52–59): **Record your answers in the spaces provided.**

52 Humans have many interactions with the environment. Briefly describe how human activities can affect the environment of organisms living 50 years from now. In your answer, be sure to:

- identify *one* human activity that could release chemicals harmful to the environment [1]
- identify the chemical released by the activity [1]
- state *one* effect the release of this chemical would most likely have on future ecosystems [1]
- state *one* way in which humans can reduce the production of this chemical to lessen its effect on future ecosystems [1]

_____

_____

_____

_____

_____

_____

_____

_____

_____

53 Plants respond to their environment in many different ways. Design an experiment to test the effects of *one* environmental factor, chosen from the list below, on plant growth.

> Acidity of precipitation
> Temperature
> Amount of water

In your answer, be sure to:
- identify the environmental factor you chose
- state *one* hypothesis the experiment would test [1]
- state how the control group would be treated differently from the experimental group [1]
- state *two* factors that must be kept the same in both the experimental and control groups [1]
- identify the independent variable in the experiment [1]
- label the columns on the data table below for the collection of data in your experiment [1]

Environmental factor: _____

_____

_____

_____

_____

_____

_____

_____

_____

_____

Data Table

Base your answer to question 54 on the article below and on your knowledge of biology.

Power plan calls for windmills off beach

The Associated Press

Several dozen windmills taller than the Statue of Liberty will crop up off Long Island — the first source of off-shore wind power outside of Europe, officials said.

The Long Island Power Authority [LIPA] expects to choose a company to build and operate between 35 and 40 windmills in the Atlantic Ocean off Jones Beach, The New York Times reported Sunday [May 2, 2004]. Cost and completion date are unknown.

Energy generated by the windmills would constitute about 2 percent of LIPA's total power use. They are expected to produce 100 to 140 megawatts, enough to power 30,000 homes. . . . .

But some Long Island residents oppose the windmills, which they fear will create noise, interfere with fishing, and mar ocean views. . . . .

Source: "Democrat and Chronicle", Rochester, NY 5/3/04

54 State *two* ways that the use of windmills to produce energy would be beneficial to the environment.   [2]

1 _____

2 _____

55 Importing a foreign species, either intentionally or by accident, can alter the balance of an ecosystem. State *one* specific example of an imported species that has altered the balance of an ecosystem and explain how it has disrupted the balance in that ecosystem.   [2]

_____

_____

_____

Base your answers to questions 56 through 59 on the passage below and on your knowledge of biology.

Avian (Bird) Flu

Avian flu virus H5N1 has been a major concern recently. Most humans have not been exposed to this strain of the virus, so they have not produced the necessary protective substances. A vaccine has been developed and is being made in large quantities. However, much more time is needed to manufacture enough vaccine to protect most of the human population of the world.

Most flu virus strains affect the upper respiratory tract, resulting in a runny nose and sore throat. However, the H5N1 virus seems to go deeper into the lungs and causes severe pneumonia, which may be fatal for people infected by this virus.

So far, this virus has not been known to spread directly from one human to another. As long as H5N1 does not change to another strain that can be transferred from one human to another, a world-wide epidemic of the virus probably will not occur.

56 State *one* difference between the effect on the human body of the usual forms of flu virus and the effect of H5N1.   [1]

_____

_____

57 Identify the type of substance produced by the human body that protects against antigens, such as the flu virus.   [1]

_____

58 State what is in a vaccine that makes the vaccine effective. [1]

_____

59 Identify *one* event that could result in the virus changing to a form able to spread from human to human. [1]

_____

PART D

Answer all questions in this part.   [13]

Directions (60–72): For those questions that are followed by four choices, record your answers in the spaces provided. For all other questions in this part, record your answers in accordance with the directions given in the question.

60 The data table below shows the number of amino acid differences in the hemoglobin molecules of several species compared with amino acids in the hemoglobin of humans.

Amino Acid Differences

Species	Number of Amino Acid Differences
human	0
frog	67
pig	10
gorilla	1
horse	26

Based on the information in the data table, write the names of the organisms from the table in their correct positions on the evolutionary tree below.   [1]

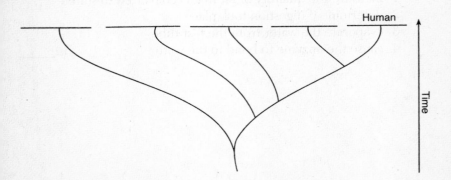

Human

Time

61 Explain why comparing the vein patterns of several leaves is a less reliable means of determining the evolutionary relationship between two plants than using gel electrophoresis. [1]

_____

_____

Base your answer to question 62 on the information and diagram below and on your knowledge of biology.

An enzyme and soluble starch were added to a test tube of water and kept at room temperature for 24 hours. Then, 10 drops of glucose indicator solution were added to the test tube, and the test tube was heated in a hot water bath for 2 minutes.

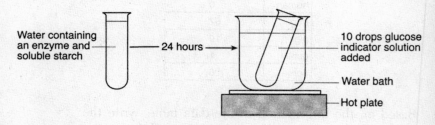

62 The test was performed in order to

  1 measure the quantity of fat that is converted to starch
  2 determine if digestion took place
  3 evaporate the water from the test tube
  4 cause the enzyme to bond to the water   62_____

63 A chromatography setup is shown below.

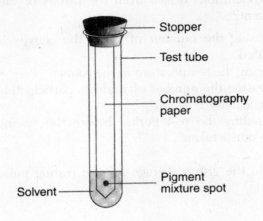

Identify *one* error in the setup.   [1]

_____

_____

Base your answers to questions 64 through 66 on the information and data table below and on your knowledge of biology.

During a laboratory activity, a group of students obtained the data shown below.

Pulse Rate Before and After Exercise

Student Tested	Pulse Rate at Rest (beats/min)	Pulse Rate After Exercise (beats/min)
A	70	97
B	74	106
C	83	120
D	60	91
E	78	122
Group Average		107

64 Which procedure would increase the validity of the conclusions drawn from the results of this experiment?

    1 increasing the number of times the activity is repeated
    2 changing the temperature in the room
    3 decreasing the number of students participating in the activity
    4 eliminating the rest period before the resting pulse rate is taken

        64_____

65 Calculate the group average for the resting pulse rate.   [1]

_____ **beats/min**

66 A change in pulse rate is related to other changes in the body. Write the name of *one* organ that is affected when a person runs a mile and describe *one* change that occurs in this organ.   [1]

Organ: _____

_____

_____

Base your answers to questions 67 through 69 on the information and diagram below and on your knowledge of biology.

A wet mount of red onion cells as seen with a compound light microscope is shown below.

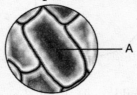

67 Which diagram best illustrates the technique that would most likely be used to add salt to these cells?

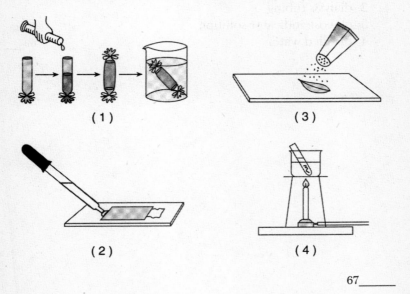

67_____

68 In the space below, sketch what cell A would look like after the addition of the salt. [1]

69 Which substance would most likely be used to return the cells to their original condition?

1 starch indicator
2 dialysis tubing
3 glucose indicator solution
4 distilled water

69_____

70 DNA electrophoresis is used to study evolutionary relationships of species. The diagram below shows the results of DNA electrophoresis for four different animal species.

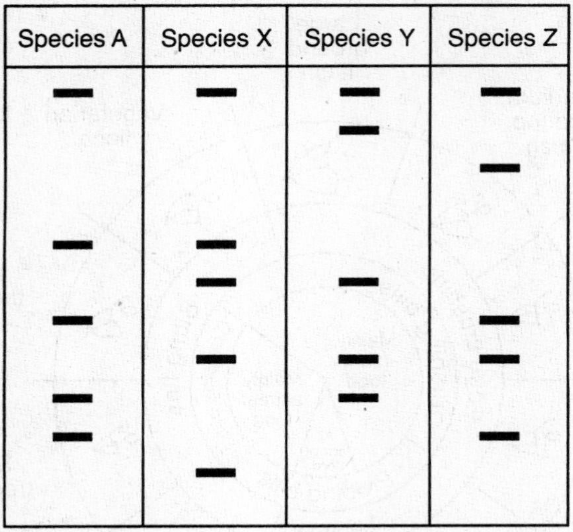

Species A	Species X	Species Y	Species Z
—	—	— —	—
			—
—	— —	—	
—	—	— —	—
— —	—		—

Which species has the most DNA in common with species A?

(1) X and Y, only
(2) Y, only
(3) Z, only
(4) X, Y, and Z

70_____

Base your answers to questions 71 and 72 on the diagram below that shows variations in the beaks of finches in the Galapagos Islands and on your knowledge of biology.

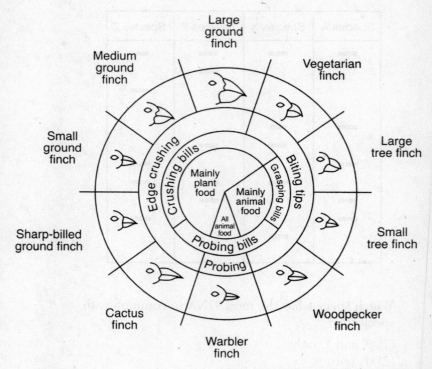

From: *Galapagos: A Natural History Guide*

71 The diversity of species seen on the Galapagos Islands is mostly due to

1 gene manipulation by scientists
2 gene changes resulting from mitotic cell division
3 natural selection
4 selective breeding

71_____

72 State *one* reason why large ground finches and large tree finches can coexist on the same island.   [1]

_____

_____

Answers
June 2009

Living Environment

Answer Key

PART A

1. 1	6. 3	11. 2	16. 2	21. 2	26. 3
2. 4	7. 3	12. 4	17. 3	22. 1	27. 4
3. 1	8. 1	13. 1	18. 4	23. 4	28. 2
4. 1	9. 1	14. 3	19. 2	24. 2	29. 4
5. 2	10. 3	15. 3	20. 1	25. 1	30. 2

PART B-1

31. 2	33. 4	35. 2	37. 4	39. 4	41. 3
32. 3	34. 4	36. 1	38. 1	40. 2	

PART B-2

42. *See* Answers Explained.
43. *See* Answers Explained.
44. *See* Answers Explained.
45. 3
46. 2
47. *See* Answers Explained.
48. *See* Answers Explained.
49. 1
50. *See* Answers Explained.
51. *See* Answers Explained.

PART C

52. *See* Answers Explained.
53. *See* Answers Explained.
54. *See* Answers Explained.
55. *See* Answers Explained.
56. *See* Answers Explained.
57. *See* Answers Explained.
58. *See* Answers Explained.
59. *See* Answers Explained.

PART D

60. *See* Answers Explained.
61. *See* Answers Explained.
62. 2
63. *See* Answers Explained.
64. 1
65. *See* Answers Explained.
66. *See* Answers Explained.
67. 2
68. *See* Answers Explained.
69. 4
70. 3
71. 3
72. *See* Answers Explained.

Answers Explained

PART A

1. 1 *The mature forest will most likely be stable over a long period of time* is the statement that best describes one of the stages represented in the diagram. The diagram illustrates stages in a typical ecological succession. Ecological succession is a term that describes the replacement of one plant community by other, progressively more complex, plant communities until a stable climax community is established. In the example given, an area populated by pioneer weeds and grasses gradually is overtaken by shrubs, which are then followed by deep-rooted trees in a young forest community. As succession continues over many years, a self-sustaining mature forest community made up of diverse plant and animal species will eventually become established.

WRONG CHOICES EXPLAINED

(2) *If all the weeds and grasses are destroyed, the number of carnivores will increase* is *not* the statement that best describes one of the stages represented in the diagram. If all the weeds and grasses are destroyed, there will be little food for the herbivorous animals in the ecosystem, causing them to die or move away. As the populations of herbivores decrease and are less available to serve as food for the carnivores, it is likely that the number of carnivores in the environment will likewise decrease, *not* increase.

(3) *As the population of the shrubs increases, it will be held in check by the mature forest community* is *not* the statement that best describes one of the stages represented in the diagram. The shrub stage of this succession is favorable for the growth and reproduction of shrubs, not mature forest. Although the shrubs will eventually be held in check, and even eliminated by invading trees, it is unlikely that the mature forest stage will be present for many years.

(4) *The young forest community will invade and take over the mature forest community* is *not* the statement that best describes one of the stages represented in the diagram. In a typical forest succession, young forest communities give way to mature forest communities. It is *not* common for young forest to invade and replace mature forest unless some intervening environmental change (e.g., flood, fire, drought, disease) occurs.

2. 4 The *circulatory* system in humans is most directly involved in the transport of oxygen. Molecular oxygen is absorbed from the air into the moist lung tissues, where it diffuses through capillary walls and into red blood cells. Red blood cells contain the protein hemoglobin that links chemically with oxygen molecules. The absorbed oxygen is then carried throughout the body within the circulatory system until it is released to tissues for use in the process of cellular respiration.

WRONG CHOICES EXPLAINED

(1) The *digestive* system in humans is *not* most directly involved in the transport of oxygen. The digestive system is responsible for the breakdown and absorption of food materials, *not* for carrying oxygen.

(2) The *nervous* system in humans is *not* most directly involved in the transport of oxygen. The nervous system is responsible for regulating the body's activities by receiving and interpreting environmental stimuli and by carrying nerve impulses from the central nervous system to effectors such as muscles and glands.

(3) The *excretory* system in humans is *not* most directly involved in the transport of oxygen. The excretory system is responsible for eliminating waste materials (e.g., carbon dioxide, urea) from the body, *not* for carrying oxygen.

3. **1** The structure labeled *1* in the diagram contains information needed for protein synthesis. Structure 1 is the nucleus of a plant cell, which contains DNA that contains the genetic codes for protein synthesis in the cell.

WRONG CHOICES EXPLAINED

(2) The structure labeled *2* in the diagram does *not* contain information needed for protein synthesis. Structure 2 is the cell membrane of a plant cell, which encloses the cell cytoplasm and regulates the passage of materials into and out of the cell.

(3) The structure labeled *3* in the diagram does *not* contain information needed for protein synthesis. Structure 3 is the food vacuole of a plant cell, which stores complex starches produced by the plant.

(4) The structure labeled *4* in the diagram does *not* contain information needed for protein synthesis. Structure 4 is the cell wall of a plant cell, which encloses the cell in a semirigid structure that gives the plant structural strength.

4. **1** Only liver cells of a human produce bile because *different cells use different parts of the genetic information they contain*. Because they develop by mitosis from a single zygote, all cells in the body contain identical genetic information. Therefore, all cells in the body contain the genes required for the production of bile. During the process of differentiation, body tissues specialize to take on specific functions. In such specialized cells, genes that are not needed to perform these specialized functions are switched "off." The genes that control bile production are switched "off" in all body cells except those cells in the liver that are responsible for the production of bile.

WRONG CHOICES EXPLAINED

(2) It is *not* true that only liver cells of a human produce bile because *cells can eliminate the genetic codes that they do not need*. All cells in the body contain the genes required for the production of bile. Unused genes can be switched "off" but cannot be eliminated by cells.

(3) It is *not* true that only liver cells of a human produce bile because *all other cells in the body lack the genes needed for the production of bile*. Because they develop by mitosis from a single zygote, all cells in the body contain identical genetic information. All cells in the body contain the genes required for the production of bile.

(4) It is *not* true that only liver cells of a human produce bile because *these cells mutated during embryonic development*. Mutations are random events, caused by mutagenic agents, which alter the genetic information in a cell. Mutations are not responsible for the production of bile by specialized liver cells in all humans.

5. **2** Although identical twins inherit exact copies of the same genes, the twins may look and act differently from each other because *the expression of genes may be modified by environmental factors*. Such environmental factors may include differences in exercise, nutrition, illness, or exposure to chemicals or radiation during critical phases of development. Twin studies have been important in the formulation of hypotheses regarding the effects of environment on gene expression.

WRONG CHOICES EXPLAINED

(1), (4) It is *not* true that, although identical twins inherit exact copies of the same genes, the twins may look and act differently from each other because *a mutation took place in the gametes that produced the twins* or because *a mutation took place in the zygote that produced the twins*. Mutations are random events, caused by mutagenic agents, which alter the genetic information in a cell. A mutation occurring either in the gametes or in the zygote would be passed on to both members of a twin pair in equal fashion and to equal effect.

(3) It is *not* true that, although identical twins inherit exact copies of the same genes, the twins may look and act differently from each other because *the expression of genes may be different in males and females*. Identical twins are genetically identical in all respects including the expression of gender. It is not possible for identical twins to have different genders; either both must be male (*XY*) or both must be female (*XX*).

6. **3** *Insulin* is the hormone that does *not* directly regulate human reproductive cycles. Insulin is produced in the pancreas and is responsible for the regulation of glucose concentrations in the blood.

WRONG CHOICES EXPLAINED

(1), (2), (4) *Testosterone*, *estrogen*, and *progesterone* are hormones that directly regulate human reproductive cycles. Testosterone (produced in the testes) regulates the production of sperm in males. Estrogen and progesterone (produced in the ovaries) work together to regulate the ovulation and gestation cycles in females.

7. **3** In making this observation, the student was most likely studying the *role of the owl in the local ecosystem*. This role is sometimes referred to as the organism's ecological niche and may encompass both the needs and the contributions of the organism to the maintenance of balance in the ecosystem.

WRONG CHOICES EXPLAINED

(1) In making this observation, the student was *not* most likely studying the *evolution of the owl*. Such a study might include observations of closely related owl species in an attempt to identify genetic similarities and differences among them that could indicate a common ancestry.

(2) In making this observation, the student was *not* most likely studying the *social structure of the owl population*. Such a study might include observations of several members of a specific owl population in an attempt to identify interactions among them in various situations.

(4) In making this observation, the student was *not* most likely studying the *life cycle of the owl*. Such a study might include observations of the mating, nesting, and rearing behaviors of several owl pairs in an attempt to identify reproductive patterns common to all members of the species.

8. **1** Sequence *1* is the sequence that best represents the relationship between DNA and the traits of an organism. In this sequence, DNA provides the code (genotype) that determines amino acid sequence by means of protein synthesis, which in turn determines the shape of the protein. The shape of the protein, including the nature of its active site, then determines the specific enzymatic function of the protein. Finally, the specific chemical reaction catalyzed by the protein (enzyme) helps to determine the trait (phenotype) displayed by the organism.

WRONG CHOICES EXPLAINED

(2), (3), (4) Sequences *2, 3,* and *4* are *not* the sequences that best represent the relationship between DNA and the traits of an organism. Each of these sequences contains one or more steps out of order. See correct answer above.

9. **1** The arrows in this sequence should be read as *leads to*. Sexual reproduction is considered to be the primary driver of genetic variation among living species on Earth. Genetic variation is known to increase the variety of living things (biodiversity) in the environment, including the roles (niches) performed by these living things. The stability of an ecosystem is enhanced by the richness of the biodiversity contained within it. Each of the elements in the sequence leads to or enhances the subsequent element.

WRONG CHOICES EXPLAINED

(2), (3), (4) The arrows in this sequence should *not* be read as *reduces, prevents,* or *simplifies*. Each of these terms implies that the elements in the sequence diminish the condition of the element following it. In fact, each element in the sequence enhances the subsequent element.

10. **3** Row *3* best describes the sequence of steps the scientists would most likely follow in this study. Scientists must logically identify the gene for production of the enzyme before removing it for insertion into the genome of the bacterium. When the bacterium has multiplied in sufficient numbers, the scientists can then extract the enzyme from the bacterial colony for use in treating the disease.

WRONG CHOICES EXPLAINED

(1), (2), (4) Rows *1, 2,* and *4* do *not* best describe the sequence of steps the scientists would most likely follow in this study. Each of these rows contains one or more steps out of logical order. See correct answer above.

11. **2** *Biological evolution* will most likely occur as a result of changes in the frequency of a gene in a particular population. Biological evolution is a term used to describe the way that new species are thought to arise from existing species by the process of natural selection. In this process, favorable genetic variations that enhance the survival of individuals become more common in the species gene pool. Over many generations, such changes in gene frequencies can result in new varieties or new species.

WRONG CHOICES EXPLAINED

(1) *Ecological succession* will *not* most likely occur as a result of changes in the frequency of a gene in a particular population. Ecological succession is a term that describes the replacement of one plant community by other, progressively more complex, plant communities until a stable climax community is established. Ecological succession is not directly affected by changes in gene frequency in particular populations.

(3) *Global warming* will *not* most likely occur as a result of changes in the frequency of a gene in a particular population. Global warming is a geological process that results in a gradual increase in the temperature of Earth due to increasing concentration of greenhouse gases in the atmosphere. Global warming is not directly affected by changes in gene frequency in particular populations.

(4) *Resource depletion* will *not* most likely occur as a result of changes in the frequency of a gene in a particular population. Resource depletion refers to the overuse of biological and mineral resources to serve human economic development. Resource depletion is not directly affected by changes in gene frequency in particular populations.

12. **4** The differences seen in this group of puppies are most likely due to *sorting and recombination of genes*. Sorting and recombination is the process by which the members of segregated allele pairs are randomly reunited in the zygote as a result of fertilization. One set of these alleles is contributed by the male, and the other set is contributed by the female. The new gene combinations that result provide the basis of genetic variations in the species that can be passed on to the next generation, in this case as a variation in coat color in a litter of puppies.

WRONG CHOICES EXPLAINED

(1) The differences seen in this group of puppies are *not* most likely due to *overproduction and selective breeding*. Overproduction is a concept that describes the natural tendency of species to produce more offspring than can possibly survive in nature. Selective breeding is a technique by which desirable traits in species are bred into offspring by careful selection of breeding pairs displaying those traits. It is possible that the variations of coat color in this group of puppies is the result of a selective breeding process.

(2) The differences seen in this group of puppies are *not* most likely due to *mutations and elimination of genes*. Mutations are random events that alter the sequence of genetic information in the cell, by either changing or eliminating genes. It is extremely unlikely that mutation could be responsible for the variations of coat color in this group of puppies.

(3) The differences seen in this group of puppies are *not* most likely due to *evolution and asexual reproduction*. Darwin's theory of evolution states that organisms best adapted to their natural environment tend to survive and pass their favorable traits on to the next generation. Asexual reproduction is a process by which new members of a species are produced from a single parent organism. It is extremely unlikely that evolution and asexual reproduction could be responsible for the variations of coat color in this group of puppies.

13. **1** The percentages of carbon dioxide and oxygen gases in the atmosphere are maintained most directly by *respiration and photosynthesis*. Photosynthesis is a series of enzyme-controlled biochemical reactions in which inorganic carbon dioxide (CO_2) and water (H_2O) molecules are combined to produce organic glucose ($C_6H_{12}O_6$) molecules and inorganic oxygen (O_2) molecules. Cellular respiration is a series of enzyme-controlled biochemical reactions in which organic glucose ($C_6H_{12}O_6$) molecules and inorganic oxygen (O_2) molecules interact to produce inorganic carbon dioxide (CO_2) and water (H_2O) molecules. These offsetting biochemical reactions recycle and replenish the supply of atmospheric carbon dioxide and oxygen gases.

WRONG CHOICES EXPLAINED

(2) The percentages of carbon dioxide and oxygen gases in the atmosphere are *not* maintained most directly by *the ozone shield*. The ozone shield is a layer of ozone gas in the atmosphere that helps to filter out harmful ultraviolet radiation found in sunlight. The ozone shield does not directly affect the percentages of carbon dioxide and oxygen gases.

(3) The percentages of carbon dioxide and oxygen gases in the atmosphere are *not* maintained most directly by *synthesis and digestion*. Synthesis is a cellular process in which complex organic molecules are built from simple subunits. Digestion is a cellular process in which complex organic molecules are broken down into simple subunits. These processes do not directly affect the percentages of carbon dioxide and oxygen gases.

(4) The percentages of carbon dioxide and oxygen gases in the atmosphere are *not* maintained most directly by *energy recycling in ecosystems*. This is a nonsense distracter. Energy does not recycle in the environment. Rather, energy enters the ecosystem from sunlight (or other sources), is passed from one trophic (feeding) level to another, and finally is dissipated into the environment as heat.

14. **3** Sequence *3* represents the order of some events in human development. A zygote is a diploid ($2n$) fertilized egg, which divides repeatedly by mitotic cell division to form differentiated cell masses known as tissues. Tissues performing similar functions group together during development to form organs and organ systems. Organ systems collectively make up the body of a developing fetus.

WRONG CHOICES EXPLAINED
(1), (2), (4) Sequences *1*, *2*, and *4* do *not* represent the order of some events in human development. Each of these sequences contains one or more steps out of correct order. See correct answer above.

[NOTE: In order to correctly answer this question, the student must interpret the arrows in the sequences as meaning "leads to." If the arrows are interpreted as meaning "derives from," then sequence 2 becomes an attractive distracter.]

15. **3** If this stem grows into a new plant, it would most likely produce *small white fruit, only*. The process described in the question is a form of vegetative propagation known as cutting. If handled properly by a competent grower, the cut stem forms a root system and other parts lost when it was severed from the parent plant, including fruits. Because the cutting contains only cells whose genetic makeup is identical to that of the parent plant, it will produce fruits identical to those produced by the parent plant, only.

WRONG CHOICES EXPLAINED
(1), (2), (4) If this stem grows into a new plant, it would *not* most likely produce *large red fruit, only*; *large pink fruit, only*; or *small red and small white fruit on the same plant*. Each of these choices describes fruits with characteristics not found in that of the parent plant.

16. **2** A mutation that can be inherited by offspring would result from *a base substitution in gametes during meiosis*. During meiosis (gamete formation), the replication of DNA may be flawed, resulting in a change in the sequence of bases (A, T, G, and C) in the resulting DNA strands. If the gamete containing DNA with this changed base sequence is involved in fertilization, the changed base sequence will be found in all the cells that develop from the zygote, ensuring that the offspring will inherit the mutation.

WRONG CHOICES EXPLAINED

(1), (3), (4) A mutation that can be inherited by offspring would *not* result from *random breakage of chromosomes in the nucleus of liver cells, abnormal lung cells produced by toxins in smoke,* or *ultraviolet radiation damage to skin cells.* Each of these choices describes a consequence of damage to the genetic structure of somatic (body) cells. Because somatic cells are not involved in the process of sexual reproduction, changes in their genetic structure cannot be passed on to offspring.

17. **3** Row 3 in the chart indicates what *A* and *B* in the boxes could represent. Proteins are complex polymeric molecules composed of many chemically linked units of amino acid. When proteins (*A*) are broken down by digestive processes, these amino acids (*B*) are released for absorption into the bloodstream.

WRONG CHOICES EXPLAINED

(1) Row *1* in the chart does *not* indicate what *A* and *B* in the boxes could represent. Starches are broken down into simple sugars, not proteins.

(2) Row *2* in the chart does *not* indicate what *A* and *B* in the boxes could represent. Starches are broken down into simple sugars, not amino acids.

(4) Row *4* in the chart does *not* indicate what *A* and *B* in the boxes could represent. Proteins are broken down into amino acids, not simple sugars.

18. **4** *The removal of wastes from both A and B involves the use of energy from ATP* is the statement that best describes the functions of those organs. In the diagram, organ *A* represents the lung, which is specialized to remove carbon dioxide from the blood. Organ *B* represents the kidney, which is specialized to remove urea, water, and salts from the blood. While carbon dioxide exits the blood tissue in the lung by simple diffusion, the expenditure of energy in the form of ATP is needed to remove urea and salts from the blood tissue in the kidney, as this movement is normally accomplished against the concentration gradient.

WRONG CHOICES EXPLAINED

(1) *B pumps blood to A for gas exchange* is *not* the statement that best describes the functions of those organs. Blood is pumped throughout the body by the heart (not shown), not by the kidney.

(2) *A and B both produce carbon dioxide, which provides nutrients for other body parts* is *not* the statement that best describes the functions of those organs. In humans, carbon dioxide is a metabolic waste, not a nutrient.

(3) *A releases antibodies in response to an infection in B* is *not* the statement that best describes the functions of those organs. Antibodies are produced and released by the blood tissue (not shown), not by the lung.

19. **2** The effect the bacteria have on humans indicates that *Salmonella* bacteria are *pathogenic organisms*. The term pathogenic means disease-causing. A pathogenic bacterium normally invades the living tissues of its host, where it multiplies rapidly and produces toxins that cause a variety of symptoms in the host. The symptoms of *Salmonella* poisoning described in the question are consistent with the concept of human disease.

WRONG CHOICES EXPLAINED
(1) It is *not* true that the effect the bacteria have on humans indicates that *Salmonella* bacteria are *predators*. Predators are multicellular animals (e.g., wolves, cougars, hawks) that kill and consume other animals for food. *Salmonella* is a unicellular bacterium, not a multicellular animal.

(3) It is *not* true that the effect the bacteria have on humans indicates that *Salmonella* bacteria are *parasitic fungi*. A parasitic fungus is an organism (e.g., ergot fungus, ringworm fungus, athlete's foot fungus) that lives in or on a host organism and derives a nutritional benefit from the association, while the host is harmed. *Salmonella* is a bacterium, not a fungus.

(4) It is *not* true that the effect the bacteria have on humans indicates that *Salmonella* bacteria are *decomposers*. A decomposer is an organism (e.g., soil bacteria, mushrooms, bracket fungi) that lives in or on the bodies of dead organisms and breaks the dead tissues down so that their chemical components can be recycled in the ecosystem. *Salmonella* lives within the bodies of living organisms, not dead organisms.

20. **1** The virus that causes AIDS is damaging to the body because it *targets cells that fight invading microbes*. The body's natural defense mechanism contains disease-fighting lymphocytes known as CD4 cells (also known as T-cells). AIDS (Acquired Immune Deficiency Syndrome) is a human disease caused by the HIV virus. When HIV infects humans, the cells it infects most often are CD4 cells. The virus becomes part of the cells, and when they multiply to fight an infection, they also make more copies of HIV that infect and destroy more CD4 cells. As the CD4 cell count drops, the body's ability to defend itself against pathogenic bacteria, fungi, and viruses diminishes, leaving the body open to multiple infections.

WRONG CHOICES EXPLAINED
(2), (3), (4) It is *not* true that the virus that causes AIDS is damaging to the body because it *attacks specific red blood cells*, *causes an abnormally high insulin level*, or *prevents the normal transmission of nerve impulses*. These effects are not associated with AIDS or the HIV virus. See correct answer above.

21. **2** In the leaf of a plant, guard cells help to *regulate oxygen and carbon dioxide levels*. The primary function regulated by guard cells is the exchange of gases between the leaf interior and the atmosphere. In regulating this function, the guard cells change shape when they produce glucose during photosynthesis and absorb water to balance their diffusion gradients, becoming turgid. In this condition, the guard cells open stomates on the leaf's surface, allowing carbon dioxide to enter and oxygen to escape. The evolutionary advantage of this regulation in land plants is to facilitate the process of photosynthesis by allowing carbon dioxide to enter the leaf at times when light is most intense and photosynthesis is occurring most rapidly.

WRONG CHOICES EXPLAINED

(1) In the leaf of a plant, guard cells do *not* help to *destroy atmospheric pollutants when they enter the plant*. This is a nonsense distracter. No known mechanism in the plant leaf is adapted to perform this function.

(3) In the leaf of a plant, guard cells do *not* help to *transport excess glucose to the roots*. This function is performed by the phloem cells found in the veins of the plant leaf, not by the guard cells.

(4) In the leaf of a plant, guard cells do *not* help to *block harmful ultraviolet rays that can disrupt chlorophyll production*. This function is performed to a limited degree by the waxy cuticle of the plant leaf, not by the guard cells.

22. **1** *They will be produced by asexual reproduction and will be more resistant to the antibiotic* is the statement that best describes future generations of these bacteria. The diagram illustrates the process of fission, a form of asexual reproduction common in bacteria. Because those bacteria with genetic variations making them more resistant to the antibiotic were more likely to survive, these surviving bacteria passed this antibiotic resistance on to their offspring, making them more resistant to the effects of the antibiotic.

WRONG CHOICES EXPLAINED

(2), (4) *They will be produced by sexual reproduction and will be more resistant to the antibiotic* and *they will be produced by sexual reproduction and will be just as susceptible to the antibiotic* are *not* the statements that best describe future generations of these bacteria. Bacteria reproduce by asexual, not sexual, means.

(3) *They will be produced by asexual reproduction and will be just as susceptible to the antibiotic* is *not* the statement that best describes future generations of these bacteria. Presumably, the gene in the bacteria for antibiotic sensitivity will be selected against, increasing the frequency of the gene for antibiotic resistance in the bacterial population and making future generations more resistant, not just as susceptible, to the antibiotic.

23. **4** The sizes of plant populations can be influenced by the *types of minerals present in the soil*. Plants absorb nutrients from soil in the form of mineral ions. The types and concentrations of minerals in the soil can have a profound effect on the populations of plants growing in a particular ecosystem, both in terms of species present and population sizes of those species. Mineral-poor soils often support only limited species of plants and sparse population sizes, whereas mineral-rich soils often support diverse plant communities with large population sizes.

WRONG CHOICES EXPLAINED

(1), (2), (3) The size of plant populations *cannot* be influenced by the *molecular structure of available oxygen*, the *size of the cells of decomposers*, or the *number of chemical bonds in a glucose molecule*. These factors are determined by the laws of chemistry and have no direct effect on the population sizes of plants in a particular environment.

24. **2** Competition between two species occurs when *chipmunks and squirrels eat sunflower seeds in a garden*. Interspecies competition occurs whenever two different species (e.g., chipmunks and squirrels) utilize the same limited resource (e.g., sunflower seeds) in an ecosystem. As the resource becomes more limited, the competition between the species becomes more intense. Intense interspecies competition can result in the elimination of one of the two species from the ecosystem.

WRONG CHOICES EXPLAINED

(1) It is *not* true that competition between two species occurs when *mold grows on a tree that has fallen in the forest*. Mold growing on and consuming a dead tree is an example of decomposition, not interspecies competition. The mold and the tree do not utilize the same limited resources in the ecosystem.

(3) It is *not* true that competition between two species occurs when *a crow feeds on the remains of a rabbit killed on a road*. A crow feeding on a dead rabbit is an example of scavenging behavior, not interspecies competition. The crow and the rabbit do not utilize the same limited resources in the ecosystem.

(4) It is *not* true that competition between two species occurs when *a lion stalks, kills, and eats an antelope*. A lion feeding on a dead antelope it has killed is an example of predation, not interspecies competition. The lion and the antelope do not utilize the same limited resources in the ecosystem.

25. **1** The arrows represented as ∿➤ most likely indicate *the energy released into the environment as heat*. Ecologists estimate that as much as 90% of the energy contained in each trophic (feeding) level of an ecosystem is converted to heat and dissipated into the environment. This leaves only 10% of that energy to be passed from one trophic level to another in a typical food chain, in this case, from seaweed, to small fish, to large fish, to sharks.

WRONG CHOICES EXPLAINED

(2) The arrows represented as ∿→ do *not* most likely indicate *oxygen produced by respiration*. This is a nonsense distracter. Oxygen is used, not produced, during respiration.

(3) The arrows represented as ∿→ do *not* most likely indicate *the absorption of energy that has been synthesized*. Energy is not synthesized in the environment but is produced and radiated from a limited number of sources in the solar system, primarily in the form of sunlight. Light energy is absorbed by green plants and converted to the chemical bond energy of glucose before being passed to animals in a food chain as food energy. The small straight arrows in the diagram may represent the transfer of this energy from organism to organism in this food chain.

(4) The arrows represented as ∿→ do *not* most likely indicate *the transport of glucose away from the organism*. Glucose is a simple sugar that is produced by green plants during photosynthesis and converted to other chemical compounds (e.g., starches, proteins, lipids) in the plants' cells. These compounds are passed from plants to animals in a food chain to be used as a source of energy and structural materials for tissue growth and repair. The small straight arrows in the diagram may represent the transfer of these compounds from organism to organism in this food chain.

26. **3** If several species of carnivores are removed from an ecosystem, the most likely effect on the ecosystem will be *a decrease in stability among populations*. In a balanced, healthy ecosystem, all the species populations present depend on each other and the niches they fill (roles they play) to maintain the overall health of the environment. Carnivores (e.g., wolves, coyotes, cougars, hawks, eagles, cormorants, northern pike) play a vital role in the ecosystem by keeping populations of prey organisms in check. Without this control, the prey species populations could easily increase in number to the point where they exceed the carrying capacity of the environment, overwhelming it and potentially causing it to collapse completely.

WRONG CHOICES EXPLAINED

(1) If several species of carnivores are removed from an ecosystem, the most likely effect on the ecosystem will *not* be *an increase in the kinds of autotrophs*. As carnivores are removed, the herbivorous prey animals will increase in number and consume autotrophs (plants), potentially eliminating plant species populations from the ecosystem. In this case, the kinds of autotrophs will be likely to decrease, not increase.

(2) If several species of carnivores are removed from an ecosystem, the most likely effect on the ecosystem will *not* be *a decrease in the number of abiotic factors*. Abiotic factors in the environment are those nonliving conditions (e.g., temperature, moisture, minerals, oxygen, sunlight) in which plant and animal populations exist in a particular environment. Abiotic factors affect,

and are affected by, the living community in a particular ecosystem. Removal of carnivorous species from an ecosystem may contribute to a change in these conditions but would not increase or decrease their number.

(4) If several species of carnivores are removed from an ecosystem, the most likely effect on the ecosystem will *not be an increase in the rate of succession*. Ecological succession is a term that describes the replacement of one plant community by other, progressively more complex, plant communities until a stable climax community is established. Succession may be set in motion by any number of events, including fire, flood, volcanic activity, or other ecological collapse. Because a potential outcome of this situation is ecological collapse, the onset of ecological succession may ultimately be triggered by the removal of carnivores from the ecosystem. Despite this likelihood, the rate at which the succession occurs once in motion will not likely be increased or decreased.

27. **4** The production and use of compost is an example of *the recycling of nutrients*. As the plant material in the compost pile decomposes through the action of bacteria and fungi, the complex organic molecules in this plant matter are broken down to simple compounds (e.g., carbon dioxide, nitrates, phosphates) that can be absorbed by, and incorporated into the structure of, living plants. Contributing to the recycling of materials by maintaining a compost pile represents a positive human activity in the environment.

WRONG CHOICES EXPLAINED

(1) The production and use of compost is *not* an example of *the introduction of natural predators*. Natural predators include animals such as wolves, coyotes, cougars, and eagles that contribute to the balance and health of the natural environment by keeping prey species populations in check. Reintroduction of natural predators represents a positive human activity in the environment. Maintenance of a compost pile does not contribute to this activity.

(2) The production and use of compost is *not* an example of *the use of fossil fuels*. Fossil fuels (e.g., oil, natural gas, coal) are naturally occurring hydrocarbon fuels that were formed millions of years ago from compression of decaying organic matter. The use of fossil fuels represents a negative human activity in the environment. Maintenance of a compost pile does not contribute to this activity.

(3) The production and use of compost is *not* an example of *the deforestation of an area*. Deforestation is the complete removal of forest trees from an area caused by human activity. Deforestation represents a negative human activity in the environment. Maintenance of a compost pile does not contribute to this activity.

28. **2** *It has genetic information contained in DNA* is the statement that best describes a chromosome. Chromosomes are organelles contained in the nucleus of the cell that control the activities of the cell through the production of enzymes and other proteins. Each of hundreds of strands of DNA in a chromosome contains a unique genetic code that is used to determine the exact chemical structure of these protein molecules. The proteins in turn determine the traits of the organism by catalyzing thousands of chemical reactions throughout the body.

WRONG CHOICES EXPLAINED

(1) *It is a gene that has thousands of different forms* is *not* the statement that best describes a chromosome. A gene is composed of one or more DNA strands that control the production of enzymes involved in determining a single genetic trait. A chromosome may contain hundreds or thousands of different genes. A chromosome is not a gene.

(3) *It is a reproductive cell that influences more than one trait* is *not* the statement that best describes a chromosome. A reproductive cell (e.g., egg cell, sperm cell) contains a nucleus housing a monoploid (*n*) set of chromosomes of the species. A chromosome is not a reproductive cell.

(4) *It contains hundreds of genetically identical DNA molecules* is *not* the statement that best describes a chromosome. A chromosome may contain hundreds or thousands of genetically different, not identical, DNA molecules.

29. **4** *Atmospheric temperature* is the environmental factor that has been most recently affected by these changes in carbon dioxide level. Increased concentrations of carbon dioxide and other greenhouse gases in the atmosphere have been linked to the phenomenon known as global warming. Greenhouse gases act to trap solar heat close to Earth's surface by preventing radiational cooling in the upper atmosphere. Scientists have demonstrated that as carbon dioxide levels have increased over time so have average air temperatures. As a result, over the past decade polar ice caps and glaciers have shrunk significantly and normal weather patterns have been disrupted with increasing frequency. Humans add tremendous quantities of this compound to the air every day, both through their metabolic processes and through their technological practices.

WRONG CHOICES EXPLAINED

(1) *Light intensity* is *not* the environmental factor that has been most recently affected by these changes in carbon dioxide level. Light intensity is influenced by natural phenomena such as solar cycles and cloud cover, as well as by human activities that produce air pollutants. Carbon dioxide is a colorless gas that does not directly affect light intensity.

(2) *Types of decomposers* is *not* the environmental factor that has been most recently affected by these changes in carbon dioxide level. Decomposers are organisms, such as bacteria and fungi, that consume the bodies of dead plants and animals. There is no scientific evidence that the types of decomposers are directly affected by changes in carbon dioxide level.

(3) *Size of consumers* is *not* the environmental factor that has been most recently affected by these changes in carbon dioxide level. Consumers are organisms, such as herbivores and carnivores, that consume other living organisms for food. There is no scientific evidence that the size of consumers is directly affected by changes in carbon dioxide level.

30. **2** One reason why people should be aware of the impact of their actions on the environment is that *the depletion of finite resources cannot be reversed*. Earth's mineral resources, such as fossil fuels (i.e., oil, coal, natural gas) and metal ores (e.g., copper, aluminum, iron, gold), were formed millions of years ago by geological processes that cannot be re-created. As such, these resources are known as nonrenewable resources. As human activities extract and use up these resources, their future availability becomes less and less certain. For this reason, it is critical that humans establish processes to recycle mineral resources and to conserve fossil fuel reserves to help extend their availability for future generations.

WRONG CHOICES EXPLAINED

(1) One reason why people should be aware of the impact of their actions on the environment is *not* that *ecosystems are never able to recover once they have been adversely affected*. An ecosystem that has been disrupted can recover over time by a process known as ecological succession. Numerous examples have demonstrated that, given sufficient time, ecosystem recovery is possible in all but the most extreme cases of environmental destruction.

(3) One reason why people should be aware of the impact of their actions on the environment is *not* that *there is a decreased need for new technology*. Humans have always depended on our ability to develop new technologies that promote our survival. In the past, we have done so without regard for the negative impact of that technology on the natural world and its diverse plant and animal communities. In the future, it is imperative for our continued survival that we develop new technologies in a manner that also promotes the health of the natural environment.

(4) One reason why people should be aware of the impact of their actions on the environment is *not* that *there is a decreased need for substances produced by natural processes*. The history of human technological development has demonstrated an increased, not a decreased, need for substances produced by natural processes. These substances include both nonrenewable and renewable resources.

PART B-1

31. **2** *The chromosome makeup of Dolly is identical to that of sheep A* is the statement about Dolly that is correct. The diagram indicates that sheep A was the donor of a single diploid (2n) nucleus from a somatic (body) cell that was inserted into the denucleated egg cell of sheep B. This process is known as cloning. The cloned egg containing the genetic information from sheep A was then implanted in the uterus of sheep C and allowed to develop into an embryo. All the cells that make up the early embryo, and ultimately make up Dolly, are derived from the cloned egg and will therefore have a chromosome makeup and contain genetic information identical to those found in the cells of sheep A.

WRONG CHOICES EXPLAINED

(1) *Gametes from sheep A and sheep B were united to produce Dolly* is *not* the statement about Dolly that is correct. The diagram indicates that a denucleated gamete (egg cell) from sheep B was used to help produce Dolly. However, the nucleus used to create the cloned egg was removed from a somatic (body) cell of sheep A, not from a gamete.

(3) *Both Dolly and sheep C have identical DNA* is *not* the statement about Dolly that is correct. In this experiment, sheep C functioned only as a surrogate mother for Dolly's embryological development. Other than genes shared in common as members of the same species, Dolly and sheep C do not have identical DNA.

(4) *Dolly contains genes from sheep B and sheep C* is *not* the statement about Dolly that is correct. In this experiment, sheep B provided a denucleated egg and sheep C functioned only as a surrogate mother for Dolly's embryological development. Dolly does not contain genes that come from either sheep B or sheep C.

32. **3** *ATP is produced most efficiently in structure C* is the statement concerning ATP and activity within the cell that is correct. Structure C in the diagram illustrates the mitochondrion, a cell organelle containing the enzymes required to carry on the process of aerobic respiration. The chemical bond energy released from each molecule of glucose oxidized by aerobic respiration is sufficient to form 36 molecules of ATP.

WRONG CHOICES EXPLAINED

(1) *The absorption of ATP occurs at structure A* is *not* the statement concerning ATP and activity within the cell that is correct. Structure A in the diagram illustrates the cell membrane, a cell organelle specialized to carry on the process of cell transport. ATP is produced in the mitochondria, not absorbed through the cell membrane.

(2) *The synthesis of ATP occurs within structure B* is *not* the statement concerning ATP and activity within the cell that is correct. Structure B in the diagram illustrates the nucleus, a cell organelle specialized to regulate cell

processes by providing the genetic codes for the synthesis of enzymes and other proteins. ATP is not synthesized within the nucleus.

(4) *The template for ATP is found in structure D* is *not* the statement concerning ATP and activity within the cell that is correct. Structure *B* in the diagram illustrates the vacuole, a cell organelle specialized to store complex food materials and metabolic wastes. The template (code) for the production of ATP is not found in the vacuole.

33. **4** *Secretions help the body to respond to changes from the normal state* is the statement that best describes events shown in the diagram. Homeostasis is defined as a state of balance, or steady state, in the body. When the homeostatic balance of the body is upset, disease or death can result. The diagram illustrates the role of the pituitary gland (often called the master gland) in regulating the secretions of other endocrine glands, which in turn regulate metabolic processes important to the maintenance of homeostasis.

WRONG CHOICES EXPLAINED

(1) *Secretions provide the energy needed for metabolism* is *not* the statement that best describes events shown in the diagram. Secretions of endocrine glands, known as hormones, have no direct role in the release of energy needed for metabolism. Cell energy is provided by the catalytic action of mitochondrial enzymes.

(2) *The raw materials for the synthesis of secretions come from nitrogen* is *not* the statement that best describes events shown in the diagram. Secretions of endocrine glands, known as hormones, are complex protein molecules synthesized from amino acid subunits. While amino acids contain nitrogen, it is incorrect to infer that hormones come from nitrogen.

(3) *The secretions of all glands speed blood circulation in the body* is *not* the statement that best describes events shown in the diagram. The secretions of endocrine glands, known as hormones, can have diverse effects on the body's metabolic processes. While some hormones help to speed blood circulation, others may have the opposite effect, or no effect, on blood circulation.

34. **4** *The catfish population helps to control the populations of water boatman and water fleas* is the statement that describes an interaction that helps to maintain the dynamic equilibrium of this ecosystem. The arrows in the diagram represent the flow of energy from organism to organism in this food web. The diagram indicates that catfish feed on (derive energy from) both water boatman and water fleas, thereby helping to control the numbers of these aquatic insects in this ecosystem.

WRONG CHOICES EXPLAINED

(1) *The frogs make energy available to the ecosystem through the process of photosynthesis* is *not* the statement that describes an interaction that helps to maintain the dynamic equilibrium of this ecosystem. Photosynthesis is an energy-capturing process carried on by producers such as the algae, not by frogs.

(2) *The algae directly provide food for both the rotifers and the catfish* is *not* the statement that describes an interaction that helps to maintain the dynamic equilibrium of this ecosystem. The diagram indicates that catfish feed on water boatman and water fleas, not on algae. The algae provide energy to the catfish indirectly via the water fleas, not directly.

(3) *The green-backed heron provides energy for the mosquito larvae* is *not* the statement that describes an interaction that helps to maintain the dynamic equilibrium of this ecosystem. The diagram indicates that the green-backed heron is a top-order consumer in this food web and so is a recipient, not a provider, of energy flowing though it. The diagram also indicates that mosquito larvae derive energy from algae, not from green-backed herons.

35. **2** Structure *B* may function in the recognition of chemical signals. Structure *B* represents a membrane-embedded protein thought to function in the recognition of hormones, neurotransmitters, and other complex substances needed by cells for the maintenance of homeostasis. These substances link with structure *B* at an active site and are transported across the cell membrane when structure *B* rotates in place.

WRONG CHOICES EXPLAINED

(1), (3), (4) It is *not* true that structures *A*, *C*, and *D* may function in the recognition of chemical signals. These structures represent different portions of the cell membrane's bilipid layer. These components of the cell membrane are thought to regulate the diffusion of simple molecules into and out of the cell but are not thought to function in the recognition of chemical signals.

36. **1** An inference that can be made from information in the diagram is that *many of the descendants of organism B became extinct*. The diagram indicates that only evolutionary lines *Q* and *S* represent contemporary (extant) species descended from ancestral species *B*, whereas all other evolutionary lines have resulted in extinction at some point in the past.

WRONG CHOICES EXPLAINED

(2) An inference that *cannot* be made from information in the diagram is that *organism B was probably much larger than any of the other organisms represented*. There is no data provided in the diagram concerning the relative sizes of the organisms in the evolutionary lines represented by the letters. For this reason, it is inappropriate to draw an inference concerning the relative sizes of these organisms.

(3) An inference that *cannot* be made from information in the diagram is that *most of the descendants of organism B successfully adapted to their environment and have survived to the present time*. In interpreting this diagram, it must be inferred that evolutionary lines represented by letters below the dashed line are no longer in existence but have died out some time in the past (are extinct). Only lines *Q* and *S* are shown above the dashed line as having survived to the present time (are extant).

(4) An inference that *cannot* be made from information in the diagram is that *the letters above organism B represent members of a single large population with much biodiversity*. This diagram is clearly captioned as representing some evolutionary pathways. This caption does not lead to the inference that the letters represent members of a single population.

37. **4** Species *D* in the chart shown is most likely to have the fastest rate of evolution. Organisms with rapid rates of reproduction are more likely than slower-breeding species to develop and display a wide variety of genetic variation due to the processes of mutation and/or segregation and recombination. If these organisms exist in an environment with changing conditions, the selection pressures on the species are likely to select certain varieties for continuation, while selecting others for elimination. Such species are known to evolve relatively rapidly.

WRONG CHOICES EXPLAINED
(1), (2), (3) Species *A*, *B*, and *C* in the chart are *not* most likely to have the fastest rate of evolution. The combination of slow reproductive rates and stable environments for these species will normally restrict their evolutionary processes to rates slower than that of species *D*.

38. **1** Species *A* would have the largest amount of available energy in this ecosystem. Species *A* in the diagram represents grass, a green plant that absorbs solar energy and converts it to the chemical bond energy of glucose in the process of photosynthesis. In any balanced ecosystem, the producer level (green plants including grass) must contain the greatest amount of energy compared to the consumer levels. As energy is transferred from one trophic (feeding) level to the next, as much as 90% of the energy in that level is lost, being dissipated into the environment in the form of heat.

WRONG CHOICES EXPLAINED
(2), (3), (4) Species *B*, *C*, and *E* would *not* have the largest amount of available energy in this ecosystem. Species *B*, *C*, and *E* represent a grasshopper, a frog, and a bird, respectively, all of which are consumers and will therefore contain less energy at their respective trophic levels than that found in species *A*.

39. **4** Organisms *C and E* are carnivores. Carnivores are animals that consume other animals for food. Organism *C* (frog) commonly consumes insects and other small animals for food. Many species of organism *E* (bird) eat worms or insects or scavenge the bodies of dead animals and are carnivores (other bird species consume fruits and seeds and are herbivores).

WRONG CHOICES EXPLAINED
(1), (2), (3) Organisms *A and B*, *A and E*, and *B and D* are *not* both carnivores. Organism *A* (grass) is a producer, not a carnivore. Organism *B* (grasshopper) is an herbivore, not a carnivore. Organisms *D* (snake) and *E* (bird) are often considered to be carnivores, although this characteristic may vary in individual species.

40. **2** *As the foxes adapted to different environments, differences in appearance evolved* is the statement that best explains how the differences between these two species came about. Most accurately, as the environment changed, favorable variations already present in the ancestral fox population were selected for continuation in the species because they provided those foxes with an adaptive advantage over others lacking those favorable variations. These favorable variations presumably led to increased rates of survival and reproduction for those foxes, which assisted in the perpetuation of the favorable variations in the two fox populations. By this process and over many generations, the kit fox population gradually became adapted to the desert environment, while the red fox population gradually became adapted to the forest environment.

WRONG CHOICES EXPLAINED
(1), (3), (4) *Different adaptations developed because the kit fox preferred hotter environments than the red fox, The foxes evolved differently to prevent overpopulation of the forest habitat,* and *The foxes evolved differently because their ancestors were trying to avoid competition* are *not* the statements that best explain how the differences between these two species came about. Each of these statements infers some purposeful action by the foxes. Species do not evolve because they wish to do so for specific purposes but rather as a matter of natural selection of favorable variations that provide adaptive advantages to some member over others.

41. **3** The organisms represented as ϵϵϵϵ are found in the area shown because of *light intensity*. These organisms are identified in the key as algae and seaweed. As producers, these organisms require sunlight in order to carry on the process of photosynthesis, so inhabit the upper layers of their water environment through which light can penetrate. This layer is known as the photic zone.

WRONG CHOICES EXPLAINED
(1) It is *not* true that the organisms represented as ϵϵϵϵ are found in the area shown because of *pH*. pH (relative acidity) is normally the same throughout a water environment, so is unlikely to attract algae and seaweed to the top of the water environment.
(2) It is *not* true that the organisms represented as ϵϵϵϵ are found in the area shown because of *sediment*. Sediments collect on the bottom of a water environment, so would be unlikely to attract algae and seaweed to the top of the water environment.
(4) It is *not* true that the organisms represented as ϵϵϵϵ are found in the area shown because of *colder temperatures*. Colder temperatures are found on the bottom of a water environment because of the increased density of cold water, so would be unlikely to attract algae and seaweed to the top of the water environment.

PART B-2

42. One credit is allowed for correctly stating what happened to the amount of dissolved oxygen and the number of fish species as the amount of sewage waste increased. Acceptable responses include: [1]

- *Both the level of dissolved oxygen and the number of fish species decreased.*
- *There is an inverse relationship between the amount of sewage waste and the level of dissolved oxygen in the ecosystem.*
- *The number of fish species that are present in the ecosystem decreased during the study, indicating that either the presence of sewage pollution or the decrease in dissolved oxygen was responsible for this decline.*
- *Both decreased.*

43. One credit is allowed for correctly marking an appropriate scale on each labeled axis. [1]

44. One credit is allowed for correctly plotting the data, surrounding each point with a small circle, and connecting the points. [1]

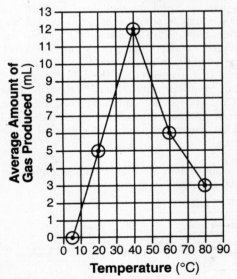

Average Amount of Gas Produced at Various Temperatures

45. **3** The maximum rate of cellular respiration in yeast occurred at *40°C*. Examination of the data in the table provided clearly indicates that the greatest level of gas production (12 ml) occurred at a temperature of 40°C.

WRONG CHOICES EXPLAINED

(1), (2), (4) The maximum rate of cellular respiration in yeast did *not* occur at *5°C, 20°C,* or *60°C*. Gas production rates at these temperatures are shown in the table to be 0 ml, 5 ml, and 6 ml, respectively, all of which are less than the 12 ml produced at 40°C.

46. **2** Compared to the other tubes at the end of 30 minutes, the tubes in group 3 probably contained the *smallest amount of glucose*. Assuming no glucose was added from other sources, the glucose present in group 3 would be less than that found in the other groups. This is true because, in order to produce larger amounts of carbon dioxide (CO_2) gas, the glucose in group 3 would have had to be used more rapidly than in the other groups.

WRONG CHOICES EXPLAINED

(1) It is *not* true that, compared to the other tubes at the end of 30 minutes, the tubes in group 3 probably contained the *smallest amount of CO_2*. In fact, the table documents that group 3 collected the greatest, not the smallest, amount of CO_2 compared to the other groups.

(3) It is *not* true that, compared to the other tubes at the end of 30 minutes, the tubes in group 3 probably contained the *smallest amount of ethyl alcohol*. The reaction equation indicates that ethyl alcohol is a product of the fermentation reaction. Since group 3 collected the greatest amount of CO_2, it would also be expected to collect the greatest, not the smallest, amount of ethyl alcohol compared to the other groups.

(4) It is *not* true that, compared to the other tubes at the end of 30 minutes, the tubes in group 3 probably contained the *same amounts of glucose, ethyl alcohol, and CO_2*. The reaction equation indicates that glucose is used up in this reaction, while CO_2 and ethyl alcohol are produced. It is likely that group 3 contained less glucose than either CO_2 or ethyl alcohol as a result of this chemical reaction.

47. One credit is allowed for placing the names of *all* the organisms in the correct locations. An acceptable response is: [1]

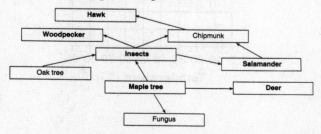

48. One credit is allowed for correctly identifying one producer recorded by the ecologist in the data table. Acceptable responses include, but are not limited to: [1]

- *Maple tree*
- *Oak tree*
- *Tree*

49. **1** *Trees absorb water as a raw material for photosynthesis* is the statement that describes how one biotic factor of the forest uses one of the abiotic factors listed in the data table. Biotic (living) factors in the environment include all the living organisms and their roles; trees are an example of one such biotic factor in the forest ecosystem. Abiotic (nonliving) factors include all the environmental conditions upon which the biotic community depends for its survival; water is an example of one such abiotic factor in the forest ecosystem. The statement identifies how a biotic factor (trees) makes use of an abiotic factor (water) in the forest ecosystem.

WRONG CHOICES EXPLAINED

(2) *Insects eat and digest the leaves of trees* is *not* the statement that describes how one biotic factor of the forest uses one of the abiotic factors listed in the data table. Both insects and trees are biotic factors in the forest ecosystem.

(3) *Erosion of sedimentary rock adds phosphorus to the soil* is *not* the statement that describes how one biotic factor of the forest uses one of the abiotic factors listed in the data table. Both sedimentary rock and phosphorus in the soil are abiotic factors in the forest ecosystem.

(4) *Fungi release oxygen from the trees back into the air* is *not* the statement that describes how one biotic factor of the forest uses one of the abiotic factors listed in the data table. This is a nonsense distracter. Fungi represent a biotic factor in a forest ecosystem, and oxygen represents an abiotic factor. However, fungi do *not* perform the role of oxygen release described in the statement.

50. Two credits are allowed for correctly filing in the blanks in parts 2 and 3 of the dichotomous key so that it contains information that could be used to identify the four animals shown. Acceptable responses include, but are not limited to: [2]

Dichotomous Key
1. a. Legs present ....Go to 2
   b. Legs not present ....Go to 3
2. a. **Four legs** ....**I (Dog)**
   b. **Eight legs** ....**II (Spider)**
3. a. **Fins present** ....**III (Fish)**
   b. **Fins not present** *or* **segments present** ...**IV (Earthworm)**

51. Four credits are allowed for correctly completing boxes 1 through 4 in the chart using the information from the diagram. An acceptable response is: [4]

Name of Structure	Letter on Diagram	Function of Structure
ovary	C	produces gametes
uterus	D	**site of internal development**
placenta	B	transports oxygen directly to the embryo

PART C

52. Four credits are allowed for correctly describing how human activities can affect the environment of organisms living 50 years from now. In your experimental design, be sure to:

- Identify *one* human activity that could release chemicals harmful to the environment. [1]
- Identify the chemical released by this activity. [1]
- State *one* effect the release of this chemical would most likely have on future ecosystems. [1]
- State *one* way in which humans can reduce the production of this chemical to lessen its effect on future ecosystems. [1]

Examples of acceptable responses: [4]

A human activity that could release harmful chemicals is driving automobiles that burn gasoline. [1] One harmful chemical released during driving gasoline cars is carbon dioxide. [1] Carbon dioxide is a greenhouse gas that is known to contribute to global warming. Global warming may result in higher temperatures and melting of the polar ice caps, both of which can hurt plants and animals around the world by drastically changing abiotic conditions. [1] One way humans could help reduce the production of carbon dioxide is to develop sources of energy for cars (such as electric motors) that do not produce this chemical. [1]

When humans dump toxic wastes into unprotected pits, harmful chemicals can be released into the environment by leaching through soil into groundwater. [1] Toxic wastes often contain heavy metals such as lead [1] that are known to disrupt reproductive cycles in animals and humans. [1] It is important for the protection of the ecosystem to develop and enforce laws that regulate the disposal of toxic wastes in a safe manner. [1]

Paper production is a human activity that can release a harmful chemical into the environment. [1] This chemical is dioxin, which results from the use of chlorine to bleach the paper. [1] Dioxin is a poison that is stored in the fatty tissues of animals and humans and is known to accumulate to harmful levels in the food chain. [1] One way to reduce dioxin produced by the paper industry is to require paper mills to catch the dioxin before it is released into the environment and chemically convert it to harmless substances. [1]

53. Five credits are allowed for correctly designing an experiment to test the effects of *one* environmental factor, chosen from the list, on plant growth. In your experimental design, be sure to:

- Identify the environmental factor you chose. [0]
- State *one* hypothesis the experiment would test. [1]
- State how the control group would be treated differently from the experimental group. [1]
- State *two* factors that must be kept the same in both the experimental and control groups. [1]
- Identify the independent variable in the experiment. [1]
- Label the columns in the data table for the collection of data in your experiment. [1]

Examples of acceptable responses: [5]

Environmental factor: *acidity of precipitation [0]*

Hypothesis: Acid rain will cause a decrease in the number of seeds that germinate. [1] In this experiment, the control group will be watered with water at pH 7, while the experimental groups will be watered with water at pH less than 7. [1] Two factors that must be kept the same in this experiment are soil used and type of plant seeds used. [1] The independent variable in this experiment is the pH of the water. [1]

Chart column headings: *pH of water, Number of seeds that germinate [1]*

Environmental factor: *temperature [0]*

The experiment's hypothesis is that colder temperature will reduce the rate of plant growth. [1] In this experiment, the control group will be grown at 65°F, while the experimental groups will be grown at temperatures of 55°F, 45°F, and 35°F. [1] Two factors that must be kept the same in this experiment are the amount of fertilizer used and the amount of water given to the plants. [1] The independent variable in this experiment is the temperature of the growing pots. [1]

Chart column headings: *Temperature of the pots, Height of the plants after 30 days [1]*

54. Two credits are allowed for correctly stating *two* ways that the use of windmills to produce energy would be beneficial to the environment. Acceptable responses include, but are not limited to: [2]

- *Windmills do not produce pollution.*
- *Use of windmills would mean that we would burn less coal and oil.*
- *Wind is a renewable resource.*

55. Two credits are allowed for correctly stating *one* specific example of an imported species that has altered the balance of the ecosystem and explaining how it has disrupted the balance in that ecosystem. Acceptable responses include, but are not limited to: [2]

- *Zebra mussels outcompete native mollusk species for food, causing native species to decline in numbers.*
- *Purple loosestrife has crowded out native plants, leaving many native animals with much less available food because they cannot eat the purple loosestrife.*
- *Rabbits imported into Australia ate much of the vegetation that previously fed many native animal species. Populations of many native species were reduced drastically, disrupting the ecosystem there.*
- *Gypsy moths imported from Europe have overpopulated in parts of the United States, eating all the leaves of many trees, causing some of the trees to die, and leaving little food for native species.*
- *Kudzu from Asia has grown over native plants in the southern United States, blocking the light and crowding the roots of native plants, causing them to die.*

56. One credit is allowed for correctly stating *one* difference between the effect on the human body of the usual forms of flu virus and the effect of H5N1. Acceptable responses include, but are not limited to: [1]

- *Most flu viruses cause a runny nose and sore throat, while the H5N1 virus can cause pneumonia.*
- *The avian flu goes deeper into the lungs and can cause severe pneumonia.*
- *The avian flu has a more severe effect on humans than most other flu viruses.*

57. One credit is allowed for correctly identifying the type of substance produced by the human body that protects against antigens such as the flu virus. An acceptable response is: [1]

- *Antibodies*

58. One credit is allowed for correctly stating what is in a vaccine that makes the vaccine effective. Acceptable responses include, but are not limited to: [1]

- *A vaccine contains dead pathogens.*
- *Weakened pathogens in a vaccine make it effective in producing immunity.*

59. One credit is allowed for correctly identifying *one* event that could result in the virus changing to a form able to spread from human to human. Acceptable responses include, but are not limited to: [1]

- *Mutation*
- *Genetic alteration*

PART D

60. One credit is allowed for writing the names of the organisms from the table in their correct positions on the evolutionary tree. An acceptable response is: [1]

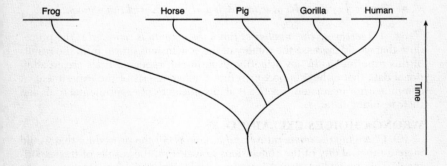

61. One credit is allowed for correctly explaining why comparing the vein patterns of several leaves is a less reliable means of determining the evolutionary relationship between two plants than using gel electrophoresis. Acceptable responses include, but are not limited to: [1]

- *Chemical similarities are more reliable than structural similarities for identifying relationships among species.*
- *Electrophoresis shows chemical similarities, which are more reliable than physical similarities.*
- *Many unrelated plants have leaves with similar vein patterns.*

62. **2** The test was performed in order to *determine if digestion took place.* Starch is a complex carbohydrate that can be converted to glucose by the process of digestion. The fact that the contents of the test tube are tested for the presence of glucose indicates that the researcher is trying to determine whether the starch has been digested by the action of the enzyme over the 24-hour duration of the experiment.

WRONG CHOICES EXPLAINED

(1) The test was *not* performed in order to *measure the quantity of fat that is converted to starch.* Fat is not identified as a variable in this experiment. Fat cannot be directly converted to starch by any known chemical reaction.

(3) The test was *not* performed in order to *evaporate the water from the test tube.* Placing the test tube in a water bath helps to ensure that water in the test tube will not be inadvertently evaporated during the experiment.

(4) The test was *not* performed in order to *cause the enzyme to bond to the water.* Enzymes bond to substrate molecules, not to water.

63. One credit is allowed for correctly identifying *one* error in the setup. Acceptable responses include, but are not limited to: [1]

- *The pigment spot is below the surface of the solvent.*
- *The level of the solvent is too high and has covered the pigment spot.*

64. **1** *Increasing the number of times the activity is repeated* is the procedure that would increase the validity of the conclusions drawn from the results of this experiment. In any scientific experiment, repeated trials create additional data that either support or refute the hypothesis of the experiment. A common error in amateur science is an insufficiency of experimental trials and their resulting data.

WRONG CHOICES EXPLAINED

(2) *Changing the temperature of the room* is *not* the procedure that would increase the validity of the conclusions drawn from the results of this experiment. This procedure would introduce additional variables into the experiment that would invalidate the results of the study. Proper experimental control would keep all variables in the experiment constant except the independent variable.

(3) *Decreasing the number of students participating in the study* is *not* the procedure that would increase the validity of the conclusions drawn from the results of this experiment. This procedure would serve to weaken the study

and invalidate any conclusions drawn from it. Good experimental procedure includes large numbers of subjects in order to reduce bias that might otherwise interfere with experimental results.

(4) *Eliminating the rest period before the resting pulse rate is taken* is not the procedure that would increase the validity of the conclusions drawn from the results of this experiment. The purpose of measuring the resting pulse is to provide accurate background data against which the experimental data are measured. Eliminating the rest period would invalidate the study by interfering with the accurate collection of this background data.

65. One credit is allowed for correctly calculating the group average for the resting pulse rate. An acceptable response is: [1]

- *73*

66. One credit is allowed for correctly writing the name of *one* organ that is affected when a person runs a mile and describing *one* change that occurs in this organ. Acceptable responses include, but are not limited to: [2]

- *heart—beats faster*
- *lungs—take in oxygen faster*
- *muscles—use energy faster*
- *skin—sweat glands open to release perspiration*
- *brain—secretes neurotransmitters that coordinate other organs*
- *intestines—reduce digestive activity*

67. **2** Diagram 2 best illustrates the technique that would most likely be used to add salt to these cells. This diagram shows a medicine dropper being used to add a drop of water containing salt to the edge of the coverslip covering the specimen on the slide. The drop of saltwater is sucked under the coverslip by capillary action and surrounds the plant tissue specimen.

WRONG CHOICES EXPLAINED
(1) Diagram *1* does *not* best illustrate the technique that would most likely be used to add salt to these cells. This diagram shows the proper laboratory technique for filling dialysis tubing for an experiment involving diffusion.

(3) Diagram *3* does *not* best illustrate the technique that would most likely be used to add salt to these cells. This diagram is a nonsense distracter that shows improper laboratory technique by using a household saltshaker to add salt to a microscope specimen.

(4) Diagram *4* does *not* best illustrate the technique that would most likely be used to add salt to these cells. This diagram shows the proper laboratory technique for heating a specimen in a water bath.

68. One credit is allowed for correctly sketching what cell A would look like after the addition of the salt. An acceptable response is: [1]

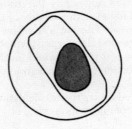

Note: Allow credit if the cell is drawn correctly but *not* shaded.

69. **4** *Distilled water* is the substance that would most likely be used to return the cells to their original condition. By adding distilled water to the specimen, the effects of the saltwater in reducing water concentration and causing water to leave the cell by osmosis would be reversed. Water concentration outside the cell would reach 100% in the distilled water bath, causing water to enter the cell rapidly by osmosis.

WRONG CHOICES EXPLAINED

(1) *Starch indicator* is *not* the substance that would most likely be used to return the cells to their original condition. Starch indicator is used to detect the presence of starch in a specimen and would not restore the cells to their original condition.

(2) *Dialysis tubing* is *not* the substance that would most likely be used to return the cells to their original condition. Dialysis tubing is used in experiments involving diffusion and osmosis and would not restore the cells to their original condition.

(3) *Glucose indicator solution* is *not* the substance that would most likely be used to return the cells to their original condition. Glucose indicator is used to detect the presence of glucose in a specimen and would not restore the cells to their original condition.

70. **3** Species Z, *only* has the most DNA in common with species A. Examination of the data indicates that species A and Z share three DNA types in common, whereas species X and Y each share only two DNA types in common with species A.

WRONG CHOICES EXPLAINED

(1), (2), (4) It is *not* true that species X *and* Y, *only*; Y, *only*; or X, Y, *and* Z have the most DNA in common with species A. Examination of the data indicates that species A and Z share three DNA types in common, whereas species X and Y each share only two DNA types in common with species A.

71. **3** The diversity of species seen on the Galapagos Islands is mostly due to *natural selection*. Natural selection is a term used to describe the way that new species are thought to arise from existing species, resulting in biological evolution. In this process, favorable genetic variations that enhance the survival of individuals become more common in the species gene pool. Over many generations, such changes in gene frequencies can result in new varieties or new species.

WRONG CHOICES EXPLAINED
    (1) The diversity of species seen on the Galapagos Islands is *not* mostly due to *gene manipulation by scientists*. Gene manipulation is a laboratory technique used by scientists to alter the genetic makeup of selected specimens for the purpose of studying the effects of genes on their genetic makeup and biological processes. Gene manipulation is not responsible for the evolution of new species.
    (2) The diversity of species seen on the Galapagos Islands is *not* mostly due to *gene changes resulting from mitotic cell division*. Mitotic cell division is a type of cell reproduction that maintains genetic information intact from one generation of cells to the next. Mitotic cell division is not normally responsible for the evolution of new species.
    (4) The diversity of species seen on the Galapagos Islands is *not* mostly due to *selective breeding*. Selective breeding is a technique by which desirable traits in species are bred into offspring by careful selection of breeding pairs displaying those traits. Selective breeding is not responsible for the evolution of new species.

72. One credit is allowed for correctly stating *one* reason why large ground finches and large tree finches can coexist on the same island. Acceptable responses include, but are not limited to: [1]

- *Large ground finches eat mainly plant food, and large tree finches eat mainly animal food.*
- *They do not compete for the same resources, so both can survive.*
- *They occupy different environmental niches.*
- *Large tree finches depend on worms and insects for their food supply, whereas large ground finches consume fruits and seeds.*
- *Because they utilize different foods, these species are less likely to compete with each other in their nutritional activities.*

STANDARDS/KEY IDEAS	JUNE 2009 QUESTION NUMBERS	NUMBER OF CORRECT RESPONSES
STANDARD 1		
Key Idea 1: The central purpose of scientific inquiry is to develop explanations of natural phenomena in a continuing and creative process.		
Key Idea 2: Beyond the use of reasoning and consensus, scientific inquiry involves the testing of proposed explanations involving the use of conventional techniques and procedures and usually requiring considerable ingenuity.	53	
Key Idea 3: The observations made while testing proposed explanations, when analyzed using conventional and invented methods, provide new insights into natural phenomena.	43, 44, 45, 46	
Laboratory Checklist	50	
STANDARD 4		
Key Idea 1: Living things are both similar to and different from each other and from nonliving things.	1, 2, 3, 17, 34, 35, 47, 48, 49, 55	
Key Idea 2: Organisms inherit genetic information in a variety of ways that result in continuity of structure and function between parents and offspring.	4, 5, 8, 10, 15, 22, 28, 31	
Key Idea 3: Individual organisms and species change over time.	11, 12, 16, 36, 37, 40, 59	
Key Idea 4: The continuity of life is sustained through reproduction and development.	6, 14, 51	
Key Idea 5: Organisms maintain a dynamic equilibrium that sustains life.	13, 18, 19, 20, 21, 32, 33, 42, 56, 57, 58	
Key Idea 6: Plants and animals depend on each other and their physical environment.	1, 7, 9, 23, 24, 25, 26, 38, 39, 41	
Key Idea 7: Human decisions and activities have a profound impact on the physical and living environment.	27, 29, 30, 52, 54	

STANDARDS/KEY IDEAS	JUNE 2009 QUESTION NUMBERS	NUMBER OF CORRECT RESPONSES
REQUIRED LABORATORIES		
Lab 1: "Relationships and Biodiversity"	60, 61, 63, 70	
Lab 2: "Making Connections"	64, 65, 66	
Lab 3: "The Beaks of Finches"	71, 72	
Lab 5: "Diffusion Through a Membrane"	62, 67, 68, 69	

ur Personal Online Test paration Tutor for the AT, PSAT, and ACT

barronstestprep.com